WITCH HUNT

ALSO BY GREGG JARRETT
The Russia Hoax:
The Illicit Scheme to Clear Hillary Clinton and Frame Donald Trump

WITCH HUNT

THE STORY OF THE GREATEST MASS DELUSION IN AMERICAN POLITICAL HISTORY

GREGG JARRETT

An Imprint of HarperCollins *Publishers*

HarperCollins books may be purchased for educational, business, or sales promotional use. For information, please email the Special Markets Department at SPsales@harpercollins.com.

Broadside Books™ and the Broadside logo are trademarks of HarperCollins Publishers.

FIRST EDITION

Names: Jarrett, Gregg, author.
Title: Witch hunt : the story of the greatest mass delusion in American political history / Gregg Jarrett.
Other titles: Story of the greatest mass delusion in American political history
Description: First edition. | New York City : HarperCollins, [2019] | Includes bibliographical references. | Summary: "Now that every detail and argument set forth in the #1 New York Times bestseller The Russia Hoax has been borne out by the Mueller report, the author is back with a hard-hitting, well-reasoned evisceration of what may be the dirtiest trick in political history"—Provided by publisher.
Identifiers: LCCN 2019031064 (print) | LCCN 2019031065 (ebook) | ISBN 9780062960092 (hardcover) | ISBN 9780062978233 (large print paperback) | ISBN 9780062960108 (epub)
Subjects: LCSH: Presidents—United States—Election—2016. | Elections—Corrupt practices—United States. | Governmental investigations—United States. | Trump, Donald, 1946– | Clinton, Hillary Rodham. | Conspiracy—United States. | Russia (Federation)—Foreign relations—United States. | United States—Foreign relations—Russia (Federation) | Hacking—Russia (Federation)
Classification: LCC E911 .J37 2019 (print) | LCC E911 (ebook) | DDC 324.973/09051—dc23
LC record available at https://lccn.loc.gov/2019031064
LC ebook record available at https://lccn.loc.gov/2019031065

19 20 21 22 23   LSC   10 9 8 7 6 5 4 3 2 1

TO CATE, GRACE AND LIV,
FOR YOUR LOVE AND ENCOURAGEMENT

This was the greatest hoax in the history of our country. And the biggest witch hunt ever.

—President Donald J. Trump, interview with the author, June 25, 2019, Oval Office, White House

CONTENTS

A MALIGNANT FORCE

If you are somebody who's being falsely accused of something, you would tend to view the investigation as a witch hunt.

—Attorney General William Barr, Senate testimony, April 10, 2019

Inventing a lie is easy. Spreading a lie is even easier. Uncovering the truth is hard. The truth will always have enemies. It is the inherent vice of corrupt men. Nothing is more dangerous to the public good than a lie contrived to usurp the reins of power.

Witch Hunt is the story of a pernicious lie that was invented and spread in a collusive effort to sabotage the election of Donald Trump. When the plot failed, the objective shifted. Newly aggressive tactics were deployed to destroy his presidency, undo his election, and drive him from office. There was never any credible evidence that Trump was a Russian agent. There were no seditious acts that he cooked up with the Kremlin. It was a damning fiction that constitutes what is surely the dirtiest political trick ever perpetrated in politics.

The most amazing part, now that the truth has been revealed, is how widely the lie was believed. How did a handful of government officials create the greatest mass delusion in American political history?

What people tend to forget about "witch hunts" is that there are no real witches, but the hunt persists because of an overwhelming and irrational desire to *believe* there must be witches. The absolute, unshakable faith in the impossible is what makes a witch hunt possible.

Trump's enemies, blinded by their own bias and hatred, argued that he could not possibly have won the presidency absent some nefarious cabal to steal the 2016 election. To them, no other explanation made sense. They could not conceive that voters had legitimately placed him in office. There *had* to be witches flying about. The lie justified their hunch.

In each chapter of this book, the reader will discover people who bent the rules or broke the law because they knew in their hearts that witches would eventually drop out of the sky to prove their thesis. Surely the tangible evidence was there. It was simply a matter of waiting for the apparition. An astonishing array of theoretically smart people convinced themselves, based on *nothing*, that Trump had committed the most noxious crime in America: a treasonous conspiracy with Russia. At the heart of the witch hunt were rogue government operators determined to stop Trump. They were tenacious and unrelenting.

The peril to democracy today is not a foreign force but a malignant force of unelected officials here at home. Armed with immense power and often lurking in the shadows, they have revealed themselves capable of uncommon corruption. Their allegiance is not to the Constitution and the rule of law but to themselves. Personal animus and a voracious appetite for authority are what motivates their zeal. Anyone like Trump, who might jeopardize their hold on power, must be immobilized by any means and at any cost. To neutralize this threat and achieve their desired ambitions, they politicized their agencies, weaponized law enforcement, and persecuted people without respect for law or process.

There is little doubt that top Obama administration officials at the FBI, CIA, Department of Justice, and other federal agencies abused their authority for political purposes. James Comey, Andrew McCabe, Peter Strzok, Lisa Page, Bruce Ohr, James Clapper, and John Brennan all regarded themselves as above the law and accountable to no one. Supremely confident in their arrogance that they were smarter than the American people, who are empowered to choose a president, their actions would serve a higher moral purpose. Or so they persuaded themselves. Audaciously, they sought to incriminate Trump for crimes he did not commit. *Witch Hunt* unravels the plot hatched by these enemies of truth, their insipid deceptions, and the extraordinary actions they took to cover up their malevolent acts.

The mainstream media, riven with bias and easily exploited, enabled the hoax and became witting accessories to the witch hunt. Their disdain for Trump, both the man and his policies, influenced their adversarial reporting on a daily basis. Their methods failed them as they covered every twist and turn in the search for wrongdoing, instead of investigating ostensible wrongdoing. Whatever position Trump adopted or action he took was reflexively opposed by the liberal intellectual quotient that inhabits newsrooms almost everywhere. Trump was their nemesis. Even the few striving for fairness in their reporting frequently tweeted only the most partisan articles to their influential followers. Television commentators set a new standard for bloviating, simply making things up, and manufacturing all manner of falsehoods. They were all thirsting to gulp the Kool-Aid. Their reporting led many Americans to swallow the big lie.

Now, of course, the country can see the last two years clearly. The illicit scheme originated during the campaign season of 2016. It was secretly financed by Hillary Clinton's campaign and Democrats, conceived by a foreign agent with a checkered past in espionage, and then brokered to solicitous collaborators at the FBI, the Justice Department, and elsewhere in government.[1] The premise of the ruse was as outlandish as the actions of those who advanced it: they posited that

Trump was a covert Russian asset who had spent many years "colluding" with the Kremlin to win the White House. This was contrary to
all reason or common sense; it was utterly preposterous.

As with most mendacious smears, it was bereft of any proof to
support it. There was nothing in the way of probable cause or reasonable suspicion to indicate that Trump had coordinated or collaborated
with Moscow. There was no *hard* evidence; indeed, there was no *soft*
evidence. The narrative patiently constructed on MSNBC and CNN
for months—Trump would get a good deal on building a Russian
hotel, Trump hires Russians to hack the DNC, and then the Russians
secretly control our foreign policy—doesn't even make sense. Trump
did not conspire or, if you like, "collude" with Russia to influence the
election. Others were colluding to frame him.

The first seeds of the canard appear to have been planted by the
CIA under the partisan stewardship of Director John Brennan, who
loathed Trump.[2] The idea slowly germinated as Clinton sprinkled her
campaign speeches with tenebrous references to her opponent's patri-
otism, suggesting a veiled fealty to Russian president Vladimir Putin.
It was rubbish, of course, but it tended to arouse the kind of suspicion
that only political bombast during an election can engender. Clinton
then took it a step further, fertilizing and harvesting the idea. Through
a series of discreetly disguised transactions that served as firewalls, her
campaign and the Democratic National Committee hired a former
British spy by the name of Christopher Steele who conjured up an
anti-Trump document known as the "dossier." In a series of seventeen
successive memos penned over a six-month period in 2016, these specious documents spun a fantastic tale of how Trump associates and
the candidate himself had consorted with sinister Kremlin officials in
an elaborate effort to pervert the election.[3]

As Steele disseminated his scribble, he confided to a high Justice
Department official by the name of Bruce Ohr that he detested the
Republican nominee and was desperate to prevent him from ascending to the presidency.[4] The end would justify the means, even if that
included falsifying or fabricating evidence against the candidate he

despised. Over several months and with remarkable stealth, Ohr propagated the information to the FBI, which was already working secretly with Steele, who had been on the Bureau payroll since early 2016.

At its core, Steele's "dossier" was based on little more than multiple hearsays from supposed Russian sources who were largely anonymous. It is quite possible that those unidentified sources never existed at all or, in the alternative, the ex-spy was fed Russian disinformation to the amusement of the Kremlin. Senior FBI officials well knew the sketchy provenance of the "dossier" and the mutable character of its author. Indeed, they had been warned.[5] They didn't care. They also knew that the accusations contained therein were so dubious and unverified that they could never be used in court or to initiate a formal investigation of its target.

Yet, undeterred by the constraints of law, the FBI used them anyway. Top officials exploited the "dossier" as a pretext to launch their investigation of Trump in direct violation of the regulations that govern initiating such a probe.[6] In essence, the outlandish rumors fueled by the partisan enemies of Trump and the lawlessness of the FBI created the biggest political "con" of all time: a deliberate deception that managed to dominate and, to some extent, disable the Trump presidency.

The FBI first got its hands on the "dossier" on July 5, 2016, the very day that Director Comey stood in front of television cameras to absolve Clinton of various crimes for mishandling classified documents. As he twisted the facts and distorted the law, one of his agents was furtively meeting with Steele in London. When the agent read the document, he was stunned and remarked, "I have to report this to headquarters."[7] Thus, on the same day Comey cleared Clinton, the "witch hunt" began in earnest.

The formal investigation was opened three weeks later on July 31, 2016. That day, rogue FBI agent Peter Strzok texted his lover, FBI lawyer Lisa Page, admitting that the Clinton case had never really mattered, but the Trump case was "momentous" and did "MAT-

TER."[8] During my interview of President Trump at the White House in June 2019, I showed him that text. He shook his head in disgust and said, "That text tells me it was all rigged from the very beginning, and it tells me that it is the worst scandal to hit the FBI."[9]

Steele did not work alone. His carefully cultivated false narrative of Trump-Russia "collusion" was engineered by hired surrogates of Clinton's campaign and the Democrats, namely Glenn Simpson, the founder of a company called Fusion GPS, which marketed itself as an opposition research firm. Simpson was aptly described by a major publication as "a mercenary for hire by anyone with fat stacks of bitcoins."[10] With clever calculation, Simpson and Steele hustled their "dossier" to the media and Clinton allies in the upper echelons of government, with the FBI as the ultimate receptacle of all phony information.[11] Simpson would later clam up by invoking his Fifth Amendment right against self-incrimination when subpoenaed to testify before Congress.

None of that impeded the FBI or President Obama's Justice Department. Weeks before the presidential election, they misused the *unverified* "dossier" as the basis of a *"verified"* warrant application from a secret surveillance court to wiretap a onetime Trump campaign adviser, boldly asserting that he was a Russian spy.[12] He was not. Vital evidence was concealed, the judges were deceived, and the court was defrauded.

The strategy to dismantle Trump's bid for the highest office was dependent on proliferating the erroneous and lurid story that he was a modern-day "Manchurian Candidate" or "Putin's puppet."[13] When it did not gain the desired traction and the Republican nominee was elected, his enemies doubled down on the scheme to portray him as a Russian stooge who would betray his nation once he set foot in the Oval Office. The complicit media were all too willing to convict Trump in the court of public opinion by perpetuating those calumnies without bothering to honestly examine the reliability of their sources or the ridiculousness of the narrative they were feeding daily.

Trump's improbable election had sent many in the press corps into a tailspin.

Equally unglued were Barack Obama's intelligence agencies. With assistance from the outgoing president's diplomats, they "unmasked" the protected names of hundreds of Americans identified in secret intelligence reports, including three senior Trump advisers.[14] One victim of the unmasking was the incoming president's pick to be national security advisor. His conversation with a Russian diplomat during the transition was secretly recorded and illegally leaked to the media to create the appearance of "collusion" so as to damage the newly elected president.[15] In the waning days of the old administration, intelligence chiefs worked furiously to enact new rules that would relax the sharing of intel, making it easier to spread scurrilous information that could prove destructive to Trump.

Shortly before inauguration day, the CIA, FBI, and the director of national intelligence (DNI) concocted a plan for Director Comey to selectively brief Trump on only the salacious part of the "dossier" while deliberately hiding the Russia "collusion" accusation and who had paid for it.[16] The discussion in the meeting was immediately leaked to journalists so that reporters would have an excuse to publish the contents of the "dossier," which they did. The new president deserved the truth about the full document, not to be smeared in the media by a false narrative underwritten by the Clinton campaign and circulated by the FBI and intelligence agencies.

Almost overnight, the Trump-Russia phantasm ignited a public firestorm. Unfounded allegations that the president was in league with Putin served as rich fodder for endless stories, commentary, and denunciations on Capitol Hill and in the compliant press. That was precisely what the progenitors of the lie had desired all along. Within three months, Trump was fed up with Comey's duplicity and misrepresentations. Privately, he was assuring the president that he was not under investigation. In public testimony before Congress he implied the opposite. Comey had also broken FBI rules in his handling of the

Clinton email case. When Trump fired the FBI director, the escalating "witch hunt" became a full-blown political maelstrom. Incensed and angry that his mentor at the Bureau had been canned, the temporary "acting" FBI director Andrew McCabe initiated a new and surreptitious investigation of Trump.[17] There was no legal justification for this. He did it because he could.

Over at the Justice Department, the emotionally overwrought Rod Rosenstein, the deputy attorney general, named Comey's longtime friend and ally Robert Mueller III as special counsel.[18] The appointment was an act of pure retribution against the president whom Rosenstein blamed for the unexpected and stinging public criticism the deputy AG had endured after recommending Comey's termination. As for Mueller, he should never have accepted the job. He had more than one disqualifying conflict of interest, including his close ties to Comey, who was a pivotal witness.[19] Even worse, Mueller had met with the president in the Oval Office the day before he accepted the assignment to investigate Trump. The appointment itself was not just perfidious; it was contrary to federal regulations, and Rosenstein must have known it. When confronted, he allegedly cowered behind his desk and blubbered, "Am I gonna get fired?"[20]

The appointment of Mueller had the insufferable Comey's dirty fingerprints all over it. When he was sacked, he took with him presidential memorandums he had purloined from the government. In contravention of the law and FBI rules, he conveyed them without authorization to a friend for the sole purpose of leaking the confidential memos to the media in order to trigger the naming of a special counsel, who just happened to be his friend and former colleague Mueller. They were the same memos Comey had concealed from the Justice Department but delivered to the media.[21] He vowed that he would never leak, and then he did. The devious plan worked, and a special counsel was appointed.

In private testimony before Congress more than a year later, Comey admitted that there was no evidence of "collusion" when the FBI had launched its investigation in July 2016, and by the time Mueller was named, "we still couldn't answer the question."[22] This means that the initial probe should never have been opened in the first place, and the appointment of a special counsel some nine months later was not authorized under the governing regulations; there must first be some articulable factual basis or evidence suggesting that a crime may have been committed.[23] There wasn't any. But the Trump resistance operation was just getting started.

In a plot that should alarm all Americans, McCabe and Rosenstein met behind closed doors to consider deposing Trump from the presidency. But how exactly could he be evicted? The deputy attorney general proposed wearing a "wire" to secretly record Trump and recruit cabinet members to remove him from office under a contorted interpretation of the Twenty-fifth Amendment.[24] When later questioned about this by Trump, Rosenstein denied the attempted coup against the duly elected president of the United States. "He said it didn't happen. He said he never said it," the president told me. "What he told other people is that he was joking. But to me, he claimed he never said it."[25] McCabe and another witness affirmed that it had been no joke.[26] It appears that Rosenstein was not telling the truth.

As Mueller ignored his own conflicts with impunity, he hired a team of partisan prosecutors, ruining the special counsel's integrity and the credibility of its investigation. Rosenstein, who was obviously a key witness in any obstruction of justice case, refused to recuse himself and continued to supervise the probe. According to Trump's lawyer, Mueller knew within a few months that there was no collusion and confessed as much during a meeting on March 5, 2018.[27] Yet he refused to say so for more than a year, likely affecting the midterm elections.

On March 22, 2019, the Mueller Report was finally submitted to the Justice Department. As expected, there was no evidence of criminal "collusion" or, as the special counsel affirmed, "the investigation did not establish that members of the Trump Campaign conspired or coordinated with the Russian government in its election interference activities."[28] After a twenty-two-month investigation, hundreds of witnesses interviewed, thousands of subpoenas issued, and more than a million documents examined, the Russia hoax was exposed for what it was: a lie. Not a single person was ever charged with a "collusion" conspiracy offense.

Though perfectly willing to render a legal judgment on "collusion," Mueller then announced that he had "determined not to make a traditional prosecutorial judgment" on obstruction.[29] His rationale was unintelligible. It got worse. He stated, "while this report does not conclude that the President committed a crime, it also does not exonerate him."[30] What? As any lawyer will tell you, it is *never* the job of a prosecutor anywhere to exonerate people. By this one act, Mueller managed to reverse the burden of proof and invert the presumption of innocence which are sacrosanct principles in American law. Instead, he spent 183 pages smearing Trump by *implying* that, under certain circumstances which did not actually exist, the facts might sustain an obstruction case.

Attorney General William Barr was admittedly baffled by Mueller's reasoning, describing it in Senate testimony as "strange" and "bizarre."[31] He and other top lawyers at the DOJ examined the report, analyzed the law and the facts, and concluded that the evidence was "not sufficient to establish that the President committed an obstruction-of-justice offense."[32] In a swipe at Mueller, Barr observed that the special counsel's legal analysis "did not reflect the views of the department" but was the product of the personal views "of a particular lawyer or lawyers."[33] The attorney general insisted that Mueller could have rendered a decision on obstruction and that no Department of Justice rules prevented him from doing so.[34]

The witch hunt ended precisely as I had argued it would in my book *The Russia Hoax*. Trump did not "collude" with Russia, but his opponent who funded and endorsed the lie did. The Clinton campaign paid for Russian disinformation in a phony "dossier." Clinton's surrogates then fed the elaborate fraud to the FBI so that it would launch a dilating investigation of Trump, and they peddled it to the media to influence the 2016 presidential election. Clinton herself incessantly advanced the false conspiracy theory that held the nation and presidency hostage for more than two years.

As the fictive "collusion" narrative persisted, the media paid no attention to what Clinton did but devoted all of its scrutiny to what Trump did *not* do. Journalists were all too eager to accept as gospel the intelligence leaks that they *assumed* were accurate and truthful. Their lack of question or curiosity was animated by their antipathy toward the president. They embraced the "dossier" as scripture, teased by hope that it might somehow, inexplicably, be proven true. There has been no shortage of media malpractice in the age of Trump. It continues to this day, as reporters bang the drum of "collusion" and obstruction, the paucity of evidence notwithstanding. This has been a perplexing phenomenon, as Attorney General Barr noted in his testimony before the Senate Judiciary Committee at the conclusion of the Mueller investigation:

> How did we get to the point where the evidence is now that the president was falsely accused of colluding with the Russians and accused of being treasonous and accused of being a Russian agent? And the evidence now is that it was without a basis.
>
> Two years of his administration have been dominated by the allegations that have now been proven false. And . . . to listen to some of the rhetoric, you would think that The Mueller Report had found the opposite.[35]

Barr was deeply disturbed by the answers he was getting when he inquired about the reasons why the Trump-Russia investigation had been initiated and the actions by government officials in pursuing a case that had proven to be utterly without merit. The known facts belied the explanations he was hearing. With stunning candor, he stated that "these counter-intelligence activities directed at the Trump Campaign, were not done in the normal course and not through the normal procedures as far as I can tell."[36] That was an understatement. The attorney general decided to launch his own investigation into potential misconduct and lawlessness. More lies and corruption will likely be exposed.

I decided to write this second book, *Witch Hunt*, because a wealth of evidence has emerged since the *The Russia Hoax* went to print in early June 2018. The inspector general at the DOJ issued a highly critical report on how the Clinton email case had been mishandled, if not rigged. More text messages between Strzok and Page surfaced, casting doubt on the legitimacy of the probe and emphasizing the bias that had contaminated the ensuing Trump investigation. The plot to overthrow the president came to light. So, too, did a myriad of details on how the Clinton campaign and Democrats, not Trump, were guilty of "collusion." Prodigious lying and spying came into sharper focus with the release of the Foreign Intelligence Surveillance Act (FISA) warrant applications and reports on the use of undercover agents. The doors were opened on heretofore closed-door testimony, as transcripts were belatedly made public. And, of course, Mueller eventually produced his magnum opus that managed to smear Trump while deflating the phony narrative that the president of the United States was a clandestine Russian agent who hijacked an election. It was folly at its best—or worst.

Attorney General Barr posed the correct question: How could it have ever happened? *Witch Hunt* uncovers the truth about the invented lies and corrupt actions of high officials who abused the power of their positions for political gain. They sought to subvert our rules of law and undermine the democratic process. By their venal acts, they damaged the institutions of American government. And they squandered the nation's trust.

WITCH HUNT

CHAPTER 1

A TALE OF TWO CASES

And damn this feels momentous. Because this matters. The other one did, too, but that was to ensure we didn't F something up. This matters because this MATTERS.

—Text message from Peter Strzok to his lover Lisa Page, comparing the closing of the Clinton case to the opening of the Trump case, July 31, 2016

That text tells me it was all rigged from the very beginning, and it tells me that it is the worst scandal to hit the FBI.

—Author's interview with President Donald J. Trump, Oval Office, White House, June 25, 2019

The United States' policy toward Russia has always been a contentious issue, often propelled by feverish electoral polemics. However, there has been a striking continuity from the Obama to the Trump administrations.[1] Both verbalized outreach and reconciliation early on, only to retreat into an adversarial posture when reality set in. If anything, the current president has demonstrated greater antagonism toward Moscow than his predecessor, who during a 2012 presidential debate dismissed Russia as "the biggest geopolitical threat."[2]

In his first two years in office, Trump imposed a series of new sanctions against Russian government officials and oligarchs, approved punitive measures targeting Moscow's defense and energy sectors, expelled dozens of diplomats, shuttered several ministerial properties, sent lethal weapons to Ukraine to defend itself against Russian aggression, authorized military force against Russian troops in Syria, and initiated withdrawal from the 1987 Intermediate-Range Nuclear Forces (INF) Treaty based on evidence that Moscow had repeatedly violated its terms.[3] He also appointed well-known Russia hawks to top-level positions in his administration.[4] These are hardly the actions of a US president who is a Kremlin sympathizer, much less a furtive Russian agent.

How did the accusation that Trump was in league with the Kremlin transcend conspiracy theorists to become the common mantra of millions of Americans?

We know it began with top officials at the FBI, in the intelligence community, and at the Department of Justice who had reason to damage or destroy Trump. As an outsider to the praetorian ways of Washington, he posed an existential threat to their positions of power. Trump, the candidate, had vowed to "drain the swamp" of those who had wielded outsized influence in government operations with little or no accountability. But the "swamp" did not want to be drained. The prospect of Trump as president represented an ignominious end to their dominion.

Power in the nation's capital can be likened to crack cocaine: it is highly addictive. Those who exert power tend to become dependent on it and crave it. They are rarely inclined to give it up without a fight. The evidence suggests that people such as CIA director John Brennan, Director of National Intelligence James Clapper, Attorney General Loretta Lynch, Justice Department official Bruce Ohr, FBI director James Comey and his phalanx of loyal lieutenants, Andrew

McCabe, James Baker, Peter Strzok, Lisa Page, and others imagined Trump as a menace to their ideas of who should control government. Clinton, by contrast, was their favored candidate. She represented the status quo—the equivalent of a third and, maybe, a fourth term of Barack Obama. The Democratic nominee signified continuity of authority and purpose. Those in the "deep state" would likely keep their jobs. The power of the entrenched would be inexorably extended under a President Clinton; it would be seriously jeopardized if her opponent prevailed.

Something had to be done to stop Trump. Remarkably, the plan nearly worked.

CLEARING CLINTON OF CRIMINALITY

To understand how the swamp normally deals with perjury, obstruction of justice, leaks, and other administrative crimes, we have to start with the mountain of compelling evidence that Hillary Clinton had committed crimes by mishandling classified documents during her four years as secretary of state. Her fate rested squarely in the hands of Director Comey's FBI and Attorney General Lynch's Justice Department. They knew that she had egregiously compromised national security and, in the process, committed a myriad of felonies under the Espionage Act and other criminal statutes.[5]

The tricky dilemma they faced was devising a way to navigate around the facts and the law to clear her of crimes.

Before she was sworn in as the nation's top diplomat, Clinton set up a private email server in the basement of her home in Chappaqua, New York. She didn't just use a personal email account; she had her emails travel through a personal server, hidden from public view by registration under a separate identity.[6] She decided to use that clandestine server to handle *all* of her electronic communications as secretary of state, including the transfer and dissemination of thousands of classified and top secret documents.[7]

State Department rules forbid this because foreign governments and cyberterrorists could readily access such materials using even rudimentary hacking techniques on an unauthorized and unprotected server. The nation's secrets would be jeopardized. Clinton knew that. She had spent eight years as a US senator. As a member of the Armed Services Committee, she had been counseled on classified documents, how to recognize them, and all the ways she must employ government-instituted safeguards to maintain their secrecy. As secretary of state, she received even more extensive indoctrination: classified records were never to be taken home or otherwise stored there.[8]

She absolutely *knew* that if all of her work-related emails were housed on the private server in her home, there would inevitably be innumerable classified documents contained therein. It would be impossible for the nation's chief diplomat to conduct extensive communications without exchanging such classified information. Nevertheless, she *intended* to establish a nongovernmental server and *intended* that it be used exclusively for all of her business as secretary of state. She *intended* that classified records be stored on and transmitted to other people via her unauthorized and vulnerable system. Such willful acts violated 18 U.S.C. § 793(d) and (e) of the Espionage Act,[9] but also a separate and more fundamental law that criminalizes the mishandling of classified documents, 18 U.S.C. § 1924:

> Whoever, being an officer of the United States . . . becomes possessed of documents or materials containing classified information of the United States, knowingly removes such documents or materials without authority and with the intent to retain such documents or materials at an unauthorized location shall be fined under this title or imprisoned for not more than five years or both.[10]

The above language is explicit. She did it *knowingly* because, by her own admission, she read the classified emails she received and sent, and she *intended* that they be retained at the unauthorized location. How many crimes were committed? At the very least, there were 110 violations of the law, representing the number of emails that were classified when Clinton sent or received them on her home system. That was made clear when Director Comey announced his findings on July 5, 2016, as follows:

> From the group of 30,000 emails returned to the State Department, 110 emails in 52 email chains have been determined by the owning agency to contain classified information at the time they were sent or received.[11]

Comey also found "about 2,000 additional emails [that] were 'up-classified,'" meaning they were not classified at the time they were sent. Under a strict reading of the law, Clinton should have been charged for mishandling those documents, too. That was emphasized by the director when he stated, "even if information is not marked 'classified' in an email, participants who know or should know that the subject matter is classified are still obligated to protect it."

Citing several email chains involving top secret communications, Comey further observed, "There is evidence to support a conclusion that any reasonable person in Secretary Clinton's position . . . should have known that an unclassified system was no place for that conversation."[12] Clinton should also have faced numerous conspiracy charges, as well, since she was acting in concert with others who, according to uncovered documents, knew she was using a private account for classified document exchanges and participated in them.[13]

Let's assume for the sake of argument (and in defiance of logic) that Clinton did not act willfully or intentionally but through unimaginable misfeasance or incompetence. She most certainly behaved with reckless disregard for the protection of classified documents. The law calls this "gross negligence," a term that is interchangeable or synonymous with "extremely careless" conduct.[14] At the very least, Clinton's mishandling of classified documents violated the "gross negligence" provision of 18 U.S.C. § 793(f) of the Espionage Act:

Whoever, being entrusted with or having lawful possession or control of any document . . . relating to national defense, (1) through gross negligence permits the same to be removed from its proper place of custody . . . or (2) having knowledge that the same has been illegally removed from its proper place of custody . . . shall be fined under this title or imprisoned not more than ten years, or both.[15]

There is no question that Clinton's mishandling of classified materials was, at the very least, grossly negligent. Indeed, her actions are the definition of reckless or extremely careless conduct. Comey grudgingly conceded that she *might* have jeopardized national security when he stated, "We assess it is *possible* that hostile actors gained access to Secretary Clinton's personal email account."[16] That was not at all accurate. It wasn't just *possible*; it was a *fact* that the FBI surely knew but tried to conceal from the public. Comey even watered down his findings when he deleted the words "reasonably likely" and substituted "possible" to describe how hostile actors might have breached the secretary's system.[17]

Sure enough, information from Clinton's server turned up on the "dark web"—a collection of encrypted websites where both criminals and rogue nations operate. Since the secretary of state was violating regulations by using an unprotected system, it was easily accessed indirectly through a source who was communicating with her.[18] That was what happened. Evidence showed that a Romanian hacker known as "Guccifer" infiltrated Clinton's emails by utilizing a server in Russia.[19] That meant that Russian intelligence likely benefited from the illegal penetration and obtained US classified material, thanks to Clinton's contempt for rules and the law. Among the hacked records was an Excel spreadsheet containing "targeting data" that would constitute top secret information. "It is inescapable that a security breach and a violation of basic server security occurred here," according to an independent review contained in FBI documents that came to light three years after Clinton was cleared of wrongdoing.[20] Yet when Comey absolved the secretary, he tried to minimize the significance of any national security breach, the very reason laws were passed criminalizing the kind of conduct engaged in by Clinton.

Having presented at a national news conference an overwhelming case of how Clinton had committed more than a hundred crimes, the FBI director offered this bizarre and incomprehensible reason why she would not be prosecuted:

> Although there is evidence of potential violations of the statutes regarding the handling of classified information, our judgment is that no reasonable prosecutor would bring such a case.[21]

COMEY TWISTS THE FACTS

Having found a plethora of evidence that laws were potentially broken, as he stated unambiguously in the first part of his sentence, Comey was duty bound to tender a criminal referral to the Justice Department that Clinton be prosecuted. The unambiguous facts, in combination with the law, demanded it. That was more than legally sufficient for presentment to a grand jury that would most certainly have rendered an expansive indictment.

However, Comey's qualifying phrase that no "reasonable prosecutor would bring such a case" amounted to pure speculation by him. It is not, and never has been, a valid legal basis for declining to levy charges. Comey invented a legal standard that does not exist. He was not the prosecutor. His job was to gather evidence through documents and witnesses. Yet he deigned to anoint himself the sole authority on whether criminal charges would be brought. In doing so, he flagrantly commandeered the power of the attorney general and violated FBI and Department of Justice regulations in the process.

Comey was also just plain wrong when he boldly declared that "no reasonable prosecutor would bring such a case." In truth, prosecutors had brought several similar cases against government officials who had mishandled classified information in much the same way that Clinton did, including military convictions. They included former national security advisor Samuel "Sandy" Berger, former CIA directors David Petraeus and John Deutch, former national security contractor Harold T. Martin III, navy engineer Bryan Nishimura, and navy sailor Kristian Saucier.[22] Comey's assertion that "we cannot find a case that would support bringing charges on these facts" was demonstrably untrue.[23]

A subsequent review of Comey's decision making was conducted by the DOJ's inspector general, Michael Horowitz, and released in June 2018. He determined that the director's unilateral actions in clearing Clinton were both "extraordinary and insubordinate," concluding that he had "usurped the authority of the Attorney General and inadequately and incompletely described the legal position of Department prosecutors." In other words, Comey had no business acting as a prosecutor who terminated the case against Clinton. "We did not find his justifications for issuing the statement to be reasonable or persuasive," wrote Horowitz.[24]

Those same reasons were cited by the Justice Department when it eventually recommended that Comey be fired ten months later on May 9, 2017: he had acted without authorization and in dereliction of his duty to follow established policies and regulations. This was a view "shared by former Attorneys General and Deputy Attorneys General from different eras and both political parties."[25] Comey's obstinate refusal to admit his errors only reinforced the need to fire him. He demeaned the work of the agency he led, damaged the integrity of the nation's premier law enforcement agency, and breached the public's trust.

Comey's hubris and bias led him to twist the facts and contort the law to absolve Clinton. He took it upon himself to assume the authority that rightly belonged to others. Or, as the inspector general found, he "engaged in his own subjective, ad hoc decision-making."[26] Comey's maladroit behavior was so acute that he should have been fired the same day he stood in front of television cameras to exonerate

Clinton. But, of course, President Obama took no such action—an obvious indication of his tacit approval of Comey's decision to clear the path for the candidate that Obama would endorse to succeed him in the Oval Office.

How do we know? In April 2016, the president sat down for an interview with *Fox News Sunday* host Chris Wallace. On national television, Obama insisted that Clinton had not jeopardized national security but had merely been "careless" in her mishandling of classified documents.[27] That was a thinly veiled directive to Comey and the FBI that the president of the United States did not want criminal charges brought against his former secretary of state, incriminating evidence notwithstanding. Obama made the same statements in two other television interviews. We now know that that message was received by Comey loud and clear. The former director confessed to it during questioning by the IG:

COMEY: [P]resident [Obama's] comments obviously weighed on me as well. You've got the President who has already said there's no there there. . . . And so all of that creates a situation where how do we get out of this without grievous damage to the institution?[28]

The salient phrase is "how we get out of this?" That is, how could Comey thread a legal needle that would disregard evidence inculpating Clinton and circumvent the law that would ordinarily result in charges, all the while maintaining some semblance of credibility at the FBI without tarnishing its vaunted reputation? With his marching orders from Obama, the FBI director proceeded to disfigure the facts and adulterate the relevant criminal statutes. He rewrote federal law and, in the process, literally rewrote his original findings that Clinton had been "grossly negligent."

The disparate treatment of the Trump-Russia investigation versus the Clinton email investigation is exemplified by Comey's admission to the IG. The director suggested that his decision was motivated by the wishes of Obama. Yet Comey never accused Obama of attempting to influence or obstruct the FBI's investigation. However, when Trump allegedly remarked that he "hoped" fired national security advisor Michael Flynn would be cleared by the Bureau, Comey later told Congress and the public, he interpreted it to be an attempt by Trump to influence or obstruct the FBI's investigation.[29] When Obama comments about a pending case, it's perfectly all right in Comey's sphere. When Trump purportedly does the same, it's a felony. Not only did the director's bias influence his views and actions toward the two presidents he served, but that same prejudice motivated him to distort the law to disadvantage one over the other.

The standard Comey and his colleagues set is that a politician must be excused for anything less than a calamitous betrayal. It would be wrong to investigate a major political figure and prosecute him or her—or any of his or her aides—without incontrovertible evidence of evil intent and significant negative consequences.

Clinton violated the law. Comey and others at the FBI and DOJ knew it. During my interview with President Trump on June 25, 2019, he spelled it out in blunt terms: "Comey kept her out of jail."[30]

COMEY SANITIZES HIS STATEMENT

One of the more stunning revelations contained in the inspector general's report is that Comey claimed he did not remember the moment he decided—and reduced to writing—that Clinton had committed crimes.

On or about May 2, 2016, Comey composed a statement summarizing Clinton's mishandling of classified documents, concluding that she had been "grossly negligent."[31] As noted earlier, those pivotal words have a distinct legal meaning because they are drawn directly from the Espionage Act. In describing Clinton's actions, Comey used the exact phrase not once but *three times* (Exhibit C in IG report). Under questioning, he readily admitted to Horowitz that he had authored the May 2 statement and penned every word of it himself. Then he offered the implausible claim that "he did not recall that his original draft used the term 'gross negligence,' and did not recall discussions about that issue."[32]

Comey's amnesia is preposterous. He participated in subsequent discussions with top officials at the FBI about Clinton's "gross negligence" and how to creatively alter the language to sidestep an indictment. Email discussions and meetings were held on the thorny subject, and contemporaneous notes and electronic evidence obtained by the IG prove that Comey was in attendance and intimately involved.[33] Those records show that although Comey had determined that Clinton had been "grossly negligent" in violation of the law, he resolved to clear her notwithstanding. To achieve that remarkable somersault and absolve the soon-to-be Democratic nominee, the legally damning terminology would have to be stricken from his statement.

Metadata show that on June 6, the FBI's lead investigator on the case, Peter Strzok, deputy assistant director of counterintelligence, sat down at his office computer to cleanse his boss's statement of the vexing term "gross negligence." With the assistance of his paramour and FBI lawyer, Lisa Page, the words "extremely careless" were substituted to make Clinton *appear* less criminally culpable.[34] Page told the IG that "to use a term that actually has a legal definition would be confusing."

It most certainly would. After all, how could Clinton be exonerated under the "gross negligence" provision of the Espionage Act if that very phrase was used to describe her behavior? The two phrases are indistinguishable and synonymous in the law, but only one appears in the statute. It was a clever feint of semantics: create the appearance that Clinton had barely skirted the law, even though she had trampled on it.

Strzok and Page also expunged from Comey's statement his reference to another statute that Clinton had plainly violated. In his original statement, Comey determined that "there is evidence of potential violations" of 18 U.S.C. § 1924 [cited earlier], which makes it a crime to retain classified information in an unauthorized place.[35] Clinton's home was not authorized to house classified records, and her private nongovernmental server was unprotected. She knew it because she'd been instructed accordingly during a comprehensive security briefing. She had affixed her signature to two documents acknowledging that she understood the law and the penalties.[36] But Comey's finding of that crime was also completely expurgated from his public announcement, likely by the same people who removed the other incriminating findings. Having cleansed the director's statement twice, the FBI wasn't yet finished with its sanitation project.

One of the more damaging conclusions drawn by Comey in his initial statement was this sentence: "The sheer volume of information that was properly classified as Secret at the time it was discussed on email . . . supports an inference that the participants were grossly negligent in their handling of that information."[37] That one sentence framed a damning indictment of Clinton. She had mishandled *so many* secret and protected documents that she *must* have known she was violating the law with impunity. However, the director saw to it that that conclusion was also purged in its entirety from his statement when he absolved Clinton.

Though this may seem to some like the distant past, it's essential to understand the standard the FBI publicly set for investigating a presidential candidate.

In a confidential email discussion with his FBI colleagues, E. W. "Bill" Priestap, the assistant director of counterintelligence, was blunt in his assessment of their predicament. He cautioned that it was "important for the Director to more fully explain why the FBI can, in good faith, recommend to DOJ that they not charge someone who has committed a crime (as defined by the letter of the law)."[38] In fact, there was no "good faith" way of explaining how someone who had broken the law could be relieved of the consequences. The scheme the FBI settled on was wrapped in bad faith: it would recast the letter of the law to achieve an unjust, but politically expedient, result.

Where did Comey and others come up with the term "extremely

careless" to dilute Clinton's felonious conduct? From the president himself. Recall that Obama had employed the term "careless" in his meticulously worded public remarks just a month earlier and in two other televised statements. That, coupled with the director's admission that the president's "comments weighed on me" provides the inevitable and unmistakable answer: Comey and his subordinates did what Obama wanted them to do. They cleared Clinton of three sets of crimes that should have amounted to more than a hundred felony charges.

Though the FBI director insisted that he had no memory of writing the words that should have indicted Clinton, he claimed to have remarkable recall of the little-known history of the Espionage Act. He informed the IG that he thought "Congress intended for there to be some level of willfulness present even to prove a 'gross negligence' violation." [39] In other words, he argued that "intent" was required under the law. But it is not. That tortured interpretation by Comey is patently and provably untrue.

If Comey had honestly read the legislative history, he would have learned that in 1948, Congress amended the original Espionage Act of 1917 to add a new "gross negligence" provision that did *not* require intent or willfulness. [40] Lawmakers plainly and deliberately omitted that. Indeed, eliminating the necessary element of "willfulness" was the whole purpose of modifying and expanding the statute. During World War II, Congress had come to realize that there were increasing instances of government and military officials becoming complacent. All too frequently, they had personally kept and handed out classified material cavalierly and to the detriment of national security. Congress sought to remedy the problem by establishing a new category of crime. Willfulness need no longer be a legal requirement of proof. Grossly negligent behavior provided a lesser alternative to willful conduct in presenting a case under the Espionage Act. Thus, Comey applied a legal standard to the Clinton case that did not exist. He just made it up. He read "intent" into the statute after Congress had removed it.

Amnesia must be contagious at the FBI. Testifying before Congress in July 2018, Strzok claimed to have no recollection of using his computer to make the critical alteration that cleared Clinton. He did, however, directly implicate the FBI director when he asserted, "Ultimately, he [Comey] made the decision to change that wording." [41] Curiously, Strzok recalled that his boss had ordered him to change words he didn't remember changing. But the critical alterations clearing Hillary were made.

THE FBI'S BLATANT BIAS IN FAVOR OF CLINTON

Comey's actions in absolving Clinton were roundly condemned by the inspector general, citing misjudgments, bias, insubordination, and unprofessionalism. But many of his top lieutenants were guilty of the same. Notes and emails show that it was a collective endeavor to sanitize the director's initial statement. Clinton was the beneficiary of what FBI deputy director Andrew McCabe described as the "HQ special"—that is, special status at the Bureau's headquarters. [42] Comey and his deputies handled the case instead of agents at the Washington, DC, field office. That departure from normal procedures allowed the case to be massaged in a way that would achieve the preconceived outcome they desired. Had the field office investigated and managed the case, Clinton almost certainly would have been indicted for her criminal acts.

Though Comey publicly maintained that the Clinton "matter," as Attorney General Lynch insisted it be called, [43] was managed in an apolitical and professional manner, private text messages exchanged between Strzok and Page contradicted this. Both individuals played pivotal roles in clearing Clinton. They were neither fair nor impartial. Their texts, uncovered by the IG, were replete with adoring compliments of the very woman they were supposed to be investigating. They lauded Clinton's nomination and stated, "God, she's an incredibly impressive woman" and "She just has to win now." [44]

But the fix was in several months before Clinton secured the nomination. As she was still being investigated and more than four months before the FBI would even interview her, Page predicted that Clinton would become president. Of course, that would have been impossible if the FBI had recommended a criminal indictment. That was not what top officials at the agency intended to do. In a text to Strzok on February 24, 2016, Page warned him that any aggressive tactics would backfire on the Bureau once they absolved Clinton and she became president:

PAGE: She might be our next president. The last thing you need [is] us going in there loaded for bear. You think she's going to remember or care that it was more DOJ than FBI?
STRZOK: Agreed.[45]

It's worth underlining how unusually naked the concern is here. It's easy to suspect that Clinton benefited from special treatment because agents feared retribution, but it's astonishing to see that someone put it in writing.

At the same time, a similar message was sent to McCabe and another FBI official. Page, Strzok, and others at the Bureau were operating their investigation on the assumption or, more likely, advance knowledge that Clinton would never be charged. That was reinforced by another text sent by Strzok on March 3, 2016, when he wrote, "God, Hillary should win 100,000,000–0."[46] Pause for a moment to consider what that message meant. Those two key FBI officials were prophesizing a Clinton presidency that could happen only if they first ensured that she would escape criminal charges.

Many members of Congress were outraged over the Strzok-Page texts when they were discovered and released by the inspector general. They demanded answers, but the FBI stonewalled efforts to force Strzok and Page to testify. Under intense pressure and repeated demands, both of them submitted to closed-door interviews in the summer of 2018. The transcripts, however, were kept private until March

2019. The record shows that Representative Trey Gowdy (R-SC) confronted Strzok as follows:

> **GOWDY**: You had her running and winning before you had concluded the investigation . . . before you even bothered to interview her. That's what we're left with.[47]

Strzok had no real answer, except to assert that his personal beliefs had not impacted the investigation. That is an implausible claim, given all of the other evidence showing that Comey and his confederates dismissed incriminating facts and rewrote the meaning of criminal statutes to reach their improbable result. In the absence of an outright admission or some other confessional document, the IG was unable to declare definitively that bias had driven the FBI to clear Clinton. Horowitz did state that it had been "potentially indicated."[48]

The confidential testimony of both Strzok and Page might have remained secret but for the decision by Representative Doug Collins (R-GA) to take matters into his own hands. As the ranking member of the House Judiciary Committee, he felt authorized to release them to the public in March 2019. The transcripts provided stunning new details of how Obama's Justice Department, with Lynch at the helm, had actively intervened to protect Clinton.

Strzok revealed that the FBI's access to Clinton's personal servers, including emails related to the Clinton Foundation, had been restricted by the DOJ during the probe. He stated, "We did not have access. My recollection is that the access to those emails were based on consent that was negotiated between the Department of Justice attorneys and counsel for Clinton."[49] To a great extent, the FBI's investigation seems to have been controlled by Clinton's allies at the DOJ and, perhaps, by the White House. Deals were struck with Clinton's lawyers, her computer server was largely kept out of reach, and immunity agreements were given out to her closest associates like party favors.

Attorney General Loretta Lynch and President Obama both

claimed they had never interfered with the FBI's investigation. But Strzok undermined those claims when he was asked about how the language in Comey's exoneration statement had been changed from "gross negligence" to "extremely careless." He said, "My recollection is attorneys brought it up, and these, of course, were DOJ attorneys." Page concurred that the DOJ had interfered when she was questioned by Representative John Ratcliffe (R-TX):

> **RATCLIFFE**: You're making it sound like it was the Department [of Justice] that told you you're not going to charge gross negligence because we're the prosecutors and we're telling you we're not going to—
> **PAGE**: That's correct.
> **RATCLIFFE**: —bring the case.[50]

It had long been suspected that Lynch, as head of the Department of Justice, had seen to it that Clinton was shielded from prosecution. On July 1, 2016, Page texted Strzok that "she [Lynch] knows no charges will be brought." [51] Clinton had not yet been interviewed by the FBI, but her exoneration was predetermined. Her answers were a superfluous exercise. The investigation was a sham.

On the day he absolved Clinton, Comey made a point of saying "I have not coordinated or reviewed this statement in any way with the Department of Justice or any other part of the government. They do not know what I am about to say." [52] Yet according to Strzok and Page, the DOJ knew exactly what he was going to say because the department had ordered Comey's FBI not to pursue a charge against Clinton under the most likely legal avenue—the "grossly negligent" provision of the Espionage Act.

The closed-door testimony of Page leaves no doubt that the FBI had made up its mind to clear Clinton long before she and most of the key witnesses surrounding her mishandling of classified documents were even interviewed. Page admitted that "Every single person on the team, whether FBI or DOJ, knew far earlier than July that we were not going to be able to make out sufficient evidence to charge a crime."[53] She indicated that that decision had been made perhaps as early as March 2016. That is a remarkable admission of how the process was corrupted. A conclusion was reached before much of the evidence was properly examined. No one could know Clinton's intent or knowledge without asking her, as well as more than a dozen other people who had been involved. Clinton's interview was a mere formality. The investigation was a farce.

Though the FBI consistently denied that political bias influenced its decision, evidence shows that the Bureau was acutely aware of the political necessity of clearing Clinton before the Democratic National Convention, which was set to begin on July 25, 2016. That was something that should never have factored into the decision-making process. Despite that, it seems to have been a guiding principle that dictated the course of the probe. At the Bureau, there was a rush by Comey and others to liberate the expected nominee of the investigation's criminal cloud, as reflected in an illuminating text that betrayed the FBI's facade of political indifference. When Trump secured the Republican nomination, the following exchange took place:

> **STRZOK**: I saw Trump won, figured it would be a bit.
> **STRZOK**: Now the pressure really starts to finish MYE [Midyear Exam].
> **PAGE**: It sure does. We need to talk about follow up call tomorrow. We still never have.[54]

"Midyear Exam" was the innocuous-sounding name the FBI had attached to the Clinton probe, but the text makes explicit that the Bureau was anxious to conclude it in time for Clinton to accept her party's nomination. That exchange of messages occurred a full two months before she was even interviewed by the FBI, along with a dozen other key witnesses in the case who had yet to be questioned.

Sure enough, within weeks the necessary changes were made in Comey's statement to clear Clinton. A brief interview with Clinton—

not under oath—was scheduled for July 2, 2016, and Comey's exoneration would be announced three days later.

In the interim, the candidate's husband, former president Bill Clinton, met furtively with Lynch on June 27, 2016, inside the attorney general's plane, which was parked on the tarmac of Sky Harbor International Airport in Phoenix, Arizona.[55] That was a scant five days before the former secretary of state was to be questioned by the FBI and thereafter cleared of any wrongdoing. There is no known written account of what exactly was said during the private discussion. Lynch refused to offer any details, and both parties dismissed it as nothing more than a "primarily social" interaction.[56]

Remember the standard the DOJ was attempting to establish here: meetings that appear nefarious should be given a pass if the participants say that nothing untoward happens. Apply that standard to the Russia hoax, and there's nothing left to investigate.

Consider the meeting in its broader context: The husband of the subject of a criminal investigation was meeting secretly with the one person who could ensure that charges would not be brought against his wife. Was the ex-president assured in this meeting that his spouse would face no legal obstacle in her quest to win the presidency? Was any pressure brought to bear to secure the equivalent of a "get out of jail free" card?

It cannot be overlooked that Lynch owed her career to Bill Clinton. She was required under federal regulations to recuse herself from the government investigation because of both a personal and a professional relationship with the spouse of its subject.[57] She refused to recuse herself, saying only that she would accept the recommendations of the FBI. In the process, she was abdicating the Justice Department's legal responsibility while improperly delegating it to Comey's FBI. Neither was ever going to allow Clinton to be prosecuted. Lynch's motivation in protecting the target of their investigation was obvious: if Clinton were to be cleared and won the presidency, the attorney general might well have maintained her coveted position of power as head of the Justice Department.

As the investigation progressed in name only, Clinton was furiously attempting to cover up what she had done by peddling a successive series of arrant deceptions. At first, she claimed, "I did not email any classified material to anyone on my email. There is no classified material."[58] When evidence surfaced to the contrary, she changed her story to say, "I never sent or received any information that was classified *at the time* it was sent or received."[59] As more facts emerged directly contradicting this unlikely claim, Clinton altered her story for a third time by asserting "I never sent or received any email that was *marked* classified."[60] That, too, proved to be untrue. Many of the emails had been marked classified. Even Comey called that statement untrue.[61]

Then Clinton altered her story yet again by proclaiming that she had not realized that the parenthetical "C" meant classified material at the confidential level when it appeared on documents.[62] As preposterous as that sounded, she was effectively arguing her own incompetence. Besides, the markings were irrelevant under the law, since the content—not the markings—made them classified. Under the law, it was no defense for Clinton to claim she had not *known* that certain matters were classified. Finally, she abandoned all of those excuses and resorted to the blanket assertion that "Everything I did was permitted by law and regulation."[63] It was not.

Clinton established an unauthorized private communications system, used it to house thousands of classified government records, left it vulnerable to theft by hacking, and jeopardized national security. That is precisely what the law forbids. There is evidence that she broke other laws, including obstruction of justice statutes, by destroying more than 30,000 documents subject to a duly authorized subpoena issued by Congress, as well as orders for the preservation of documents.[64] If anyone else had done that, he or she would most assuredly have been prosecuted for crimes committed.

THE FBI'S BIAS AGAINST TRUMP

Once Clinton was absolved, the FBI immediately turned its full attention to the political opponent who was the only remaining obstacle in her path to the White House. That was the moment the Bureau began its Trump-Russia investigation in earnest. The pretext for the probe was an unfounded accusation that Trump was "colluding" with Russia to win the presidential election. Thus, as one case ended, the other began. It literally happened on the same day that the FBI made its stunning announcement that the soon-to-be Democratic nominee was free and clear.

As Comey stood in front of television cameras on July 5, 2016, his FBI was meeting secretly in London with Christopher Steele, the author of the fictitious anti-Trump "dossier" that was funded by Clinton's campaign and Democrats. That document, together with many other unfounded accusations, would be exploited in a malicious attempt to frame Trump for unidentified crimes he did not commit.

The genesis of the new investigation will be explained in greater detail in the next chapter. But within two weeks, Strzok was on a plane to London to mine intelligence information and sources there, including confidential informants. Armed with a new case that might damage or destroy the man he so openly loathed in his text messages, he seemed giddy with excitement at the prospect. In a message to Page on July 31, 2016, the day papers were signed at the FBI to officially launch the case against Trump, he compared the dismissed Clinton case to the burgeoning Trump case:

> Damn this feels momentous. Because this matters. The other one did, too, but that was to ensure that we didn't F something up. This matters because this MATTERS.[65]

That one text, more than any other, exemplifies the way the FBI's personal and political opinions of Trump innervated the Bureau's desire to pursue him, even in the absence of any plausible evidence that he had done something wrong. In her private testimony, Page admitted that at that point the FBI had had almost nothing to go on.[66] Her lover, Strzok, didn't care. He was determined to use the full force and power of the FBI to investigate Trump. No resource would be spared. More than nine months later, after the special counsel was appointed, Page confessed that evidence of "collusion" with Russia was still unproven. "We still couldn't answer the question," she said.[67] Comey also confirmed that startling admission when he told Congress in private sessions that there had been almost no evidence of "collusion" when his investigation began and little had changed by the time he was fired and Mueller was appointed.[68]

But to Strzok and others at the FBI, only the Trump case really mattered. The Clinton case, by comparison, didn't. This is hard to comprehend inasmuch as Clinton had actually jeopardized the nation's security secrets by her flagrant mishandling of thousands of classified documents—materials that might have included military and economic secrets, terrorism, crime, energy security, and cybersecurity. Foreign powers likely gained that valuable information. There was overwhelming proof that Clinton had done so in violation of several felony laws. Nonetheless, Strzok seemed convinced that that somehow paled in comparison with his obsessive belief in a nonexistent plot that Trump had "colluded" with the Kremlin.

To some extent, it is understandable how Strzok and others never considered the Clinton "matter" a real and legitimate investigation. How could they? The decision was made well in advance that she would never face prosecution. But the Trump investigation was something that "MATTERS." Fueled by partisan hostility, stopping Trump became their top priority. Covering their tracks after the wrongful exoneration of Clinton was all that mattered. The inspector general would later conclude that he "did not have confidence that Strzok's decision to prioritize the Russia investigation . . . was free from bias."[69]

When he testified before Congress, Horowitz averred that Strzok's texts had "clearly showed a biased state of mind" and that that was "antithetical to the core values of the department and extremely serious." He added, "I can't imagine FBI agents even suggesting that they would use their powers to investigate any candidate for office."[70] But that was precisely what happened.

Strzok wasn't just *suggesting* he could use his powers against Trump, he was vowing to do so. Having solved the Clinton conundrum by cleansing Comey's statement, he was rewarded with a promotion as he took command of the Trump case. A week after it was formally launched, the two paramours exchanged a message that crystallized their intent to bring down Trump, whom they referred to as a "menace":

> **PAGE**: And maybe you're meant to stay where you are because you're meant to protect the country from that menace.
> **STRZOK**: Thanks. And of course I'll try and approach it that way. I just know it will be tough at times. I can protect our country at many levels, not sure if that helps.[71]

The newly empowered Strzok envisioned himself as the FBI's superagent who would "protect" the country from a dangerous and harmful Trump presidency at all costs. In their private interviews with Congress, Strzok brazenly denied that their reference to "menace" had meant Trump. Page, the one who had employed the word in her text, was more forthright. She admitted they had been conversing about Trump. He was the perceived menace.

Two days later, Page became distressed at Trump's rising status in the polls and precipitated another text exchange about the possibility that their favored candidate, Clinton, might lose her bid to become president:

> **PAGE**: He's not ever going to become president, right? Right?!
> **STRZOK**: No. No he's not. We'll stop it.[72]

Strzok was now the lead FBI agent in charge of the one-week-old Trump-Russia "collusion" investigation. When asked about what he had meant, he told Congress that he did "not recall" composing the infamous text vowing to stop Trump from being elected president.[73] But then he tendered this explanation: "What I can tell you is that text in no way suggested that I or the FBI would take any action to influence the candidacy." That was a remarkably dexterous justification for something he did not remember doing. When confronted with dozens of other messages extolling Clinton and disparaging Trump, Strzok had the temerity to say, "I do not have bias." Later, he claimed, "Those text messages are not indicative of bias."[74]

No one could read through the multitude of incendiary Strzok-Page messages without recognizing their strident political agenda and personal bias against the very man they were investigating.

In those texts we see the mind-set that led to the greatest mass delusion in American political history. Without any factual basis, without anything more than a perverted kind of wishful thinking, two people who should have known better decided that only a vast international criminal conspiracy could have led to President Trump's election.

Consider the cryptic text of this August 15 message to Page just two weeks after Strzok signed papers formally launching the probe:

STRZOK: I want to believe the path you threw out for consideration in Andy's office that there's no way he gets elected— but I'm afraid we can't take that risk. It's like an insurance policy in the unlikely event you die before you're 40.[75]

"Andy" was Andrew McCabe, the deputy director of the FBI. He, too, suffered an acute bout of memory loss and claimed he had no recollection of the meeting in his office.[76] Page, on the other hand, was more scrupulous and candid. The "insurance policy," she confirmed, was the FBI's then-secret Russian "collusion" investigation of Trump.[77] Under their plan, it would be quietly investigated by the

Bureau but held in abeyance unless and until the "unlikely" event occurred—the election of Trump. That was especially devious, since she admitted that there was a "paucity" of evidence, meaning the Bureau had almost nothing at all.[78]

In truth, there wasn't even sufficient evidence to open an investigation under FBI guidelines, as will be explained later.[79] One was launched anyway. At the time, the phony Steele "dossier" was about all the agents had to go on. It was utterly unverified and suspect on its face. But that didn't seem to matter, because the FBI assumed that Trump would never win. If he defied all expectations and prevailed on election day, it would then kick its probe into overdrive. The spurious allegation that Trump was a Russian asset would be pursued with vigor. In the minds of officials there, Trump was a threat only if he became president. The chimera of "collusion" might then be used against him to undo his presidency. It was the FBI's version of an "insurance policy" against the risk of Trump.

Numerous texts between Strzok and Page show a stunning hostility toward the man they were investigating. They called him "awful," "loathsome," a "disaster, a "f***ing idiot," an "enormous do*che," and other disparaging names that were laced with incandescent profanity. Republican supporters were smugly branded as "retarded," "the crazies," and "ignorant hillbillies" who "SMELL."[80] Once Trump assumed office, Strzok wrote that his investigation of the president could be used to impeach him.[81] Strzok and Page used FBI-issued phones to exchange more than 50,000 texts to each other, many of them while they were at work. Those toxic rants underscored an abiding enmity toward Trump that poisoned any chance that the FBI's investigation would be neutral, objective, and fair. Attorney General William Barr agreed:

> Well it's hard to read some of the texts and not feel that there was gross bias at work and they're appalling. . . . Those were appalling. And on their face they were very damning and I think if the shoe was on the other foot we could be hearing a lot about it.[82]

Just as Strzok and Page played a pivotal role in driving the "collusion" case against Trump, they were also a driving force in clearing Clinton. Their text missives were replete with adoring compliments of the Democratic candidate, lauding her accomplishments and predicting how she would coast to victory in the election.[83] Just two weeks after the FBI exonerated Clinton, Page was celebrating her nomination when she wrote, "Congrats on a woman nominated for President in a major party! About damn time! Many, many more returns of the day!" Later, she warned, "We do not want this election stolen from us."[84]

That text shows how just how deeply invested the FBI was in helping Clinton evade prosecution. Comey, Strzok, Page, and others had worked hard to bury evidence and shield her from legal jeopardy. They did not want an outsider like Trump to sabotage their labors by "stealing" the election from them. They had a greater attachment to their own power than to the public good. Trump was a threat to them.

In the summer of 2016, the FBI officials who had overseen both cases *assumed* that Clinton would win the presidency, and their actions were dictated accordingly. However, on September 26, 2016, a communication from a New York FBI agent to Washington headquarters threatened to derail that goal. Hundreds of thousands of emails had been discovered on the laptop of former representative Anthony Weiner, who was married to Clinton's close aide Huma Abedin.

The New York office immediately advised Comey's deputy, Andrew McCabe, that many classified documents were among them. The Clinton case would have to be reexamined. Instead of taking prompt action to reopen the case based on the new evidence, the Washington headquarters of the FBI ignored it entirely and did nothing. In his self-serving book *The Threat: How the FBI Protects America in the Age of Terror and Trump*, McCabe said he *assumed* that others in the Bureau's counterintelligence unit would take control of the laptop and examine the emails.[85] Right. In his subsequent report, the inspector general seemed baffled as to why no immediate action had been taken. Under questioning, agents and officials offered both conflicting and nonsensical explanations. It appears that senior leadership tried to bury the problem, hoping it would somehow disappear. A month later, and only under repeated pressure from the New York office, Comey belatedly and reluctantly reopened the Clinton investigation.[86]

The IG concluded that Strzok and others at FBI headquarters no longer cared about the Clinton case they'd dismissed because pursuing Trump was their main priority.[87] Having been forced to reopen the investigation of Clinton's emails, Comey moved at breakneck speed to close it. Naturally, Strzok was the point man. He cut off all contact with the New York FBI and took control.

According to an in-depth investigation by journalist Paul Sperry, the FBI never actually examined the vast majority of the emails found on Weiner's laptop.[88] This despite the assurance Comey gave in writing to Congress that the Bureau had "reviewed all of the communications."[89] That was untrue. Sperry found that "Only 3,077 of the 694,000 emails were directly reviewed for classified or incriminating information."[90] Among those, *classified* documents were discovered. Once again, Comey and Strzok shuttered the case against Clinton, nine days after reopening it. The director claimed that no new evidence had emerged. In truth, Comey could not possibly have known that because only a fraction of the newly discovered material had been studied. Abedin, of course, was never charged.

It was fatuous for Strzok to adamantly deny bias in the face of such graphic and overwhelming evidence to the contrary. In the conclusion of his five-hundred-page report, the inspector general stated that the Strzok-Page communications "are not only indicative of a biased state of mind but imply a willingness to take official action to impact a presidential candidate's electoral process."[91] Although Horowitz said he could find no *direct* or *testimonial* evidence that such pervasively severe bias had affected investigation decisions, that seems to have been a deflection of responsibility. No one puts into writing a smoking-gun admission such as "Let's clear Clinton for political reasons even though she committed crimes" or "We should contrive a false case against Trump to influence voters." The proof was self-evident.

THE BOTCHED CLINTON INVESTIGATION

The totality of evidence and the disparate handling of the Clinton and Trump cases provides more than sufficient evidence that unconstrained bias drove Comey and his cadre of aides to treat Clinton with extraordinary deference but to target Trump with a vengeance. Consider the titles they chose for their respective probes. Clinton's was innocuously called "Midyear Exam." Trump's was labeled with the ominous-sounding "Crossfire Hurricane."[92] One outcome was predetermined; the other was preconceived.

Other notable departures from normal procedures served up glaring red flags of selective prosecution and unequal justice. Five people close to Clinton were given immunity agreements in exchange for nothing of any value. They seemed to have been given a free pass from any legal jeopardy for crimes that some, if not all, of them certainly appeared to have committed.[93] In return, those individuals provided no incriminating evidence that resulted in anyone else being prosecuted, which is the customary practice when granting immunity.

Two of the immunized fact witnesses, Clinton's chief of staff, Cheryl Mills, and senior adviser, Heather Samuelson, were even allowed to accompany Clinton during her remarkably brief July 2, 2016, interview with the FBI. That was not only irregular but highly improper, unethical, and probably illegal.[94] Clinton's interview lasted all of three and a half hours, during which time the former secretary of state replied, "I do not recall" or "I don't remember" some thirty-nine times.[95] She was not placed under oath and was cleared by Comey three days later, solidifying the charade. Strzok participated in the softball questioning.

Juxtapose with that the treatment of Trump associates. Several were prosecuted for relatively minor "process crimes." Instead of immunity deals, they were threatened with prison unless they said something incriminating about Trump. Teams of FBI agents swarmed their homes with guns drawn, seized their property, arrested them on the spot, and charged them with making false statements. Some were charged with more serious offenses unrelated to Trump or his campaign.

Clinton associates who were suspected of lying were treated with favor. Huma Abedin and Cheryl Mills, two of Clinton's closest advisers, told FBI agents that they had *never known* their boss was using a private, unsecured server until *after* she had left office.[96] That was demonstrably untrue. Emails show that they certainly knew and even discussed her unauthorized computer system at the time it was in use.[97] Justin Cooper, Clinton's technology expert who set up her secret server, testified that Abedin had approached him to do the work and helped him create the system.[98] Additionally, in her FBI interview, Abedin denied knowing anything about hacking attempts on her boss's server.[99] Yet emails show her reacting with alarm when Cooper informed her of such attempts, writing the response "omg" (shorthand for "Oh, my God!").[100] Without a doubt, Abedin knew about the multiple hacking efforts; there were simply too many emails that belied her claims. Her statements to the FBI were untrue. She had also "dawdled" for twelve hours before notifying Mills and other staff members not to use the compromised server for sensitive communications.[101] Naturally, neither Abedin nor Mills was ever charged with making materially false statements, unlike Trump associates.

Federal judge Royce C. Lamberth, presiding over a lawsuit involving Clinton's server, offered this scathing critique of Mills:

I was actually dumbfounded when I found out, in reading the I-G report, that Cheryl Mills had been given immunity because . . . I had myself found that Mills had committed perjury and lied under oath in a published opinion I had issued in a Judicial Watch case where I found her unworthy of belief, and I was quite shocked to find out she had been given immunity by the Justice Department.[102]

In a different email case involving Mills, Judge Lamberth described her conduct as "loathsome."[103]

Why were no charges brought against Abedin, Mills, and three other Clinton aides who appear to have committed crimes? When asked that question by Congress, Comey chalked it up to failed recollections that should be forgiven.[104] No one ever extended such benefit of the doubt or forgiveness to a Trump associate. Indeed, it appears that the FBI elected to *help* those close to HRC by "agreeing to destroy records and laptops of Clinton associates after reviewing them."[105] Since when does the FBI engage in the business of destroying evidence in a case that was subject to review by the Justice Department and was, at the same time, being investigated by Congress? In essence, the Bureau became an active participant in a conspiracy to obstruct justice and obstruct Congress. All of that happened under the toxic stewardship of James Comey.

It can be no coincidence that it was the very same small set of agents and officials that investigated both the Clinton and Trump cases. The normal practice of having field offices handle the probes was subverted. FBI headquarters, under Comey's direction, commandeered all of the investigations. Was that because he knew that most of the FBI was incorruptible—that the vast majority of agents were good and honorable people who would have processed the cases impartially and without favor?

Had field office agents been given the chance to gather the evidence and interview the witnesses in the Clinton case, she would surely have been referred for criminal prosecution, given the abundance of evidence. Conversely, given the paucity of credible evidence that Trump or anyone in his orbit had "colluded" with Russia, it is likely that any preliminary investigation by a field office would have crumbled and vanished for lack of probable cause or, as FBI regulations require, an "articulable factual basis for the investigation."[106] Regular agents would never have launched a formal investigation in search of a crime, reversing and bastardizing the legal process.

In exonerating Clinton, the FBI abandoned traditional investigative practices and misconstrued the law. FBI general counsel James Baker, the Bureau's top lawyer who worked closely with Comey, admitted during his closed-door testimony before House investigators that he had originally believed that Clinton should have been criminally charged. Rather than adhering to his legally correct conviction, he allowed Comey's confederates to talk him out of it when they argued erroneously that Clinton had not had the "intent necessary to violate [the law]."[107]

Baker likely knew that "intent" was not required. Still, he eventually relented under pressure "pretty late in the process, because we were arguing about it, I think, up until the end."[108] He acquiesced even though it appears that he was convinced that Clinton had broken the law by repeatedly mishandling classified documents. "I thought that it was alarming, appalling, whatever words I said, and argued with others about why they thought she shouldn't be charged," he recounted.[109]

Baker's legal instincts were correct. But the fix was in. Comey and others were hell bent on clearing Clinton. Joseph diGenova, a former US Attorney for the District of Columbia who also served as an independent counsel, called it a fake criminal investigation of Clinton:

> This story is about a brazen plot to exonerate Hillary Clinton from a clear violation of the law. Absolutely a crime, absolutely a felony. [The FBI] followed none of the regular rules, gave her every break in the book, immunized all kinds of people, allowed the destruction of evidence, with no grand jury, no subpoenas, no search warrants. That's not an investigation. That's a Potemkin village. It's a farce.[110]

DiGenova's reference to Clinton's destruction of documents is another aspect of how the FBI ignored or overlooked criminal behavior that would have subjected anyone else, especially Trump associates, to a charge of obstruction of justice. Under the Federal Records Act and the Foreign Affairs Manual, all of Clinton's emails were required to be captured and preserved by the State Department.[111] While serving as secretary of state, Clinton refused to comply. Those records were not her personal property. Every single one of her work-related emails was owned by the US government as an official federal record.[112] In fundamental terms, Clinton was stealing government documents. Converting such records for personal use is a felony under 18 U.S.C. § 641.[113] It's called theft.

More than a year after she left office, Congress learned that Clinton's State Department email account was empty. Only under legal pressure did she relent and return some, but not all, of her emails to the department over which she had once presided. Tens of thousands of her emails were destroyed while her server was wiped clean by using file-deleting software.[114] Although Clinton claimed that only personal emails had been made to disappear, the FBI discovered that that was not true. Thousands of relevant work documents were never turned over and had been destroyed. Her mobile devices had been broken in half or demolished with hammers. A special computer program called "BleachBit" was used to wipe clean any trace of the deleted files.[115] All of that violated another statute making it unlawful to willfully destroy government documents, 18 U.S.C. § 2071(b).[116]

Clinton had been instructed by Congress to preserve all of her documents.[117] In response, she promised to cooperate fully in its demand to protect and produce the records. A follow-up subpoena was sent to her lawyers. By her own admission, she directed that all of her emails be deleted. Her lawyer, David Kendall, sent a letter to Congress confirming that both personal and business emails had been removed from his client's server and backup systems.[118]

Destroying documents under those circumstances creates a convincing case of obstruction of justice, which makes it a crime to act "corruptly" by "withholding, concealing, altering, or destroying a document or other information" in a congressional investigation.[119] There is no evidence to indicate that the FBI ever seriously considered charging Clinton on that basis. In his book, McCabe described how the FBI spent many months, significant manpower, and untold taxpayer money attempting to recover or reconstruct the emails that Clinton willfully destroyed.[120] That recovery effort underscores the importance of such documents to both the congressional and FBI investigations. Yet McCabe never bothered to mention that that deliberate destruction of evidence almost surely constituted obstruction of justice. Had Trump destroyed emails and documents, he would have been accused of obstruction in an instant.

Nor did the Bureau contemplate bringing a criminal case against Clinton for conveying classified documents to people who did not have security clearance to receive them. Comey was asked during a House Oversight Committee hearing, "Did Hillary Clinton give non-cleared people access to classified information?" The director's response was unequivocal, "Yes, yes."[121] Clinton gave "up to ten people" access to classified information, according to Comey.[122] None of them was permitted under law to have such protected materials, and therefore it is a criminal act for those classified materials to have been conveyed to them. It is confounding and inexplicable that Clinton was never charged for such a blatant violation of the law.

Three years after Clinton and others were absolved by Comey, the US State Department was forced to divulge that there had been, at minimum, thirty separate security breaches of Clinton's unauthorized email system.[123] In a letter to Senator Charles Grassley, the department "assessed culpability to 15 individuals, some of whom were culpable in multiple security incidents. DS has issued 23 violations and 7 infractions."[124] Another trove of documents released by the FBI revealed that "Clinton aides were shocked at apparent attempts to hack her private email servers."[125] Some hacks were successful. Chinese intelligence reportedly gained access to Clinton's server housing classified documents and "inserted code that forwarded them a copy of virtually every email she sent or received" in real time.[126] That was discovered in 2015 by the intelligence community inspector general (ICIG), who immediately drove over to the FBI Building in Washington to alert the FBI. Peter Strzok was the agent who was warned by the ICIG.[127] Yet Strzok was an instrumental part of the FBI team that cleared Clinton of any wrongdoing the next year.

Consider what such a foreign intrusion meant. On a daily basis, our adversaries in Beijing were reading the United States' national security secrets, courtesy of Clinton. If the Chinese could so easily hack the secretary of state's unprotected system housed in her home, so could the Russians and many other enemy states. As noted earlier, the infamous hacker known as "Guccifer" was using a Russian server to penetrate Clinton's system, which means that Moscow's intelligence agents obtained everything he stole. During the secretary of state's tenure, it is fair to assume that the United States had no national security secrets. So much for Obama's claim that Clinton never jeopardized national security. Few people believed it at the time, although the FBI attempted to peddle the myth as part of its excuse for not prosecuting Clinton.

By contrast, the FBI and Special Counsel Robert Mueller pursued with seeming vengeance many of the people connected, however tangentially, to Trump. Cases of false statements of the type that had been overlooked in the Clinton case were brought with intense vigor during the Trump investigation. Selective prosecution and unequal justice became the hallmarks of how both investigations were handled. It wasn't just Comey who allowed personal animus and political bias to infect those cases. Representative Doug Collins, who released to the public many of the private interview transcripts, referred to what he called a "corrupt triumvirate of Page, McCabe and Strzok." [128] But it was the self-imagined "superagent" Strzok who seemed to drive the malignant force that animated corrupt acts within the FBI, according to Collins:

> Peter Strzok was a man who thought he was untouchable. He
> became a hero in his mind's eye in thinking that he was going

to be able to control and sometimes maybe showing off for his mistress at the time, Lisa Page. . . .

Peter Strzok was central, going back to the [Clinton] e-mail investigation, into the Russia investigation, into what became the Mueller investigation. . . .

The transcripts reveal the bias.[129]

The extramarital affair between Strzok and Page was a microcosm of the FBI dysfunction that was endemic in the upper echelons of the agency. As insipid as their pro-Clinton and anti-Trump text messages were, they illustrate how pervasive bias contaminated both the Clinton and Trump investigations. Strzok, Page, and others considered themselves above the law because they *were* the law. They could do as they pleased, even violate rules and regulations that warned against such intimate personal relationships because it could make them "susceptible or vulnerable to recruitment" or blackmail by a foreign adversary.[130] At one point, Bill Priestap, the assistant director of the counterintelligence division, confronted them and warned that their relationship could compromise both them and the Bureau by creating a dangerous security risk.[131] Unbelievably, Priestap did nothing to stop it because, he told Congress, "I'm not the morality police."[132] The FBI didn't care that their conduct might jeopardize an important and sensitive counterintelligence case.

Instead of removing or demoting them from the case, they were promoted. After leading the FBI's "collusion" investigation of Trump during the ten months before Mueller was appointed special counsel, Strzok was elevated to the newly commissioned probe in the role of lead investigator there, as well. Page was rewarded with an assignment to the special counsel team. Strzok told the House Judiciary Committee that he had carried over all of the information he had gathered while at the FBI.[133] To the extent that the first probe was tainted by his anti-Trump prejudices, that taint was then incorporated into the second probe, infecting it with the same bias.

When Special Counsel Mueller was advised in August 2017 of

all the incendiary texts exposing Strzok's abiding and pervasive bias, he quietly removed him from the special counsel team but kept it a secret. Congress inquired why the lead investigator had been removed, but Mueller never explained. Perhaps he was hoping to conceal from the American public the likelihood that the integrity and credibility of his probe had been horribly corrupted.

Yet, according to Strzok's testimony, when the special counsel called him into his office to question him about the explicitly biased texts, he never bothered to ask him whether his Trump-hating opinions had influenced the Russia investigation and any decisions made.[134] No one else on the team made such an inquiry either, said Strzok.[135] How was that possible? Maybe the special counsel didn't care. Perhaps he feared a truthful answer. In either case, Mueller's failure to pose such a vital question suggests an unwillingness to even consider that his investigation of Trump had already been incurably contaminated by Strzok's actions and the misconduct of so many others at the FBI. As one *Washington Post* columnist observed, "Nobody did more damage to Robert Mueller than Peter Strzok." [136]

When Mueller fired Strzok, he did not ask that the agent hand over his infamous cell phone as potential evidence in the ensuing FBI investigation of his conduct. It had been issued by the special counsel, and yet Mueller did not confiscate the device to preserve the text messages. The data on the iPhones of both Strzok and Page were stripped of their contents and text messages vanished to the consternation of the inspector general.[137] It is hard to comprehend how a prosecutor with Mueller's experience would not automatically take steps to seize and secure evidence. As journalist Byron York wrote, "One might think that, given the reason for the removal, Mueller might want to check the content of the Strzok and Page iPhones." [138] Unless, of course, the special counsel feared even worse consequences to his probe if more incendiary texts came to light.

Despite the capacious nature of Mueller's investigation, there was never any plausible evidence to support the incessant accusations that Trump had "colluded" with Russia to steal the election. Again, it's important to think about the Clinton email scandal as a precedent for how this FBI and DOJ looked at political investigations. In the Clinton case, they were willing to overlook on-the-record lying, suspicious meetings, improper public statements, and obstructions of the investigation, because it was better to let the voters decide. They did so despite the fact that crimes had clearly been committed.

So how did the same group of people start the Trump case with no evidence of a crime and then convince the country an unparalleled investigation was necessary?

Attorney General Barr said it best when he observed, "Mueller has spent two and a half years and the fact is there is no evidence of a conspiracy. So it was bogus, this whole idea that Trump was in cahoots with the Russians is bogus." [139]

CHAPTER 2

CLINTON COLLUSION

Russian regime has been cultivating, supporting and assisting Trump for at least 5 years. Further evidence of extensive conspiracy between Trump's campaign team and Kremlin, sanctioned at highest levels.

According to several knowledgeable sources, his conduct in Moscow has included perverted sexual acts which have been arranged/monitored by the FSB.

—"Dossier" from anonymous Russian sources
commissioned by the Clinton campaign

I had nothing to do with Russia and the election. I don't know any Russians. The idea that I was conspiring with them is ridiculous. Prostitutes peeing on a bed? C'mon . . . where's the evidence of that?

—Author's interview with President Donald J. Trump,
September 17, 2017, Bedminster, New Jersey

magine if Donald Trump's campaign had secretly paid a foreign spy to obtain Russian disinformation about Hillary Clinton and then fed it to the media, the FBI, the Justice Department, and the State Department, all to unduly influence the 2016 presidential campaign. Trump would immediately have been accused of "colluding" with Russia. Demands for his indictment and impeachment would have reverberated through the halls of Congress, in newsrooms across America, on television airwaves, and on social media websites everywhere.

Trump did not do that to Clinton. Clinton did it to Trump. However, the president is the one who was hounded by the constant condemnation of having "colluded" with Moscow to win the presidency, even though he did no such thing and there was no credible evidence that he did. On an almost daily basis, the media declared him guilty in the court of public opinion. He was saddled with multiple investigations by the FBI, Congress, and a special counsel. Clinton was not. It is hard to make sense of it until you consider the virulent bias that animated the actions of those who sought to destroy Trump and undo his election.

The unadulterated truth is that Trump never did get any opposition research from Russia, but the Clinton campaign and the Democratic National Committee did. They paid for Russian phony information and then fed it to the FBI and the media to damage Trump's candidacy and influence the election. When that failed, the effort was kicked into overdrive to exploit the same bogus material to destroy Trump and undo his election. In sum, Clinton's campaign "colluded" with Russia to falsely accuse Trump of "colluding" with Russia. It was so far-fetched and so harebrained that you could not make it up. Yet it happened.

In March 2016, Clinton told a California gathering that with Trump in the White House "it will be like Christmas in the Kremlin."[1] In a June 2 speech in San Diego, Clinton connected Trump to Putin four times. "If Donald gets his way, they'll be celebrating in the Kremlin," she declared. She accused him of having a "bizarre fascination with dictators and strongmen," mentioning the Russian president by name.[2] Thunderous applause greeted her when she said her next line, "I'll leave it to the psychiatrists to explain his affection for tyrants."[3] Was Clinton insinuating that Trump was under the influence of Putin and acting as a clandestine agent for Russia? The implication was most certainly there. The audience reaction showed that it was a well-received, winning line of attack.

Clinton's devoted coterie of disciples also alluded to an illicit Trump-Russia partnership. Clinton's campaign chairman, John Podesta, labeled it a "bromance," while her senior adviser Jake Sullivan called it "a national security issue."[4] Clinton's website was more trenchant, posting accusations in the form of queries such as "Why does Trump surround himself with advisers with links to the Kremlin?" The site all but stated that Trump had "ties to Russian oligarchs" and raised the specter that he was tangled in a Russian plot to "interfere in our election."[5] It escalated to a point where Clinton herself openly predicted that if Trump won the presidency, he would be a "puppet" of Putin.[6]

All the campaign needed now was a series of leaks from the intelligence community that would appear to back up her claims.

THE CLINTON CAMPAIGN
COMMISSIONED THE "DOSSIER"

The Clinton campaign and the Democratic National Committee (DNC) commissioned the series of documents that became known as the "dossier." The money was secretly funneled through a Washington, DC, law firm called Perkins Coie and a lawyer representing the campaign and Democrats, Marc E. Elias. The firm received between $9.2 million and $12.4 million in legal and consulting fees, of which $1.02 million was then given to Fusion GPS and its founder, Glenn Simpson, for any negative information they could drum up against Clinton's opponent, Trump.[7]

One of the myths associated with the "dossier" is that it was initiated and funded by a conservative website called The Washington Free Beacon.[8] This is not true, as Simpson explained in his testimony before the Senate Judiciary Committee on August 22, 2017. The website, funded substantially by Republican donor and hedge fund billionaire Paul Singer, had retained Fusion GPS months earlier to conduct standard opposition research on Trump that was derived from open-source information found in public records, corporate and financial data, and online material. When it became clear that Trump would likely win the GOP nomination, both the conservative funding and Fusion's work against Trump ceased. Only thereafter, when the Clinton campaign and the DNC hired Simpson, did Simpson first contact a former British intelligence service agent by the name of Christopher Steele in London.[9] This sequence of events was also confirmed by The Washington Free Beacon, which stated that its work had ended before Steele produced the "dossier."[10]

For years, Fusion GPS founder Glenn Simpson had been a reporter for the *Wall Street Journal*. He had penned several articles about Russian organized crime and other Russia-related subjects. He fancied himself an expert, even though he did not speak Russian and had never once visited the country. In his subsequent congressional testimony, he seemed to throw around Russian buzzwords such as *kompromat* and "kleptocracy," as if that somehow established his bona fides.[11] Simpson was rather vague about why he had first begun to think that Trump might have questionable ties to Russia, but in June 2016, he contacted Steele, who had specialized in Russian intelligence for MI6 and was now operating his own firm called Orbis Business Intelligence. Commensurate with Clinton's public accusations against her opponent, Simpson asked Steele to investigate any relationship Trump might have had with Russian businesses or the Russian government. Steele and his company pocketed $168,000 for their work.[12]

Just how this former spook went about compiling his dubious "dossier" remains largely a mystery since he has resisted attempts to be questioned by Congress. The FBI, which had many contacts and conversations with him, has remained mum, except to confirm that he was fired for leaking to the media and lying about it.[13] Even though Steele had not set foot in Russia in more than a decade, he managed to compose the first of seventeen "dossier" memos in a remarkably short period—a couple of weeks, if that. That was next to impossible. Bear in mind that this kind of detailed intelligence gathering from a foreign power would normally demand months of intensive culling of information from confidential human sources and an extensive process of attempting to corroborate its reliability. Steele accomplished it almost overnight, like magic. He never ventured beyond the shores of Great Britain when he did it. It's as if he had picked up a telephone, dialed a couple of numbers at the Kremlin, and was happily handed incriminating information that should have been both sensitive and highly guarded—quite an accomplishment for a guy who had been out of the spy game for many years.

Steele's initial document was dated June 20, 2016, and titled "Republican Candidate Donald Trump's Activities in Russia and Compromising Relationship with the Kremlin."[14] Two qualities immediately stand out when examining the document. First, it seems utterly preposterous on its face. Second, no direct sources are identified. It is based entirely on multiple hearsay from anonymous individuals, who are referred to cryptically as "sources A, B, and C." This opening "dossier" memorandum leveled three main charges against Trump:

- "Russian regime has been cultivating, supporting and assisting TRUMP for at least 5 years."
- "So far TRUMP has declined various sweetener real estate business deals offered him in Russia in order to further the Kremlin's cultivation of him."
- "TRUMP's (perverted) conduct in Moscow included hiring the presidential suite of the Ritz Carlton Hotel, where he knew President and Mrs. Obama (whom he hated) had stayed on one of their official trips to Russia, and defiling the bed where they had slept by employing a number of

prostitutes to perform 'golden showers' (urination) shows in front of him." [15]

No evidence was offered, and there was never any proof that it was true. No video has ever surfaced, although it was widely reported to exist. Either this first "dossier" was invented out of whole cloth, or Steele was fed preposterous disinformation by Russian sources, who must have laughed at their handiwork and the gullibility of the recipient. Interestingly, the term "trusted compatriot" was used in the memo. This suggests that it was written, at least in part, by a Russian determined to invent a collection of lies to create mischief.[16]

It is possible that the Trump-Russia "collusion" fiction arose from Steele's relationship with one of his clients at Orbis, a Russian oligarch named Oleg Deripaska, who had hired the ex–British spy as a subcontractor in 2012 to conduct research for a lawsuit that had been filed against the aluminum magnate.[17] Deripaska's business interests included properties in Ukraine, where he came in contact with Paul Manafort, Jr., who briefly served as Trump's campaign manager in the summer of 2016. Later memos in the "dossier" accused Manafort of acting as an intermediary in the supposed Trump-Russia conspiracy. That false story about Manafort also explains why the FBI interviewed Deripaska in New York in September 2016 and allegedly tried to recruit him as an undercover informant.[18] He refused.

Simpson, whom the *Los Angeles Times* aptly called "a mercenary for hire by anyone with fat stacks of bitcoins," [19] was also obsessed with Manafort. As far back as 2007, he and his wife, Mary Jacoby, had coauthored a story for the *Wall Street Journal* that identified Manafort as representing Ukrainian and Russian business interests, including those of Deripaska.[20] That nine-year-old story bears an odd and striking resemblance to several of the subsequent "dossier" memos authored by Steele and, perhaps, Simpson. Jacoby's name would surface again in a Facebook post on June 24, 2017, which has since vanished. The journalist Lee Smith reported that he had seen screenshots of the post in which Jacoby had given her husband fu

credit for creating the "collusion" narrative: "It's come to my attention
that some people still don't realize what Glenn's role was in exposing
Putin's control of Donald Trump. Let's be clear. Glenn conducted
the investigation. Glenn hired Chris Steele. Chris Steele worked for
Glenn."[21] According to Smith, Jacoby never replied to his repeated
requests for comment.

In his Senate testimony, Simpson said that Manafort's role in the
Trump campaign made him suspicious that the Republican candidate might somehow have improper ties to Russia, which was what
prompted him to hire Steele, who was a subcontractor for Deripaska.[22]
If all of this seems rather circular and incestuous, it was. As he was
trashing Trump as a suspected agent of the Kremlin by peddling the
phony "dossier," Simpson was also working on behalf of a Russian
real estate firm by the name of Prevezon Holdings, which is operated
by an oligarch family with a deep allegiance to Putin.[23] According to
Senator Charles Grassley (R-IA), Simpson's assignment was to discredit a critic of the Russian president through the same kind of salacious "opposition research" or smear campaign that he employed
against Trump:

> There are public reports that the FBI used the dossier to kick
> start its Russia investigation. Did the FBI know that Fusion
> pitched Russian propaganda for another client as it pushed the
> dirty Trump dossier? What would that say about the reliability
> of the information?[24]

The evidence is compelling that it was Simpson who was actually
"colluding" with Russia through his surrogate or intermediary, Steele,
while falsely accusing Trump of doing the same. Moreover, Simpson
was not registered as a foreign agent under the Foreign Agents Registration Act (FARA).

But that's not all. Simpson seems to have played a crucial role in the
controversial Trump Tower meeting in June 2016 between Russian
lawyer Natalia Veselnitskaya and Donald Trump, Jr., also attended

by several campaign staffers. The lawyer purportedly had negative information on Clinton. However, those who attended the meeting all concurred that she passed no such information along. Simpson, who worked with Veselnitskaya on the Prevezon case, admitted he had met with her immediately before and after the Trump Tower meeting.[25] In an interview with NBC News, Veselnitskaya stated that it was Simpson who had supplied her with "supposedly incriminating information" about Clinton that was supposed to be passed along to the Trump campaign.[26] In an interview with Fox News, she insisted she had not provided such information.[27] Based on those statements, one can reasonably conclude that the Trump campaign was set up by none other than Simpson and the lawyer. "They were baiting a trap for the Trump campaign to make it appear as if they were colluding with Russian officials," as one editorial described it.[28] Simpson denied knowing anything about the meeting, which means that one of them is not telling the truth. This may account for why Simpson later invoked the Fifth Amendment and clammed up.

The lightning speed with which Steele collected his anti-Trump fables may be explained in another way: cold, hard cash might have lubricated the dissemination of disinformation. According to Michael Morell, who twice served as acting director of the CIA, Steele seems to have accumulated his dubious "dossier" material from former agents of the FSB (Federal Security Service, the successor of the KBG) by lining the pockets of intermediaries who, in turn, doled out payments to the supposed sources:

> If you're paying somebody, particularly former FSB officers, they are going to tell you truth and innuendo and rumor, and they're going to call you up and say, "Hey, let's have another meeting, I have more information for you," because they want to get paid some more.[29]

In British court documents, Steele laid bare that his "dossier" was not worth the paper on which it was written. He readily confessed that the accusations against Trump were "unverifiable" and derived from "limited intelligence."[30] He also called it "raw intelligence," meaning it had not been checked for its authenticity. It was so poorly sourced that it was inherently untrustworthy. Further admissions by him cast genuine doubt that any of it reflected so much as a thread of truth.

In his videotaped deposition, Steele was even blunter in discrediting his own work. He said his "dossier" incorporated what he had thought was a story on CNN's website, not realizing that it was "nothing more than any random person posting things on the internet."[31] He conceded that he had been trolling the internet for information to insert into his "dossier." He dismissed it as "research" for a private company, not verified or reliable information. In other words, his anti-Trump memos were uncorroborated junk.

Steele's former boss at MI6, Sir John Scarlett, appears to have been quite aware that his onetime agent had composed a dodgy document that would never meet the standards of credible intelligence. Scarlett rejected it as "unverified and overrated"—nothing more than a "commercial" enterprise to make money.[32] When asked what working with Steele had been like, all that was offered was the curt reply "I'm not going to comment." His silence spoke volumes. Others who worked with Steele were caustic and unkind in their assessments of both his skill and intelligence. The current head of MI6 "is said to be livid" at Steele for "causing worldwide embarrassment to British secret services."[33]

A congressional investigation found that Clinton associates were also feeding Steele bogus information as he compiled his "dossier."[34] One of the more salacious allegations he described as "fifty-fifty" in its validity. It appeared to have come from a Belarusan-born American businessman named Sergei Millian, who was regarded by Simpson as "a big talker."[35] In a veiled reference to the partisan motivations behind his report, Steele cautioned that "its content must be critically viewed in light of the purpose for and circumstances in which the information was collected."[36] In plain language, it was an untruthful political attack intended to vilify an opposing candidate, Donald Trump.

As for Simpson, he grudgingly admitted in his congressional testimony that he had taken no steps to vet or verify any of the contents of the "dossier" or even ask Steele about its veracity.[37] He had accepted it on blind faith because, as he explained, Steele was the Russian expert, and he trusted his competence. It did not matter to Simpson that the allegations against Trump were offensive, outrageous, and even ludicrous. He appeared to be more interested in the political havoc he could wreak on the Republican nominee than in truth or accuracy. In other testimony, Simpson said that Steele's dossier had been elicited from "human source information." He then added, "And humans sometimes lie, and more frequently they just get it wrong."[38] Thus, both men knew that their anti-Trump memos might well contain faulty intelligence or, more plainly, lies. That did not deter them at all.

Steele and Simpson were determined to utilize the "dossier" to nullify Trump in his bid to be the next president of the United States. That was not a new tactic for Fusion GPS. The firm and its founder had an alarming reputation for smear campaigns based on "prepared dossiers containing false information" and "carefully placed slanderous news items," as Senate investigators learned from several of its victims.[39] Simpson was at it again. This time, he had a much bigger target in his sights, and the damage he could incite would be momentous.

Simpson and Steele devised a comprehensive strategy to disseminate their "dossier." Part one was to circulate the document to sympathetic officials at the FBI and Department of Justice, who might then launch an investigation of Trump. Part two was to share it with certain anti-Trump members of the press, who would serve as an instrumental part of the plan to destroy the GOP nominee with "collusion" stories in publications, on television, and on social media. Facts were irrelevant as long as Simpson and Steele could sell a salacious story to those who were predisposed to believe it, even in the absence of real evidence. They would also profit by it financially.

As soon as Steele composed his June 20 memo, he notified Simpson, revealed its contents, and sent him a copy via an encrypted communication. In several ensuing telephone conversations, the two men solidified their blueprint for propagating the "dossier": feed it to the FBI and circulate it among the rapacious American media. In later testimony before the Senate Judiciary Committee, Simpson claimed that Steele had alerted the FBI during the first week in July 2016.[40] This may not be entirely accurate. There is some evidence that Steele delivered the document to the FBI within days of its completion either by sending it through a secure communication to an agent in Rome or by hand delivering it to him after catching a flight to the Italian capital.[41] Regardless, a *formal* meeting between Steele and the FBI agent was scheduled for early July in London. It happened on the very day that Comey cleared Hillary Clinton in her email scandal.

THE PIVOTAL DAY: JULY 5, 2016

It was 11:00 a.m. on Tuesday, July 5, 2016, when James Comey stood before television cameras and microphones at a nationally watched news conference. Mangling the law and contorting the facts, he announced that he was exonerating Clinton of any crimes for her mishandling of thousands of classified documents. As far as Comey was concerned, that was the end of the Clinton case.

While Comey was speaking, Americans had no idea that crucial steps were being taken to simultaneously convert the FBI's immense power and investigative resources from one presidential candidate to the other. As the Clinton case concluded, the Trump case was just beginning. Had Americans known, they would have been both bewildered and outraged that such a Machiavellian dynamic was unfolding on two continents concurrently.

Some 3,660 miles away from Washington, DC, Comey's FBI was meeting at Orbis headquarters for the first time with Steele since he had composed his initial "dossier" memo dated June 20, 2016. The agent, identified as Michael Gaeta, had flown to London from Rome, where he was an attaché at the US Embassy.[42] Victoria Nuland, a top Obama State Department official, authorized the meeting, although FBI headquarters would certainly have been well aware of it.[43] During the encounter, Steele conveyed the contents of his initial "dossier" to Gaeta, as well as other unsubstantiated allegations against Trump that Steele had been preparing for four additional July memos. Gaeta then passed the information to Nuland, who delivered it to the FBI.[44] Those memos were equally outrageous in their accusations and, as before, derived from either fictitious Russian sources or unreliable multiple hearsay:

- Trump associates Paul Manafort and Carter Page were "intermediaries" in the Trump-Russia conspiracy to influence the presidential election.
- Page held "secret meetings in Moscow" with top Kremlin officials and promised to have Trump lift sanctions against Russia in exchange for a 19 percent stake in the oil company Rosneft (further detailed in a later "dossier" memo).
- "TRUMP and senior members of his campaign" were involved in the hacking by Russia of Democratic National Committee (DNC) emails then published by WikiLeaks.
- Michael Cohen, Trump's lawyer, held "secret discussions with Kremlin representatives" in Prague.
- Instead of consummating real estate deals in Russia, "TRUMP had had to settle for the use of extensive sexual services there from local prostitutes rather than business success."
- "An intelligence exchange had been running" between the Trump team and the Kremlin "for at least 8 years."[45]

All of those accusations were little more than fantasy. Each of them could be disproved, had the FBI investigated at the outset. Page had not held secret meetings in Moscow. No evidence surfaced to the contrary. He had traveled there to deliver a speech at the New Economic School, where Barack Obama had given a speech years earlier.[46] Page's address was well publicized in advance, which is likely how Steele and Simpson gained their information—from newspapers and/or the internet. Similarly, the hacking of the DNC and Clinton campaign emails was already known to the public by the time Steele authored his "dossier" memo on the subject, and Russia was the prime suspect.[47] There was never any evidence that the Trump campaign was a participant. Finally, Steele's contention of an eight-year-long relationship of intelligence sharing between the "Trump team" and the Kremlin defied common sense and logic. It was a silly claim that undermined the "dossier" as a whole.

But the most comical assertion was that Page would pocket a 19 percent stake in one of Russia's most valuable assets, the state-controlled oil company Rosneft, in exchange for the prospect of lifting Western sanctions. In dollars, the payout would approximate $11 billion.[48] It is absurd to think that Moscow would pay an enormous bribe to a volunteer "junior, unpaid adviser" who had never met the candidate and had few real connections to the campaign.[49] That, of course, did not stop Democrats like Representative Adam Schiff (D-CA) of the House Intelligence Committee from advancing this screwball conspiracy during hearings on Capitol Hill.[50] Page vigorously denied any involvement in such a plot. Despite his having been wiretapped by the FBI for more than a year, no incriminating evidence was ever found and no charges were ever brought against him. With their "dossier," Simpson and Steele managed to victimize an innocent man.

As will be discussed in a later chapter, Manafort was convicted of financial crimes that predated by many years his joining the Trump campaign. The charges had nothing whatsoever to do with Trump-Russia "collusion," and no evidence was ever presented by prosecutors alleging a Kremlin conspiracy to throw the presidential election. Like Page, Manafort was never an "intermediary" working at the behest of Russia. A later iteration of the "dossier" claimed that Trump's lawyer Michael Cohen had met secretly in Prague with a Kremlin official to arrange cash payments and devise a cover-up operation.[51] Had the FBI bothered to look, records showed that Cohen had been in Los Angeles and New York during the stated time frame. No documents placed him in Prague. During his congressional testimony in February 2019, Cohen stated under oath that he had never so much as visited the Czech Republic.[52]

There was nothing intelligent about the so-called intelligence document created by Steele and Simpson. With seed money from the Clinton campaign and Democrats, they composed a collection of fables designed to defame and discredit Trump and those associated with him. For a while, it seemed to work. The FBI eagerly appropriated the "dossier" and, without first verifying any of the claims, launched a dilating investigation of Trump that began in earnest on the same day Comey cleared Clinton.

As the document gained currency in the FBI, agent Peter Strzok boarded a plane at the end of July or early August and headed to London, from which the "dossier" had originated. Text messages he exchanged with his lover, Page, were redacted in key places by the FBI and DOJ, making it difficult to discern the identities of individuals with whom he met in the United Kingdom. But his visit appears to have been profitable in furthering the "collusion" hoax, since Strzok described a particular meeting as productive.[53] A discussion took place with Page about how to conceal information from future Freedom of Information Act (FOIA) requests in the event the trip became known. References in those emails make it clear that the CIA was involved and "the White House is running this."[54] But Strzok was operating in the shadows.

FEEDING THE "DOSSIER" TO THE DEPARTMENT OF JUSTICE

Steele and Simpson were not merely satisfied with dumping the "dossier" into the lap of the FBI. They wanted to ensure that it received a wider audience in US law enforcement, inasmuch as the presidential election was fast approaching. Something had to be done about Trump. He was behind in the polls, but nothing could be left to chance. If the coconspirators could trigger a criminal investigation of the candidate and leak the information to the press, Clinton's waltz to the White House would be all but assured.

Steele flew to Washington, DC, to meet with a high-ranking Justice Department official he had known for years, Bruce Ohr. On July 30, 2016, at 9:00 a.m., the ex–British spy met at the May-flower Hotel with Ohr, who held the number four position at DOJ as associate deputy attorney general.[55] Steele conveyed the contents of the "dossier" to him but added that the FBI already had some of the memos in its possession.[56] Of course it did; Steele had shared the documents with a Bureau agent on July 5. This, and other evidence, clearly undermined the long-standing contention by Democratic representative Adam Schiff that Steele's information "did not reach the counterintelligence team investigating Russia at FBI headquarters until mid-September 2016."[57] In truth, the FBI team had it in July and perhaps as early as June, when the first memo was written. Either Schiff knew that, or he was badly misinformed.

Also attending the Mayflower Hotel meeting with Steele was Ohr's wife. Not coincidentally, she had been hired months earlier by Simpson as a contractor for Fusion GPS. Her assignment was to develop anti-Trump research that could be used for the benefit of Fusion's clients, the Clinton campaign, and the DNC. In her private testimony before the House Judiciary Committee, Nellie Ohr stated that her specific task was to develop evidence that might show how Trump and his family had engaged in illicit dealings with the Russians.[58] She told Congress that she investigated Trump's children, Ivanka Trump and Donald Trump Jr., and even the candidate's wife, Melania Trump.[59] Naturally, that was intended to help support the "dossier." She admitted that her political views had influenced her decision to target Trump, "Because I favored Hillary Clinton as a presidential candidate."[60]

Just how much anti-Trump information Nellie Ohr shared with

her husband is unclear, because she refused to answer many key questions by invoking the spousal privilege that protects one spouse from being incriminated by the testimony of the other.[61] When Bruce Ohr appeared behind closed doors for his testimony, he insisted that he had never read the opposition research his wife had prepared against Trump. He also stated that he had not read any of the documents that Steele had given him. Instead, he said, he had decided to hand them over to the FBI without examining their contents. Hard to believe, but that's what he stated.

There is now substantial evidence that Nellie Ohr provided false testimony when she appeared before the two House committees conducting a joint investigation.[62] Under questioning, she insisted that she had shared her research only with Fusion GPS, Steele, and her husband. This was not true. Three hundred thirty-nine pages of emails show she sent it to at least three other prosecutors at the Justice Department.[63] Those same emails prove that she was well aware of the DOJ's Russia investigation, even though she claimed otherwise when she testified. A criminal referral on Nellie Ohr was promptly sent to the Justice Department.

While Nellie Ohr was working behind the scenes to discredit and damage Trump, her husband at the DOJ continued to serve as a "conduit" between Steele/Simpson and the FBI to feed the "dossier" beast. He would later deliver to the Bureau two USB flash drives or "memory sticks" containing derogatory information about Trump, including the "dossier," that had been given to him separately by his wife and Simpson.[64] Ohr's cloak-and-dagger role as a covert intermediary is curious, if not suspect, since Steele had already met with FBI agent Gaeta in London three weeks earlier and had been on the FBI's payroll as a confidential human resource (CHR) since at least February 2016.[65] Steele wasn't working for just one client; he was working for three. He was collecting money from the FBI, the DNC, and the Clinton campaign. He and his benefactors, who funneled the cash through Simpson's Fusion GPS, all had a concomitant interest: trashing Trump.

Whether Ohr knew it or not, he was being manipulated by Simpson and Steele with unverified and phony information. Predictably, Ohr contacted a top official at the FBI immediately following the breakfast meeting with Steele. The desired result was achieved. The very next day, July 31, 2016, the Bureau executed papers to formally launch its investigation of Trump, nicknamed "Crossfire Hurricane"—a description not nearly as benign as Clinton's "Midyear Exam." [66] Not surprisingly, it was the Trump-hating FBI agent Peter Strzok who signed the documents that would metastasize into a full-blown campaign to harass and torment Trump for the better part of the next three years. [67] It was an unprecedented action. A presidential administration had launched a counterintelligence operation against the campaign of a candidate from the opposing party. Ohr never advised his superiors about his many actions in feeding the FBI the anti-Trump material nor his clandestine meeting with Steele and others of the rendezvous that followed. [68] Instead, shortly after the July 30 debriefing by Steele on the "dossier," Ohr convened a meeting with FBI deputy director Andrew McCabe and the FBI lawyer who worked for him directly, Lisa Page. Their get-together occurred in McCabe's office. No one seems to remember the exact date. Although given the urgency that Ohr said he had felt, it appears to have happened directly following the July 30 breakfast with Steele. After detailing the "dossier," Ohr testified that he had specifically warned McCabe and Page that the information in the document was highly dubious and driven by a biased author who despised Trump. Ohr also advised that it had been commissioned by Fusion GPS, where his wife worked, because "I wanted the FBI to be aware of any possible bias." [69]

QUESTION: Do you recall whether it was [to] you that Chris Steele said he was desperate that Donald Trump not win?
OHR: I think I said that to the FBI, yes. . . . I don't recall the exact words. I definitely had a very strong impression that he did not want Donald Trump to win . . . [70]

Later in his testimony, Ohr reiterated all of the warnings he had given to the FBI that Steele might not be a trustworthy source and that the information in his "dossier" might be unreliable, if not bogus:

OHR: When I spoke with the FBI, I told them my wife was working for Fusion GPS. I told them Fusion GPS was doing research on Donald Trump. . . . I told them this is the information I had gotten from Chris Steele. . . . I told them that Steele was desperate that Donald Trump not get elected. So those are all facts that I provided the FBI.[71]

When questioned later by the FBI investigators, Ohr was even more emphatic about Steele's flagrant bias. Official Bureau files note that Steele had stated he "was desperate that Donald Trump not get elected and was passionate about him not being president."[72] Ohr insisted that he repeated that warning to the FBI on more than one occasion. He pointedly informed it that Trump's political rival, the Clinton campaign, was financially underwriting the "dossier," which would call into question its veracity because the campaign had a motive to distort or fabricate in order to damage its opponent. In his congressional testimony, Ohr insisted that he had cautioned the FBI, "These guys were hired by somebody relating to—who's related to the Clinton campaign, and be aware . . ."[73] The funding wasn't merely "related to" the campaign; it *was* the campaign, along with the Democratic National Committee.

The testimony of Ohr disproved another story that Representative Schiff kept selling to the American public—that Ohr had never told the FBI anything about Steele's "dossier" until *after* the presidential election "in late November."[74] That was utterly untrue, and Schiff surely knew it. Ohr, McCabe, and Page all confirmed the July 30 meeting and the "dossier" discussion. FBI records attested to it. Those records also corroborate all the warnings issued by Ohr. Either Schiff didn't care to do his homework, or, more likely, he sought to misrepresent the facts. Many in the media uncritically accepted Schiff's version and propagated it as truth. It was not.

MORE MEETINGS, MORE WARNINGS

Given Steele's acute bias and motivation to lie, given that his "dossier" seemed preposterous on its face and was completely unverified, and given that it was financed by the Clinton campaign and Democrats, the FBI should have approached the "dossier" with trepidation. Members of the intelligence community specializing in Russian affairs should have vetted the information, verifying what could be true. After all, it's not possible that freelancers could have uncovered a conspiracy of that scale that US intelligence had not also run across.

That failure may be the most important clue in how the mass delusion spread. A key part of any conspiracy theory is that what the public can see must be the tip of the iceberg. From reporters to pundits to the average voter, anyone who bought into the scam believed that the FBI wouldn't have started all of it without having more than rumors to go on. That was one point that, until the Mueller Report dropped, both sides could agree on: it would be unprecedented, horrifying, and borderline insane for the FBI to have pushed the investigation for two years based on nothing at all. But the opposite happened; armed with the "dossier," the FBI used it as a pretext to pursue Trump relentlessly. Credible and verified evidence didn't seem to matter.

As Steele and Simpson continued to push their inflammatory "dossier," Ohr decided to disseminate it to others within the FBI and Justice Department. "I subsequently met with Lisa Page, Peter Strzok, and eventually Joe Pientka at the FBI," said Ohr.[75] Although he concealed from his superiors at the DOJ the fact that he was secretly acting as a conduit between Steele and the FBI, Ohr also elected to give the improbable intelligence to three prosecutors at the Justice Department in yet another meeting. That occurred within a few days of the FBI formally opening the Trump-Russia investigation on July 31.

Ohr's surprising admission came in the middle of his congressional testimony when he mentioned—almost as an aside—"I also provided this information to people in the criminal division, specifically Bruce Swartz, Zainab Ahmad, and Andrew Weissmann." [76]

Two of those individuals, Ahmad and Weissmann, were later hired by Robert Mueller to be part of his assembled team of special counsel prosecutors that escalated the investigation of Trump beyond the FBI and DOJ. They were also informed that the Clinton campaign and Democrats had paid for the Steele "dossier" through the company Fusion GPS and its founder, Glenn Simpson, who also employed Ohr's wife. Here is Ohr's testimony:

> **QUESTION**: So, the record is clear, what the Department of Justice and the FBI was aware of . . . was your relationship with Christopher Steele and Glenn Simpson, your wife's relationship with Christopher Steele and Glenn Simpson, Mr. Steele's bias against Donald Trump, Mr. Simpson's bias against Donald Trump, your wife's compensation for work for Glenn Simpson and Fusion GPS, correct?
> (Discussion off the record.)
> **OHR**: Right. [77]

They were also made aware that the Clinton campaign and the DNC had paid for the phony Russian information (or disinformation) that was intended to influence the election. Yet instead of investigating Clinton and her confederates for conspiring with foreigners to defraud the US government, the FBI used the Clinton-Russian "dossier" to target Trump despite a dearth of evidence that it was authentic. Thus, four members of Mueller's special counsel team—Weissmann, Ahmad, Strzok, and Page—all knew as early as August 2016 when the investigation of Trump was initiated by the FBI that it was riven with prejudice and driven by a defective document. They deliberately disregarded the source of Steele's funding and sought to exploit his "dossier."

Amazingly, Ohr was peddling the uncorroborated Steele document while benefiting from it financially. Since his wife was working on the anti-Trump "opposition research" documents, she was being paid by Fusion GPS, which was being subsidized by the Clinton campaign and Democrats. That money was being deposited into the Ohrs' joint bank account at the same time Bruce Ohr was using his position at the DOJ to advance the false intelligence document. That created a disqualifying conflict of interest for Ohr. He was legally obligated under Justice Department regulations to recuse himself from any investigation in which his wife was involved. Ohr did not seek a waiver of that conflict. Instead, he omitted the information on all relevant forms and continued to act as the "conduit."

Upon joining the DOJ, he had signed an agreement stating that he would be fired if he violated its rules. Inexplicably, he was not terminated, which only reinforces the impression that impropriety and concealment continued at the highest levels of the department. Not only did Ohr fail to disclose that Fusion GPS was paying his wife, but it appears he did not fully report the nature of the work performed in financial disclosure reports as required under regulations.[78] Willfully filing a false government report constitutes a crime under federal law—specifically 8 U.S.C. § 1001.[79] It is also against the law to use a public office for personal financial gain. This runs afoul of the federal bribery and gratuity statutes, as well as the honest services fraud statute.[80] Yet there appears to have been no known investigation of Ohr by the Justice Department, where he is still employed, although he was demoted when his advocacy in the "dossier" fiasco was made public. Amazingly, he was "awarded a $28,000 performance bonus while the Russia probe was ongoing," according to DOJ records.[81]

Perhaps motivated by money, Ohr moved aggressively in his role as facilitator. He agreed to meet in person with Glenn Simpson on August 22, 2016.[82] At that point, the FBI's investigation was not even a month old. Simpson pretended to be confident that Steele's document was authentic and reliable, even though he had provided not a shred of proof or corroboration. He vouched for the credibility of

Steele, not knowing that the former spy had been "admonished" by the FBI months earlier for reasons the Bureau has consistently refused to disclose.[83] Steele would soon be fired by the FBI for leaking to the media and lying about it. He was declared "not suitable for use as a CHS [confidential human source]." His FBI "handler" advised that Steele should henceforth "not operate to obtain any intelligence whatsoever on behalf of the FBI."[84] Steele had proven to be deceitful and, hence, radioactive to the Bureau. Despite that, the FBI kept going back to Steele even after he was fired. (Also, even as the FBI was leaking like a sieve over the "dossier," somehow Steele's standing within the organization failed to leak.)

During the August 22 meeting, Ohr listened as Simpson wove a tall tale of how Paul Manafort and Carter Page *must* be Russian assets engaged in an illicit scheme to rig the election. He couldn't prove it, but the dots were there. All the FBI had to do was connect them. Simpson implored Ohr to pressure the FBI into taking swift action against Trump. If nothing else, the candidate was somehow guilty by association—of what, exactly, was left unstated.

When Simpson testified before the House Intelligence Committee in November 2017, he denied that he had ever met with Ohr before the election:

QUESTION: You've never heard from anyone in the U.S. Government in relation to those matters [the "dossier"], either the FBI or the Department of Justice?

SIMPSON: After the election. I mean, during the election, no.[85]

Simpson's testimony was inaccurate and untruthful. Ohr's testimony and his notes, including an email on the date of their liaison, August 22, show that he had met with Simpson more than two months prior to the election. Either the Fusion GPS founder forgot about such an important meeting, during which he had disclosed the "dossier's" contents, or he lied. That may further explain why Simpson subsequently invoked the Fifth Amendment when asked to

appear again before Congress to explain his deceptive account of interactions with Ohr.[86]

That was not the only instance in which Simpson is believed to
have given false testimony. Senator Charles Grassley, chairman of the
Senate Judiciary Committee, recited another statement that Grassley
described as "extremely misleading, if not an outright lie."[87] Citing
the actions of Simpson, Fusion GPS, and Steele, Grassley and Senator
Lindsey Graham (R-SC) sent a criminal referral to the Justice Department on January 4, 2018, demanding a comprehensive investigation into the apparent lies.[88] (The standard here seems to be that false
testimony related to Russia is a crime and a scandal when a Trump
supporter does it but not when anyone else does.)

A month after Ohr met with Simpson, Ohr met for a second time
with Christopher Steele over breakfast at 8:00 a.m. at the Capital Hil-
ton Hotel in Washington, DC. The date was September 23.[89] The ex–
British spy had been exceedingly busy. Days earlier, on September 19,
he had flown to Rome to deliver even more memos in his "dossier"
to his FBI "handler," special agent Michael Gaeta. Gaeta, in turn,
had forwarded them to the Bureau in Washington. During one meeting with Gaeta, attended by four agents flown in from Washington
(in either September or October), Steele was offered a jaw-dropping
$50,000 bonus if only he could verify with "solid corroboration"
the information in his "dossier" that Trump was "colluding" with
Russia to influence the election.[90] That presented an insurmountable dilemma. The ex-spy could not possibly verify something that
had never occurred. His information was derived from anonymous
sources—assuming they even existed—and was based on multiple
hearsays. In the end, Steele did not collect the bonus. But the offer of
$50K by the FBI shows how desperate the Bureau was to verify the
unverifiable and to damage Trump's candidacy in the face of the fast-
approaching presidential election. That failure to verify, however, did
not seem to discourage either the FBI or Steele in the least.

With every new "dossier" memo accusing Trump of "collusion,"
Steele made doubly sure that they were passed on to both the FBI and

Ohr at the DOJ. By late September, there was a total of eleven such memos. As before, Ohr promptly conveyed the latest Steele memos to FBI agent Joe Pientka, who was now assigned as Ohr's "handler." Pientka then dispatched the material to his Bureau partner, Peter Strzok, who was overseeing the Trump-Russia investigation.

During the September 23 visit to the nation's capital, Steele met with someone else in the US government: Jonathan Winer, a State Department official. When that surreptitious meeting was later exposed in the media, Winer penned a bizarre and nauseatingly self-aggrandizing op-ed in the *Washington Post*. He seemed to brag about his role as a facilitator of anti-Trump material.[91] He described how it had worked: like a four-person Olympic relay team, Steele had dropped a digest of the "dossier" gossip off to Winer, who had handed it off to Nuland, who had then passed the baton on to Secretary of State John Kerry, who had apparently completed the final leg by racing it over the finish line to the FBI.[92] Of course, the Bureau already had it, courtesy of Simpson, Ohr, and Gaeta.

But Winer took it a step further; he also met with longtime Hillary Clinton friend and confidant Sidney Blumenthal, who had his hands on what became known as the "second dossier."[93] That equally dubious document peddled many of the same sordid stories about Trump's purported "collusion" with Russia. It had been created, according to Winer, by Clinton ally Cody Shearer. At that point, a reverse relay was enacted as the information in the "second dossier" made its way to the FBI: Shearer to Blumenthal to Winer to Steele.[94] It is quite likely that some of Shearer's accusations against Trump were transcribed into the last few Steele memos, which were then passed on to the FBI.

The comical stupidity of this scenario would make the fictional detective Sam Spade blush. It was inane. And so was the second treatment of "collusion," which echoed some of the identical fantastic allegations as the original Steele-Simpson version—equally bereft of evidence or corroboration. It constituted a collection of rumors, innuendo, and wild conjecture. Special emphasis was placed on the notion that Trump had magically engineered the hacking of Democrats' email accounts, including Blumenthal's. Speculation supplanted facts. None of the accusations against Trump was authentic.

This invites the question, just who is Cody Shearer? Mark Hemingway, a senior writer at *The Weekly Standard*, has followed the activities of Shearer and offered this assessment:

> It's worth noting that Shearer is one of the most disreputable characters in Washington, and has been frequently connected to the most scandalous acts of the Clintons' political careers. If Steele passed on information and/or allegations from Shearer to the FBI, and that information was acted on, it raises serious concerns about the impartiality and judgment of Steele and the FBI.[95]

Steele did give Shearer's "second dossier" to the FBI. Hemingway was also unsparing in his appraisal of Blumenthal, writing that "Blumenthal is a known liar and rumormonger so disreputable that the Obama White House put their foot down and nixed her [Clinton's] attempt [to hire] Blumenthal at the State Department."[96]

Blumenthal and Shearer were not the only Clinton allies circulating the "collusion" canard. Perkins Coie lawyers for the DNC and Clinton campaign, who had hired Simpson to dig up dirt on Trump, were now passing information along to the FBI. Michael Sussmann, a partner at the firm, met with FBI general counsel James Baker on September 19, the first of several meetings. According to Baker's testimony, he received a stack of papers and an electronic "disk" from Sussmann that constituted purported evidence of secret communications between Trump and Russians involving Alfa Bank in Moscow.[97] Baker delivered the materials to Comey, McCabe, and Strzok. Like many of Steele's other accusations, that one was subsequently proven to be false, but not before it had been fed to the media, including the *New York Times* and Slate, which published stories.[98]

THE MEDIA PUBLICIZE THE "DOSSIER"

The scheme hatched by Simpson and Steele to distribute their phony "dossier" to the FBI and Justice Department was only the first part of their overarching strategy to damage Trump and buttress the candidacy of his opponent. To have an impact on the election, the document would need to be fed to the ravenous anti-Trump media in advance of the election. Simpson, a former reporter, orchestrated the ensuing meetings with numerous journalists he knew. Steele, as an ex–intelligence agent, attended many of them to add the imprimatur of legitimacy to the supposed "intelligence" they were selling. The journalists, predisposed to run with negative stories on Trump, were informed that government law enforcement was already in possession of the same document, giving the reporters an excuse to author articles on the accusations contained therein even though none of them had been verified as true. That crafty calculus paid off as stories began to emerge alleging that Trump was "colluding" with Russia to win the presidency.

Steele met with Michael Isikoff of Yahoo! News, who published an article on September 23, 2016, repeating an allegation that Carter Page had "opened up private communications with Senior Russian officials," and discussed the lifting of economic sanctions.[99] The story stated that the Trump foreign policy adviser had met with two Russians who were mentioned in Steele's memos and connected Page to "suspected efforts by Moscow to influence the presidential election."[100] The Clinton campaign called it a "bombshell report" and sought widespread distribution. The campaign published a thumping screed on Trump-Russian "collusion" while concealing from everyone that it had secretly funded the fabrications. Blasting it out to millions of voters, the campaign accused her rival of "an action that could directly enrich both Trump and Page while undermining American interests."[101] Without using the word "treason," Clinton all but accused Trump of being a traitor. It was not true, of course. That did not stop the FBI and DOJ from using Isikoff's story to obtain a warrant to wiretap Page, even though the reporter's source was the same source—Steele—that was cited for the bulk of the warrant application. The judges were never informed of that and were led to believe that there were separate sources to justify the wiretap. All this will be discussed in the next chapter.

The Isikoff story was just the opening salvo in an all-out assault against Trump by the corrupt triumvirate of Simpson, Steele, and the media. The two men behind the imaginary "dossier" also briefed CNN, *Mother Jones*, BuzzFeed News, ABC News, the *New York Times*, the *Washington Post*, *The New Yorker*, Politico, and other news organizations that saw huge traffic numbers in allegations about a presidential nominee from extremely trustworthy, high-level sources. Simpson and Steele weren't that, which was why it was so important to be able to say the FBI was investigating the matter. No credible journalist would report on the "dossier" as truth. What was needed were respected journalists who would write a story saying that respectable people were talking about the "dossier."

One of the more damaging stories on Trump was published by David Corn of *Mother Jones*, who had been given a copy of Steele's "dossier." Corn's October 31, 2016, article gained widespread attention just one week before Americans went to the polls to cast their ballots. He asked, "Does this mean that the FBI is investigating whether Russian intelligence has attempted to develop a secret relationship with Trump or cultivate him as an asset?"[102] Corn identified his source as "a former senior intelligence officer for a Western country who specialized in Russian counterintelligence." That, of course, was Steele. Corn described him as "a credible source with a proven record of providing reliable, sensitive, and important information to the US government."[103]

Corn's story was the basis of the fever dream that would possess anti-Trump forces for more than two years. It covered rumors instead of substance; it portrayed unreliable people as models of morality; it included hedges about whether or not the whole thing was true; and it hinted that where there was smoke, there must be fire. Corn quoted directly from Steele's "dossier" that portrayed Trump as a Russian asset who was being controlled through "blackmail" and also associated him with the hacking of the DNC and the Clinton campaign email server. The FBI was in "shock and horror" at what had been uncovered, according to Corn. There was concern that the election was being "undermined" and "delegitimized." [104] Any American who read the story and believed it would surely cast his or her vote for someone other than Trump. It was a devastating indictment. Never mind that it was untrue. Near the end of the article—almost as an afterthought—Corn penned a rather feeble disclaimer: "There's no way to tell whether the FBI has confirmed or debunked any of the allegations contained in the former spy's memos." [105] Thus, after maligning Trump as a probable traitor with depraved intent, Corn belatedly inserted the caveat that it might all be a lie. It was the quintessential journalistic hit job.

Corn didn't stop there. He also provided a copy of Steele's dossier to the FBI's general counsel, James Baker, who dutifully distributed it to the agents and officials investigating Trump-Russia "collusion." [106] The extent to which the two men discussed the document and the FBI's investigation is not fully known because Baker, under criminal investigation by the DOJ for potential violation of FBI regulations, was directed by his lawyer not to answer such questions.[107] However, by the time they met, the Bureau was receiving multiple copies of every successive memo invented by the former British spy. The distribution chain was the definition of overkill. But Simpson and Steele were taking no chances. They planned to inundate the FBI, the DOJ, and the media with the spurious accusations that Trump was in bed with Putin to manipulate the presidential election results.

Corn and Isikoff, it should be noted, subsequently teamed up to coauthor a book, *Russian Roulette: The Inside Story of Putin's War on America and the Election of Donald Trump*, that was inspired by their initial reporting. They described it as "the story of political skull-duggery unprecedented in American history." [108] Weaving together a circumstantial argument that *implied* that Trump and/or his campaign was involved in Russian efforts to throw the election his way, Corn and Isikoff seemed to take credit for the "collusion" scandal that, in the end, proved unfounded and inauthentic. No matter; the book became a number-one *New York Times* best seller. The *Times*, which was itself wholly committed to the "collusion" delusion, lauded it as "the most thorough and riveting account so far." [109] Ironically, Isikoff's previous best seller with Corn, *Hubris: The Inside Story of Spin, Scandal, and the Selling of the Iraq War,* had exposed the faulty and fraudulent intelligence leading to the Iraq War. If only they'd brought the same skepticism to this project.

Nine months after *Russian Roulette* was published, Isikoff told an interviewer that "Steele was clearly onto something." But then, in a cathartic moment, he grudgingly admitted that when "you actually get into the details of the Steele dossier, the specific allegations, we have not seen the evidence to support them, and, in fact, there's good grounds to think that some of the more sensational allegations will never be proven and are likely false." [110] His remarks made headlines. So in a follow-up tweet, he performed a perfect reverse somersault by declaring that Trump had "aided and abetted Moscow's attack on American democracy." [111] Three months later, Special Counsel Robert Mueller delivered his report concluding that "the investigation did not establish that members of the Trump Campaign conspired or co-ordinated with the Russian government in its election interference activities." [112]

JOHN BRENNAN, THE INSTIGATOR OF THE HOAX

President Obama's CIA director, John Brennan, was instrumental in proliferating the "dossier." But even before the Clinton campaign and Democrats funded Simpson and Steele's project to smear Trump with the "collusion" hoax, the seeds of the narrative were germinated by none other than Brennan.

A close confidant of President Obama and a strident supporter of Clinton's candidacy, Brennan was determined to do what he could to place a bull's-eye on the back of her political rival. His reasons were both political and personal: if Clinton succeeded Obama, Brennan stood an excellent chance of maintaining the reins of power as the United States' chief spook at the CIA. He had been close to President Bill Clinton, delivering daily intelligence briefings. He was a lifelong liberal who admitted that he had once voted for the Communist Party. Anyone like Trump who espoused conservative policies was viewed by Brennan as a villain.

It is impossible to overestimate the immense power that resides in the hands of a CIA director. His actions are cloaked in stealth and secrecy. He hides behind legal protections afforded to clandestine operators who are supposed to be acting for the benefit of Americans and against foreign enemies that are dedicated to causing mischief and harm. Technically, he is subordinate to the director of national intelligence; during Brennan's tenure, the DNI was James Clapper. In reality, the CIA director operates quite independently and is answerable to no one, except the president who appointed him or her. Sometimes even the president is kept in the dark. Against that enigmatic backdrop, Brennan's machinations in the "collusion" hoax are difficult to unspool.

There is evidence that Brennan began formulating the Trump-Russia odyssey sometime in the latter part of 2015, long before the Clinton campaign and Democrats commissioned the "dossier" from Steele. According to the British publication *The Guardian*, "Britain's spy agencies played a crucial role in alerting their counterparts in Washington to contacts between members of Donald Trump's campaign team and Russian intelligence operatives." [113] Though this may sound ominous, the purported Russian "contacts" by associates of the campaign appear to have been either innocuous or nonexistent. No evidence of an election conspiracy or coordination was picked up by what was described as "routine surveillance" of Russians.[114] If that sketchy information was, indeed, flagged by foreign intelligence, Brennan would have been the natural recipient in the United States.

The CIA is prohibited by law from monitoring or spying on its own citizens.[115] Brennan certainly obtained information from somewhere, and the vaunted intelligence service of the United States' closest European ally seems logical. The CIA director boasted about how he had been the first to alert the FBI about "collusion" when he testified before the House Intelligence Committee in May 2017:

> I was aware of intelligence and information about contacts between Russian officials and U.S. persons that raised concerns in my mind about whether or not those individuals were cooperating with the Russians, either in a witting or unwitting fashion, and it served as the basis for the FBI investigation to determine whether such collusion—cooperation occurred.[116]

Brennan was just getting started. As he exerted uncommon pressure on the FBI to pursue a counterintelligence probe on Trump, he resolved to help spread the false allegations to Congress and the media. Simpson and Steele were already hounding reporters with their "dossier" when Brennan took a trip to Capitol Hill to tell top members of Congress of his "information indicating that Russia was working to help elect Donald J. Trump president." [117] Brennan argued that Trump was in on the conspiracy. He politicized phony intelligence and instigated the fraudulent case against Clinton's opponent.

Among the House and Senate members whom Brennan briefed were the so-called "Gang of Eight." [118] These are the leaders of both parties, as well as the chairmen and ranking members of the respective intelligence committees. Brennan held separate meetings with each of them, detailing the uncorroborated information on "collusion" that was based largely on the phony "dossier." Although Brennan testified before lawmakers in May 2017 that he had been unaware of Steele's document and/or its contents until after the election, that was patently untrue.[119] Brennan also denied, under oath, that the "dossier" had been a part of the intelligence review of Russian interference in the election and therefore could not have influenced the published report. That was also untrue, as evidenced by documents later uncovered.[120] Lying to Congress is a crime.

Brennan's next move was truly Machiavellian. Armed with the "dossier" itself or, at the very least, its allegations, he briefed Senate majority leader Harry Reid (D-NV) on August 25.[121] The plan was for Reid to send a letter to Comey demanding an investigation. The correspondence would be leaked to the media, giving journalists an excuse to publish stories about Trump-Russian "collusion." Reid crafted the letter, which all but indicted Trump. Two days later, on August 27, Reid dispatched his correspondence to Comey accusing Trump of being an "agent of Russia and the Kremlin" in an effort to "undermine our free and fair elections."[122] The accusations were straight out of Steele's "dossier"—almost word for word. Trump was accused of participating in the hacking of Democratic emails and a Trump adviser was accused of being bribed to effectuate future Trump policies toward Russia. It was all nonsense, but that didn't matter. Obviously, Brennan had delivered the contents of Steele's salacious document to Reid, and it was being put to destructive use. Naturally, Reid demanded an investigation, although he and Brennan certainly knew that one had already been officially opened. The letter and its allegations were all for show. The media needed a hook to justify running the story.

Like Pavlov's dogs, journalists capitulated. The *New York Times* and the *Washington Post* reacted as predicted, publishing articles suggesting that Trump was a Russian asset. Other news organizations jumped on board and reported much the same. On August 29, the *Times* connected Trump advisers to "Russian leadership."[123] In an August 30 column, Josh Rogan of the *Post* employed the word "collusion" in the lead sentence of his piece.[124] That seems to have been the genesis of a term that was laden with criminal overtones and a descriptive pejorative that would become the fountainhead of the

Russian hoax. Hillary Clinton seized on Reid's letter and publicly demanded that Trump be investigated. Her supporters cheered the fake news.

It is difficult to know just how many millions of Americans accepted that propaganda, but it did not gain the result that Brennan and Reid desired. The Republican nominee did not sink in the polls. A second "sub rosa" attack would have to be engineered. It happened on October 30, when Reid fired off another letter to Comey that repeated his accusation of "explosive information about close ties and coordination between Donald Trump . . . and the Russian government." [125] As before, it was immediately leaked to the media, which were now working full-time to debilitate Trump's chances in the election that was just days away. The *Washington Post*, which had coined the invidious term "collusion," repeated the word in a story the next day.[126] So did other news outlets as they papered the nation with Reid's incendiary letter and fabricated accusations. From there the word seemed to take on a life all its own. For the next two years and beyond, it would be misused and misunderstood in a deliberate effort to falsely accuse Trump of crimes he did not commit. The media, which never bothered to examine the legal meaning of "collusion," embraced it as their cudgel to bludgeon Trump in the press and on the airwaves.

As befitting common crooks who get caught and turn on each other, Brennan and Reid later gave conflicting accounts of what they'd done. Brennan claimed he had urged Reid not to send Comey his letters, while Reid claimed he had acted as Brennan wanted.[127] Neither one of them has an ounce of credibility, so it's difficult to know which one is lying. Reid has a notable and remorseless history of spreading a lie to damage a Republican candidate for president.[128] Brennan was trained in deception, duplicity, and chicanery. He has a long and distinguished record of dishonesty that has been well chronicled over the years.[129] Input the words "Brennan lies" into any conventional search engine on the internet, and your computer will explode. Having repeatedly demonstrated that he cannot be trusted with classified or top secret information, the former spook constituted a one-man national security risk.

We expect CIA directors to be nonpartisan and apolitical. Brennan was not. We expect CIA directors to be circumspect, knowledgable, and concerned about the reputation of their organization. Brennan was not. We expect CIA directors to be focused on genuine national security threats and getting to the bottom of things. Brennan was not. He misused classified information and disseminated false intelligence. Instead of collecting and analyzing intelligence to protect Americans from harm, Brennan misappropriated it to harm them, including Trump. He weaponized his powers for political reasons and personal gain. In August 2018, Trump announced that Brennan's security clearance would be revoked, but it never happened.[130] It was too late; the CIA director had already wrought enormous damage.

"MAVERICK" JOHN MCCAIN AND DAVID KRAMER GIVE THE "DOSSIER" REPUBLICAN CRED

Senator John McCain (R-AZ) seemed to revel in his moniker, "Maverick." Like the lead character in the movie *Top Gun*, he was a rebel and a nonconformist who chaffed at party doctrine and often embraced unorthodox positions for a Republican. He saw himself as independent minded. Others viewed him as disruptive, reckless, and inept. As the GOP nominee for president in 2008, he had run a miserable losing campaign that was notable mostly for his chronic mistakes and misjudgments. *Newsweek* called it "aimless and chaotic," wondering in a headline if it was the "worst ever" campaign.[131] That was a generous assessment.

It was no wonder that McCain loathed Trump. The two men had traded insults for years, but in July 2015, while campaigning in Iowa, Trump had raised the level of rancor by disparaging McCain's status as a war hero.[132] The Arizona senator was livid and never seemed to set aside the invective that Trump had hurled at him. Two years later,

McCain groused to a *60 Minutes* interviewer that the president had failed to apologize.[133] The incident helps explain why the senator may have been more than eager to seek retribution when the opportunity presented itself immediately after the presidential election in November 2016.

Simpson and Steele had been industrious and relentless in proliferating their "dossier." They had given it to the media, the FBI, the DOJ, the State Department, the CIA, and other intelligence agencies, as well as the British government. Sir Andrew Wood, who had served as the United Kingdom's ambassador to Russia and had ties to Steele's company, Orbis Business Intelligence, had been briefed on the document's contents.[134] He would become a key player. It was Wood who decided to tip off David Kramer, McCain's longtime associate, while they attended a security conference together in Halifax, Nova Scotia. So alarmed (or excited) was Kramer that he scheduled an impromptu summit. Wood, Kramer, and McCain met privately on or about November 19, 2016, whereupon the ambassador began spinning the tall tale of how Trump was in league with Putin and the Russian government. The incontrovertible evidence was contained in memos by Steele. Preposterous as the story was, McCain was anxious to exploit it against his nemesis. Kramer was dispatched to London to meet with Steele, and his document was thereafter obtained.[135]

Once McCain had his hands on the "dossier," the dissemination process commenced. The senator met with Comey on December 9 and gave the document to the director.[136] McCain would later deny having done that, only to eventually recant. Of course, the FBI already possessed the "dossier," having received it from various sources in periodic installments over the previous five months. Simultaneously, Kramer began a comprehensive campaign to blanket the media with the false but provocative allegations. He met with or spoke to a dozen reporters, including from CNN, ABC News, *The Guardian*, NPR, McClatchy, *Mother Jones*, the *Washington Post*, and the *Wall Street Journal*.[137] Yet no one was informed that the "dossier" had been financed by Hillary Clinton's campaign and Democrats. Conveniently, Steele's disqualifying anti-Trump bias and subsequent termination by the FBI for leaking and lying were also concealed. It did not seem to matter to anyone that the document was suspect on its face, as was the motivation of its creators, Simpson and Steele. It was all conveniently overlooked. Or studiously buried.

McCain's role was necessary and vital. By virtue of his senior status as a US senator and a Republican to boot, he lent the document the kind of credibility and stature that Simpson and Steele had long craved. The Democratic origins of the "dossier" could be obscured if a respected official from the same party as Trump stamped it with a hearty endorsement. It had to be taken seriously, they reasoned, since it was now countenanced by the esteemed John McCain. The senator's abiding antipathy toward the new president was dexterously overlooked or forgiven.

Among the journalists to whom Kramer delivered the "dossier" was Ken Bensinger, a reporter for BuzzFeed News. In his deposition, Kramer told a ridiculous story of stepping out of the journalist's office for a period of time and leaving the "dossier" behind, not realizing that Bensinger would photograph it with his iPhone.[138] Right. Kramer later confessed that that account was untrue and that he had, in fact, given the reporter a copy. Kramer admitted to another false statement he had made to Steele.[139] Hence, Kramer is a completely unreliable witness, except for the fact that he obtained the "dossier" and gave it to journalists. BuzzFeed published the document for the first time on January 10, 2017, just ten days before Trump would be inaugurated. CNN aired its story about it the same day. It had been leaked to the network by the director of national intelligence, James Clapper, who proceeded to condemn leaks in his television appearances.

The story was about to explode, as newspapers and television networks raced to implicate Trump as an agent of Russia who had stolen the election from Clinton. The witch hunt was on. It would encumber his presidency from the outset, impair his ability to govern, and torment him with false accusations for the next two years. Congress would investigate and harass. Comey's FBI would scrutinize his every action. A special counsel would pick up the baton and launch an un-warranted probe laden with unfairness. And the mainstream media would convict Trump in the court of public opinion on an almost daily basis. Treason would be the stated and unstated truncheon that would haunt the president until the day Robert Mueller would reveal that there was no evidence of a "collusion conspiracy."

It is nearly impossible to conceive how two nefarious people, Steele and Simpson, could hold a nation and a presidency hostage for more than two long years over nothing more than a conjured hoax. They had assistance, to be sure. Funding by Clinton's campaign and Dem-ocrats nourished the con. Unscrupulous officials at the FBI and DOJ drove it to excess, abusing their powers and defiling the rule of law. Journalists became witting and fervent partners throughout the witch hunt. They never flagged in their devotion to destroy Trump and undo his election.

NO, THE "DOSSIER" WAS NEVER "MOSTLY TRUE"

One of the most prominent and enduring falsehoods promoted by the mainstream media in their "collusion" narrative was that the "dossier" was mostly true. That was a constant refrain for more than two years. It was a clever sleight of hand by journalists who became deeply in-vested in the story that Trump had conspired with Russia to unduly influence the presidential election. It turned out that the only por-tions of the "dossier" that were true were not at all related to Trump campaign "collusion" but gleaned from readily available records and public reporting. The parts alleging a Trump-Russia conspiracy were false. Yet the media could never admit that they'd been played. They yearned for the "dossier" to be true in all respects, so they kept repeat-ing the "mostly true" mantra in print and on television as if saying it was so somehow made it so.

BuzzFeed, which was the first to print Steele's memos in their entirety in January 2017, was especially committed to justifying what it had done. The website was severely criticized for publishing the outlandish allegations that had not been verified. Nearly two years later, in December 2018, BuzzFeed's editor in chief, Ben Smith, gave an interview to *Vanity Fair*. The magazine shrewdly cherry-picked one of Smith's isolated remarks and adopted it for the article's bold headline, " 'The Broad Outline of What Steele Was Writing Is Unquestionably True.' "[140] However, if you dig through both the story and the attached transcribed interview, that brief declaration by Smith is the only comment he made about the purported veracity of the "dossier." He offered nothing more. He furnished no explanation of what was "unquestionably true" and provided no evidence to back up his assertion. The article's author, Joe Pompeo (who presumably posed the questions), did not bother to ask the natural and inevitable question "What exactly in the dossier is 'unquestionably true'?" His complete lack of curiosity suggests that Pompeo himself had accepted the document as the equivalent of gospel written in stone. Hence, readers were left with the distinct impression that Steele's document *must* be true. Period. No challenge was made, and no criticism was offered. It was simply recognized and sanctioned on faith alone.

Consistently on CNN, MSNBC, and several broadcast networks, anchors and reporters repeated the theme that the "dossier" was mostly true. So did their guests. Ex–CIA director John Brennan and former director of national intelligence James Clapper, both of whom had served under Obama, were particularly adept at leveling the fiction without enumerating details. That extended to their print interviews, as well.

In a question-and-answer session with Salon that was picked up by other news outlets, Clapper proclaimed, "Some of what was in the dossier was actually corroborated—but separately—in our intelligence community assessment, from other sources that we were confident in."[141] He tendered no explanation of what parts had been corroborated. Instead, he claimed, "We did not use the dossier as a

source for the intelligence community assessment, that's point one." [142]
Later he added, "As time has gone on more and more of it has been
corroborated, but I can't actually give you a percentage." [143]

That may be the most important moment in the entire mass delusion. How could the average citizen, watching the media frenzy, figure
out what had actually happened? The easiest answer was to listen to
what the former head of intelligence said, and here he was saying that
intelligence agents looked into it and most of it was true. If Trump
had personally coordinated with KGB hackers to break into voting
machines, that was what you would have expected Clapper to say.

Clapper's contention that the "dossier" was never used as part of
the intelligence community assessment (ICA) report issued in January 2019 was repeated by Brennan when he told Congress, under
oath, that the document was "not in any way used as the basis for
the intelligence community's assessment" that Moscow had interfered
in the election for the benefit of Trump. [144] However, Brennan and
Clapper were spinning a deception. A prominent colleague contradicted them and produced documents as proof that they were not
telling the truth. In a classified letter to Congress, National Security
Agency director Michael Rogers disclosed that the uncorroborated
document "did factor into the ICA" report. [145] Rogers confirmed to
the chairman of the House Intelligence Committee that a summary
of Steele's document had even been added as an appendix to a draft
version of the ICA that had contributed to its final analysis. [146] Having been caught in a falsehood, Clapper then repudiated his earlier
statement by telling CNN that "some" of the "dossier" had been used
for the intelligence assessment report. Was Clapper lying in his first
statement or his second statement? Brennan continued to deny all of
it, the contrary evidence notwithstanding.

None of that seemed to concern CNN. The network hired Clapper as a "national security analyst." [147] His pal Brennan was retained as
a paid "intelligence analyst" for MSNBC and NBC News. [148] The two
superspooks launched an all-out attack on Trump, exploiting their
new television platforms to advance the toxic fiction that the president

was a secret Russian asset who had "colluded" with Putin. It didn't matter to CNN that a House Intelligence Committee report determined that it had been Clapper who had leaked news of the phony "dossier" to the network before Trump had ever taken office.[149]

At first, Clapper lied during his private testimony with congressional investigators by "flatly denying" the leak. He eventually confessed to the committee that he had told CNN host Jake Tapper about the "dossier" in early January 2016.[150] Ironically, Clapper persisted in publicly condemning the leaking of information while he covertly leaked to other news organizations beyond CNN.[151] The network didn't care about its new hire's culpability or the false and sleazy document he had propagated. It happily collected a prestigious award for its Clapper-fed report on what turned out to be a nonexistent conspiracy with Moscow. Nor did it matter to NBC/MSNBC that Brennan had been the instrumental "instigator" of the Russian hoax and had delivered less-than-honest testimony before Congress on numerous subjects. He was allowed to use its TV pulpit to condemn Trump with every appearance and freely accuse the president of a death penalty offense, treason. The humiliating level of media malpractice in the age of Trump will be explored in depth in a later chapter of this book.

The persistent fairy tale that the "dossier" was "mostly true" is one of the more confounding aspects of the entire Trump-Russia faux scandal. The media, as well as Democrats, routinely pointed to select parts of Steele's memos as justification for their assertion. An example of this is a January 2019 online story by CNN, published two months before the special counsel ended its probe, which told its readers and viewers that "many of the allegations that form the bulk of the intelligence memos have held up over time, or have proven to be at least partially true."[152] The network then explained what parts were true or "partially true." It turned out that all of those earth-shattering parts were simply a restatement by Steele of what had already been a matter of public record before Steele ever composed his memos. In other words, the ex–British spy mined old information and sold it as new. He appears to have gathered some of it from the internet, newspapers, or television stories. The CNN story is worth examining because it proves the point and is illustrative of what other journalists contended. The reporters identified four sections of the "dossier" that they claimed were true.

First, CNN cited Steele's claim that there had been "Russian meddling in the 2016 election." Yes, that is an obvious truth. But it has been true for decades and well chronicled and documented by various intelligence agencies throughout the years, as noted in the ICA report.[153] An intelligence agent such as Steele would have been well aware of that. The Russians have attempted to sow discord in Western elections in much the same way that the US government has spread false information to influence elections in other nations of the world.[154] It is a frequent practice. There is nothing *new* in this. It's old news. Yet the media sold it as a stunning revelation. But Steele added a new dimension to it by claiming that Trump had been involved in the Russian interference. There was no evidence of it when the "dossier" was written, nor is there any evidence of it now.

Second, CNN repeated Steele's claim that there had been "secret contacts between Trump's team and Russians" during the campaign, adding that they had involved "at least 16 Trump associates." The genesis of that yarn came from a February 14, 2017, story in the *New York Times* that cited unnamed officials as its source.[155] McCabe reportedly shot down the story, stating "It's total bullshit . . . it's not true." [156] Comey told Congress under oath that the article "in the main, was not true." [157] Yet the media did not care that the FBI had debunked it. Journalists continued to perpetuate the fiction without attempting to establish even a semblance of balance by reporting that there had actually been more known contacts between the Hillary Clinton campaign and Russians.[158] The mainstream media largely ignored evidence from the Ukrainian Embassy that during the 2016 campaign a contractor for the Democratic National Committee "solicited dirt on Donald Trump's campaign chairman and even tried to enlist the country's president to help" in an effort to boost Clinton's chances against her opponent.[159]

Both Republicans and Democrats had numerous conversations with Russian officials before and during the campaign. Many of them were public knowledge, while others were not. So what? Foreign policy is always a relevant election issue, and no law is violated by engaging in constitutionally protected speech. This includes discussions or communications with foreigners and foreign governments. Absent some evidence of a conspiracy to defraud an election, it is not a crime to meet or talk with a Russian. Once again, Steele's "dossier" took otherwise legal and permissive "contacts" and contorted them into a plot to rig the presidential election result. Proof of such a scheme never emerged. Steele's account of it was nothing more than a fiction that was then exaggerated and animated by the media.

Third, CNN recited Steele's claim of "Trump's real estate dealings in Russia." No real estate "deals" were ever consummated by Trump, despite periodic efforts to explore potential development projects and licensing agreements in both Moscow and Saint Petersburg in an effort to expand the Trump brand. Dating back to 1987, before the fall of the Soviet Union, Trump had begun scouting prospective partners and government permission to discuss the construction of hotels, office buildings, and mixed-use commercial and residential towers.[160] He had made several visits to Russia. The ventures were often publicized and well known. That was likely how Steele gained the information. From time to time, plans were drawn and discussions were held. However, nothing ever came to fruition. A year before the presidential election, Trump signed a "letter of intent" to *discuss* the construction of a Trump tower along the Moscow River.[161] As before, the talks ended without an agreement in mid-2016. No contracts were signed, and nothing was built or licensed. Given the worldwide presence of Trump Organization projects, it would have been extraordinary if he had not at least explored the idea of a real estate venture in Russia. Simply put, there was nothing illegal or nefarious in a real estate developer seeking developments in real estate. But the media ran with it as evidence of "collusion."

Finally, CNN regurgitated Steele's claim that retired army general

Michael Flynn, who later became an adviser to Trump, had traveled to Moscow in December 2015 to deliver a paid speech to a state-run news organization called Russia Today (RT). That was most certainly true. A businessman and private citizen at the time who was entitled to earn an income, Flynn nevertheless met with Pentagon officials before the speech to make them aware of his upcoming visit, and he debriefed them upon his return.[162] But Flynn was not alone in his sojourn to Russia. Jill Stein, who became the 2016 Green Party presidential nominee, also attended the dinner.[163] They were both seated at a table with Putin. Photographs were taken, and the event received some publicity. Steele seems to have incorporated that public information into his "dossier" and contorted it into some kind of an odious criminal enterprise. It was not.

The only truthful information in Steele's "dossier" was derived from preexisting and inconsequential public information that could easily have been obtained by almost anyone. Moreover, none of the so-called truths implicated Trump or his campaign in wrongdoing or a grand conspiracy to steal the presidential election. Yet the media gave sustenance to Steele's document by misrepresenting that it was "mostly true." They implied that "collusion" must be true because noncollusion statements were true. Steele had been clever; he had carefully woven a handful of true statements into his many untruthful memos. In the alternative, he had been spoon-fed a combination of lies and truths by sources in Moscow. This is a typical tactic employed by Russian intelligence operatives in covert actions. By mixing truths with untruths, "it encourages you to believe the falsehoods," according to Daniel Hoffman, a former CIA station chief.[164] Had journalists been honest and forthright, they would have said, "None of the *collusion* allegations in the Steele dossier has been corroborated or proven true." That they did not do.

The media practice of vouching for the credibility of Steele's "dossier" continued up through, and including, the day Attorney General William Barr revealed, "The Special Counsel's investigation did not find that the Trump campaign or anyone associated with it conspired or coordinated with Russia in its efforts to influence the 2016 U.S. presidential election." [165] Suddenly journalists stopped arguing that the "dossier" was "mostly true." They pretended that the "collusion" narrative driven by Steele's document had never happened and that they had never been complicit in advancing its many false stories.

Those in the US intelligence community who had been the chief proponents of the "dossier" suddenly became deathly allergic to it. In his tiresome book, the tired James Clapper dismissed it as "a collection of seventeen 'pseudo-intelligence' reports created by a private company, which I first learned about from John Brennan." [166] Hmm. Then why did Clapper leak it to the media in January 2017? Is he now blaming Brennan? Is he trying to erase or redact his own role in peddling a phony document that created a national nightmare? The answer is yes.

For his part, Brennan now pretends he knew all along that the "dossier" was inauthentic. Yet for more than two years he championed the document as evidence of Trump-Russia "collusion" and accused the president of treason. He even insisted that the "dossier" be included in the classified intelligence community report on Russian interference.[167] Then came the special counsel report. In a Jekyll-Hyde transformation, the sheepish Brennan conceded, "I don't know if I received bad information, but I think I suspected there was more than there actually was." [168] That's quite a metamorphosis for a guy who had enthusiastically endorsed the "dossier" and who had been quoted as saying that "it was in line" with his own CIA sources, in which he "had great confidence." [169] Really? What sources? Steele and Simpson? Clinton?

And then there is James Comey. He did not hesitate to exploit the "dossier" to fuel the launch of the Trump-Russia investigation. He later admitted to the president and Congress that he had known it was "unverified." But that had not stopped him from representing to a court that it had been "verified" in order to obtain a warrant to spy on the campaign. This will be explored at length in the next chapter.

Former federal prosecutor Andrew McCarthy observed that none of that prevented Trump's enemies from making the document, funded by the Clinton campaign and Democrats, the centerpiece of their drive to frame him:

> By any objective measure, Steele's dossier is a shoddy piece of work. Its stories are preposterous—the "pee tape," the grandiose Trump-Russia espionage conspiracy, the closely coordinating Trump emissaries who turned out not even to know each other, the trips and meetings that never happened, the hub of conspiratorial activity that did not actually exist.[170]

The "collusion" narrative was a conspiracy in and of itself, fabricated as a political instrument and then weaponized by unscrupulous government officials. Attorney General William Barr likened their actions to those of the ancient Praetorian Guard, an elite unit established to protect Roman emperors but that instead often plotted to overthrow them:

> Republics have fallen because of Praetorian Guard mentality where government officials get very arrogant, they identify the national interest with their own political preferences and they feel that anyone who has a different opinion, you know, is somehow an enemy of the state. And you know, there is that tendency that they know better. . . . That can easily translate into essentially supervening the will of the majority and getting your own way as a government official.[171]

CHAPTER 3

LYING AND SPYING

I think spying did occur.

—Attorney General William Barr, Senate testimony, April 10, 2019

I have not gotten answers that are satisfactory. And some of the facts I've learned don't hang together with the official explanations of what happened. Things are just not jiving.

—Attorney General William Barr,
interview with CBS News, May 31, 2019

Media madness and Democrat delirium reached a new level of irrational frenzy the moment Attorney General William Barr uttered five simple words while testifying before a Senate appropriations subcommittee on April 10, 2019. In observing "I think spying did occur," the attorney general was stating the obvious. Any reasonable person who had been following the "collusion" hoax and ensuing witch hunt was well aware that the FBI and Department of Justice not only had wiretapped a Trump campaign associate and gained access to his electronic communications but had retained one or more undercover informants to covertly infiltrate the campaign to gather incriminating evidence that was never found. It was classic spying, to be sure. Yet Trump's opponents fumed incessantly over how Barr had misappropriated what they regarded as such a crass term to describe the truth of what had occurred.

The attorney general is an experienced and learned man. He did not recklessly employ terminology to mislead or misrepresent. He had been asked a simple question and chose his words carefully to give an even simpler answer. The controversy was sparked when Barr informed senators that he would examine diligently why the FBI had initiated its counterintelligence investigation and "the conduct of intelligence activities directed at the Trump campaign" because "I think spying on a political campaign is a big deal, it's a big deal."[1] Indeed it is. The somewhat stunned and disbelieving Senator Jeanne Shaheen (D-NH) later posed a follow-up question in the negative, the answer to which she seemed not to expect:

> **SHAHEEN:** So, you're not—you're not suggesting though that spying occurred?
> **BARR:** I don't—well, I guess you could—I—I think spying did occur. Yes, I think spying did occur.
> **SHAHEEN:** Well let me . . .
> **BARR:** But the question was whether it was predicated—adequately predicated and I'm not suggesting it wasn't adequately predicated but I need to explore that. I think it's my obligation. Congress is usually very concerned about intelligence agencies and law enforcement agencies staying in their proper lane and I want to make sure that happened.[2]

What did Barr mean by "adequately predicated"? He intended to find out whether the spying had been conducted *lawfully*, and he seemed skeptical. At one point in his testimony, he ventured that "there was probably a failure among a group of leaders" at the FBI, and he felt "an obligation to make sure that government power was not abused." By any reasonable standard, Barr's statements made perfect sense. His attitude is precisely what any law-abiding American would want from their attorney general. As the chief lawyer for the federal government, his role is to ensure that everyone upholds laws without exception and to bring prosecutions where warranted. This necessarily includes any violations of law by those who serve in government—even law enforcement. If people within the FBI and DOJ abused their positions of trust and committed illegal acts, Barr would be duty bound to hold them accountable. A lawful warrant, for example, that was obtained by unlawful means—such as lying to a court—would constitute criminal behavior.

Once Barr uttered the sacrilegious S-word, he was universally condemned by the media elite and Democrats predisposed to traduce anyone who dared to be fair to Trump or other persons whose rights might have been violated. The attorney general, who had a distinguished record and reputation as a nonpolitical "lawyer's lawyer" was suddenly denounced as a legal heretic who had succumbed to Trump's presidential skullduggery.[3] House speaker Nancy Pelosi (D-CA) decided that Barr had gone "off the rails," while Senate minority leader Charles Schumer (D-NY) insisted that the AG was "perpetuating conspiracy theories."[4] NBC News's Chuck Todd replicated the "conspiracy theory" meme by insisting that there was "zero factual basis" to say that spying had occurred.[5] The *Meet the Press* moderator must have overlooked the eavesdropping warrant and slept through the multitude of stories on how an undercover agent for the FBI had insinuated himself into the Trump campaign. Tim O'Brien of Bloomberg likened Barr to a "ruthless and sleazy attack dog" in the mold of the infamous lawyer Roy Cohen.[6] *Washington Post* columnist Jennifer Rubin wrote that lawmakers should demand the AG's resignation for being "Trump's toady."[7] Not to be outdone by anyone, CNN's chief legal analyst, Jeffrey Toobin, described Barr's terminology as "paranoid lunacy of the right wing."[8] The attorney general's impressive résumé of accomplishments and sterling reputation for integrity should have disabused any person of the notion that he is paranoid or prone to lunatic embroidery. By employing the word *spying*, he spoke in plain language that was both accurate and apropos to the circumstance. By describing what had transpired, he was telling the uncomfortable truth. The unreasoned reaction it provoked only made it seem more so.

The verb *spy* encompasses a broad meaning. The *Oxford Dictionary* defines it this way: "work for a government or other organization by secretly collecting information about enemies or competitors."[9] Let's apply this to the known facts. The FBI and Justice Department sought and secured a warrant to wiretap Carter Page, who had served for a few months as a Trump campaign foreign policy adviser. FBI director James Comey, who signed the first warrant application, asserted without equivocation that Page was a Russian agent and hence a treasonous enemy of the United States. That turned out to be completely untrue, of course. But the Foreign Intelligence Surveillance Court judge *trusted* that Comey was telling the truth and that the evidence he had submitted had been duly verified, as FBI rules and FISC regulations required.

Once the warrant was issued on October 21, 2016, the FBI began a wiretap on Page's communications devices and accessed all his electronic data, including emails and texts. For up to a year, Bureau agents secretly listened to his conversations and collected a variety of information without his knowledge. Although Page had left the Trump campaign by the time the surveillance of him commenced, the warrant allowed the FBI to retroactively access and read his communications with campaign officials and staff during the six months he had been an unpaid adviser.[10] It was a way to spy on him going both forward and backward. In the absence of the court order, that extreme intrusion would have been an otherwise impermissible violation of the Fourth Amendment rights of Page to be secure from unreasonable search and seizure by the government. But the issuance of the judicial order did not render the term *spying* inapplicable or irrelevant. There is *legal* spying and *illegal* spying. Both are a proper use of the terminology.

It was confounding and silly for the media and Democrats to take such vociferous exception to Barr's use of the term *spying*. They were quibbling over semantics and drawing a distinction where none existed. The FBI's conduct matched all of the definitional elements of the word. That is, the *government* was the active operator, *information* was *collected*, it was accomplished through *secrecy*, and the target was alleged to have been an *enemy* of the country. Thus, when Attorney General Barr stated, "I think spying did occur," he was delivering an honest description of what had happened. He openly acknowledged reality, to the shock of Democrats and the liberal media, who are unaccustomed to candor. The facts appear indisputable because a redacted version of the wiretap warrant application was subsequently made public for all to read, and FBI officials have testified that Page was surveilled. Columnist Byron York distilled the question to its bare essence when he asked, "Is a wiretap 'spying'? It is hard to imagine a practice, whether approved by a court or not, more associated with spying." [11]

The AG's statement was supported by the established evidence and was unquestionably accurate. Barr then explained that it was a vital part of his job as attorney general to examine whether the spying had been "adequately predicated." If the warrant application was dishonest and defective, then it had not been adequately predicated. If Comey and others who had prepared the application willfully misrepresented incriminating facts and concealed exculpatory evidence, a fraud on the court had been perpetrated. A firestorm erupted over the unvarnished truth. It was not at all "stunning or scary," as former director of national intelligence James Clapper called it, unless he was referring to alarmingly illegal behavior by the FBI.[12] He was not. He seemed to be *stunned* and *scared* that government malfeasance might be investigated and exposed. As former Justice Department lawyer Hans von Spakovsky pointed out, "Simply ignoring the issue of whether the spying against the Trump campaign was justified would be irresponsible and a dereliction of duty by the attorney general." [13] In a democracy, the government of one party should never be permitted to misuse its immense powers in law enforcement and intelligence gathering to target the candidate of an opposing party. That was what the attorney general vowed to investigate. He expressed his suspicions.

Barr found it bewildering that the FBI had never bothered to notify Trump or his top aides that Russians might be trying to infiltrate his campaign. "That is one of the questions I have, that I feel normally the campaign would have been advised of this," said the attorney general.[14] Former New Jersey governor Chris Christie and former New York mayor Rudy Giuliani were senior campaign officials with national security experience as former federal prosecutors. According to Barr, they would be the kind of people the FBI would have typically alerted. Nor had congressional leaders been fully and candidly briefed on the counterintelligence case. Comey's FBI had deliberately kept them in the dark, sidestepping traditional agency protocols. Kimberley Strassel of the *Wall Street Journal* offered the most plausible explanation for the secrecy:

Mr. Comey and his crew have also testified that they were all convinced Mrs. Clinton would win the election. That would have meant that no politician other than the incoming Democratic president would have known the FBI had spied on the Trump team. Nor would the public. A Clinton presidency would have ensured no accountability.[15]

When the unexpected happened and Trump was elected president, the FBI and intelligence officials such as Clapper and CIA director John Brennan were forced to change tactics and reverse course. They concocted a plan to leak the anti-Trump "dossier" to the media by first telling Trump about it. Evidence shows that it was Clapper's idea and Comey agreed to it. They met with Trump on January 6, 2017, in a conference room at Trump Tower in Manhattan. Adhering to their prepared script, Comey stayed behind to selectively brief Trump on the "salacious and unverified" material in the "dossier."[16] The director later said that Trump had been shocked. Comey also advised the president-elect that he was not personally the subject of their counterintelligence investigation.[17] This was untrue. Trump was most certainly the main target of the probe.[18] News of the Trump debriefing on the "dossier" was then leaked to journalists by Clapper and others. That gave journalists the pretext or excuse they needed to air and publish stories, reasoning that since the incoming president had been made aware of the "dossier," they were free to run with the story even though the collection of memos was unverified and might well be false.

Was President Obama a party to the scheme? Consider the sequence of events and the known facts. On January 5, 2017, the day before the Trump briefing, Obama held a meeting at the White House with Comey, Clapper, Vice President Joe Biden, acting attorney general Sally Yates and national security advisor Susan Rice. They deliberated ways to keep Trump in the dark about the counterintelligence operation in which he was the prime target.[19] The next day, Comey delivered an incomplete, if not deceptive, briefing to the president-elect. Thereafter, the media was fed the "dossier" designed to damage him. The incoming president was blindsided by the outgoing president's collaborators.

It was a clever and insidious subterfuge. And it worked perfectly. Within days, CNN and BuzzFeed were running full tilt with the narrative that Trump had "colluded" with Russia. Those news organizations and others that repeated the story had corroborated none of the accusations against the president-elect. Neither had the FBI. Simpson and Steele, with financial backing from the Clinton campaign and the Democratic National Committee, had finally accomplished their devious goal, albeit belatedly. They had not stopped Trump from ascending to the highest office in the land, but they were now poised to drive him from that office and undo the election result.

It is revealing that Comey did not tell the president-elect about the rest of the "dossier," which now totaled thirty-five pages in seventeen separately dated memos that had been composed by Steele over the previous six months. Trump was not advised that the document accused him of "colluding" with the Kremlin to win the election. He wasn't informed that Clinton's campaign and Democrats had paid for it. Nor did Comey divulge that the FBI had been spying on former campaign adviser Carter Page for three months, eavesdropping on his conversations and accessing all of his electronic communications forward and backward. Trump was not told that the FBI had declared in court papers that Page was a Russian spy. Consistent with the Obama meeting at the White House, the director deliberately hid all of that information from the incoming president. Those were lies by omission. Remember, the purpose of a counterintelligence investigation is to collect intelligence involving a foreign threat to be provided to the president. Instead, Comey was using unsubstantiated information under the guise of a counterintelligence operation to investigate Trump and his campaign, all the while actively concealing it from the president-elect.

Comey later claimed that he had decided to brief Trump on the "dossier" only because journalists were poised to publicly report the information contained therein.[20] This is untrue according to the Senate Judiciary Committee, which determined that "the media generally had found the dossier's unverified allegations unreportable, and CNN only broke the story on the dossier because Mr. Comey briefed the President-Elect about it."[21] Given the way the media tied their stories to the Comey-Trump meeting, the FBI director's account seems remarkably similar to many of the other deceptions that have severely blemished his reputation as a fair broker of the truth.[22] In truth, honesty was never Comey's strong suit. For example, when Barr ventured that "spying did occur" against the Trump campaign, the fired FBI director pretended to be dumbfounded during an interview and sputtered, "I don't understand what the heck he's talking about."[23] Comey knew what Barr was talking about because Comey was the one who did it. He signed the first warrant application to spy on Carter Page, and then he signed two more renewals. But he subsequently feigned ignorance, claiming that "the FBI and the Department of Justice conduct court-ordered electronic surveillance. I have never thought of that as spying."[24] Tell that to Page, who had no idea that FBI agents had accessed all of his electronic communications and were eaves-dropping on his telephone conversations. Having interviewed him, I can attest that he certainly considers himself the victim of govern-ment "spying," regardless of whether a court sanctioned it or not.[25] Someone then illegally leaked it to the media, which reported that Page was under surveillance as a suspected spy.

Semantics aside, there is compelling evidence that the *secret* FISA court that issued the *secret* order to *secretly* surveil Page was also a vic-tim of *secret* lies by top officials at the FBI and DOJ.

THE SECRET FISA STAR CHAMBER

The term *Star Chamber* is a pejorative that can be traced to late-fifteenth-century England, where a high court would convene to dis-pense "justice" in secret. Without public accountability, judges would dispense harsh rulings arbitrarily and in contravention of ordinary rules meant to ensure fairness. Secrecy invited corruption and abuse. The Star Chamber was so anathema to the fundamental rights of man to be judged impartially and in open court that the Habeas Cor-pus Act abolished it in 1640–1641.[26]

When the Framers crafted our Constitution more than a cen-tury later, they were acutely aware and fearful of the kinds of judicial perversions that had taken place during such secret proceedings in their antecedent country. That was the primary reason they devised the guarantee of "a speedy and *public* trial, by an impartial jury" embodied in the Sixth Amendment and the protection against self-incrimination in the Fifth Amendment.[27] The Supreme Court recog-nized that openness and transparency in public judicial proceedings were instrumental to democracy when it stated, "The Star Chamber has, for centuries, symbolized disregard of basic individual rights."[28]

Notwithstanding historical concerns that secret courts tend to breed corrupt acts and corrosive abuses of power, in 1978 Congress passed the Foreign Intelligence Surveillance Act (FISA), establishing a secret court.[29] At the time, it was considered a necessity to counter the prevalence of foreign spies during the Cold War. Privacy rights, it was felt, would have to be sacrificed for the sake of security. Though this specially constituted tribunal does not preside over trials as the Star Chamber did centuries ago, the FISC considers and grants government requests to conduct electronic surveillance on individuals for foreign intelligence purposes. There is a total of eleven judges, all of whom are appointed by the chief justice of the US Supreme Court. However, they do not sit "en banc" or all at once. Instead, a single judge, who normally presides in a district court somewhere in the federal system, rotates into Washington, DC, for a week at a time and occupies a secure courtroom in the US District Courthouse on Constitution Avenue.

Not only are the proceedings of the FISC secret, but they are typically conducted "ex parte," meaning that only the government's side is presented. It is a lopsided process.[30] In other words, there is no one of an adversarial nature in attendance to test or remonstrate against the truthfulness of what applicants from the FBI and Justice Department are representing to the court. To a great extent, the judges must *trust* that government actors are honest, forthright, and fair in presenting evidence in support of "probable cause" that a Fourth Amendment intrusion should be allowed. Statistical evidence suggests that the government is given extraordinary deference. This is the inherent flaw in the FISC.

It is extremely difficult for a judge to discern whether entirely truthful information is being conveyed. Evidence can be slanted or even fabricated. Unsupportable accusations can be made without objection or opposition. Exculpatory material can be concealed, and almost no one would know. In essence, the system is vulnerable to deception because there is no adverse party there to challenge the authenticity and veracity of the documents submitted. A conscientious judge can try to do so, but it is impossible to know how frequently this occurs. The judges' various interactions with federal prosecutors are completely hidden from public view to protect what is purported to be classified national security information. What *is* known is that the FISC approves roughly 99 percent of surveillance requests.[31] In the first thirty-three years of the court's existence, it considered more than 33,900 applications, approving *all* of them except eleven.[32] That constitutes a rejection rate of a mere .03 percent.[33] These startling figures alone create a suspicion of abuse.

It is tempting to conclude that the judges on the FISC simply rubber-stamp whatever the government submits to them. This impression was fueled when documents produced by the Justice Department in September 2018 revealed that no formal court hearings had been held for any of the four successive surveillance warrants that were sought to spy on Carter Page.[34] As a consequence, no stenographic record exists of what questions, if any, were asked by the judges or the verbal representations that were made in court by the FBI and DOJ. However, this does not mean that there were no interactions beyond the paper submissions to the FISC. Formal hearings are not required. Sometimes the judge or a staff member will ask questions informally during in-person conversations or over a secured telephone line and seek additional information.[35] On other occasions, there might be no interaction whatsoever, and the warrant application is approved. If a judge signals that he or she is reluctant to sign off on surveillance, the government will often request a formal hearing.[36]

In the case of the Page warrants, there were no such hearings. Any other interactions that may have happened have not been made public and likely never will be. So it is difficult to assess whether the judges applied some level of scrutiny or, perhaps, none at all. It may have been the case that the FISC was overwhelmed. In 2016, the court had to process more than twenty-eight applications, on average, every week.[37] It is hard to envision how a single judge can handle so many requests in such a short period and give each of them the deliberate and careful examination it deserves. However, the Page warrants were not the customary or conventional fare before the FISC. This was, after all, spying on an associate of a presidential candidate and accessing his communications with the campaign. The political alchemy was glaring. The government of a Democratic administration was seeking permission to surveil the presidential nominee of the Republican Party.[38] That elevated electronic eavesdropping to a whole new and combustible level that demanded a serious and sober examination of the legal justification. Did any of the FISC judges who considered the initial warrant and three renewal warrants say to themselves or their staffs, "Gee . . . we should probably hold a formal and extensive hearing on the record before I affix my signature to these extraordinary documents granting the power to spy on a presidential campaign."

Had such a hearing been convened, surely each of the judges would have asked government lawyers where they had obtained the document (the "dossier") upon which they so wholly relied in their application to spy. Who composed it? Is this person (Steele) as reliable and credible as you contend? Has he expressed an intense political bias? Was this opposition research? Who paid for it? Who exactly are the sources cited in the document? Have *their* accusations been verified or corroborated? Is there any exculpatory information about Page of which we should be made aware? Did you notify the candidate that Russians might have tried to infiltrate or influence his campaign? Further, one wonders whether any of the FISC judges ever requested to examine the "dossier" itself from which the surveillance application was largely drawn. If they had, it would surely have been tossed into the nearest trash can, along with the warrant, for lack of "probable cause." These are the kinds of legitimate inquiries that the unique circumstances compelled. This is especially true given Page's history of *helping* the FBI prosecute Russian agents. That critical fact seems to have been conveniently omitted in the documents presented to the court.

THE PERSECUTION OF CARTER PAGE

Carter Page was probably one of the most insignificant individuals involved in the Trump campaign in 2016. He never met the candidate or spoke with him.[39] There were no electronic communications between the two men. To put it bluntly, Page was a political ornament. He was so peripheral and invisible that few people in the hierarchy of the campaign organization, including Trump, could have picked him out of a lineup. He was hastily selected, as many others were, to serve as an uncompensated volunteer on the Trump foreign policy advisory council. The group was assembled almost overnight by Sam Clovis, who would later become cochair of the campaign. There was apparently very little deliberation and scant vetting of anyone. It was a do-nothing job that was intended to mollify critics who claimed that the presumptive GOP nominee was inexperienced in international relations and deficient in his US foreign policy credentials. The board was window dressing. It literally became a photo op at one point, and Page wasn't even there the day cameras were clicking.

Every campaign tends to engage in such a charade. It cobbles together an economics council or a health care board or a tax cut committee. If the candidate at some juncture needs a talking point or a rebuttal to a thorny issue, an adviser is occasionally consulted. Periodically, the members will convene and talk. The candidate is rarely in attendance. Some "advisers" write position papers that almost no one reads. Their reports get deep-sixed in a file cabinet in a back office. Sometimes, they will help edit a speech that focuses on their specialty. If the candidate wins, the same individuals typically vie for positions in the new administration. Some are picked, while others are thanked for their service and unceremoniously shown the door with the faux-sincere promise "We'll be in touch." It is quintessential politics—image over substance.

Trump was reluctant to play the customary game, but eventually he relented and capitulated. In March 2016, he decided to burnish his foreign policy bona fides by naming less than a dozen people to his board. One of them was Carter Page. Another was George Papadopoulos, a young policy consultant in oil and gas. Like Page, he had never met Trump. In his testimony before Congress, Page described himself as "a junior, unpaid adviser."[40] The record reflects that not much advice was ever given or taken.

If Page can be blamed for making a mistake in retrospect, it rests squarely on his fateful decision to accept an invitation from the New Economic School (NES) in Moscow to serve as one of its commencement speakers on July 8, 2016. The NES had been founded years earlier with American and Western support. Page was flattered to be asked and did not consider his acceptance to be problematic. Why would it be? He knew that President Barack Obama had spoken at the school's commencement on July 7, 2009, just six months after taking office. No one had batted an eye back then or accused him of being a Russian spy. The *New York Times* had even printed a text of Obama's speech for its American readers.[41] The president had lavished praise on the "new Russia," spoken of a global partnership with other nations, and called for a revitalized bilateral effort at cooperation and goodwill. He had reiterated his previous desire for "a 'reset' in relations between the United States and Russia."[42] Obama's proposed rapprochement was strikingly similar to what candidate Trump would advocate seven years later while on the campaign trail to succeed him. Obama's words had been heralded as visionary. Trump's words earned him accusations of being a Russian sympathizer.

Page surmised that the NES had been interested primarily in inviting someone associated, however tangentially, with the upcoming 2016 presidential election.[43] He alerted the Trump team of his speaking engagement and received permission to proceed, as long as it was understood that he was not in any way representing the Trump campaign during his visit.[44] Page agreed and accepted the invitation, never envisioning that such an innocuous trip would soon engulf him in a controversy that would ruin his business, deprive him of income, and make him a target (and victim) of a yearlong surveillance operation by the FBI. As Page would subsequently lament, "It's just so outrageous, preposterous. Where do you even begin?"[45] How could he possibly defend himself against the full force of the FBI, DOJ, and other government actors who branded him, without evidence, as a traitor? How often would his protestations of innocence fall on deaf ears?

The speech Page delivered at the NES was nowhere near as ingratiating toward Moscow as Obama's message. There was nothing polemic or apologetic about the unpaid adviser's words; they were rather banal. And there was nothing clandestine in his appearance at the venue, since his speech was televised locally in Russia and can still be seen on YouTube.[46] Since the NES was supported by the Kremlin, numerous Russian leaders were in attendance. When Obama spoke, former president Mikhail Gorbachev had been seated in the audience, along with then president Dmitry Medvedev. When Page spoke, other Russian leaders were present, although not as high ranking or distinguished. Page would later testify that he had never met privately with any of those officials before or after his speech. No evidence was ever produced to the contrary. Other than shaking a few hands and saying "Hello," his recorded appearance at the school itself was the sum total of all interactions with Kremlin officials.

Upon returning to the United States, Page sent to the campaign what he characterized as a "readout" or synopsis of his visit to Russia.[47] It might appear from his analysis that he had gathered information from Russians directly, perhaps during private conversations. In that regard, Page was probably guilty of an "unfortunate habit of self-puffery," as one writer described it.[48] However, the observations in his "readout" were derived mainly from listening to other speeches, reading Russian newspapers, and watching Russian television. The details of his synopsis were excruciatingly boring, and the totality of his insights was inconsequential. The information had not, as Page would later testify, been acquired during surreptitious meetings with Russian leaders who were hoping to influence the 2016 campaign.

But that was not what Christopher Steele alleged. In his "dossier," he seized on Page's visit to Moscow to accuse him of conspiring with Russian officials to unduly influence the election in favor of Trump. Page is identified several times in the document. On page 9 (the July 19, 2016, memo) it is alleged that Page held "secret meetings" with Igor Sechin, the chief executive officer of the Russian energy giant Rosneft, and Igor Diveykin, a senior Kremlin internal affairs official. The "dossier" claims that the "lifting of western sanctions" was discussed and the Russian officials allegedly "hinted" that they had "kompromat" (compromising information) "on Trump which the latter should bear in mind in his dealings with them."[49] The document noted that "Page was non-committal in response."

The only sin committed by Page was traveling to Russia to deliver a rather mundane speech.

Steele and his employer, Glenn Simpson at Fusion GPS, saw to it that Page's nonexistent secret meetings with Russians were leaked to the media and delivered to the FBI. The *Wall Street Journal* contacted him on July 26, 2016, and inquired about the alleged secret meetings with Kremlin officials that were cited in the "dossier." Page said it was all "ridiculous" and had never happened.[50] Other journalists started pestering him. They had been furnished the "dossier" by Simpson and Steele. No reporter specifically identified Page by name in a story until Michael Isikoff of Yahoo! News fell for the hoax. His story was published on September 23, 2016, and alleged that US intelligence officials were monitoring and investigating Page over secret talks with Russians.[51] When Page read it, he was stunned. The article cited a statement by Senator Dianne Feinstein (D-CA) and Representative Adam Schiff that seemed to tie him to a suspected Russian scheme "intended to influence the outcomes of the election."[52]

Isikoff's story was a near replica of Steele's fictitious account of meetings with Sechin and Divyekin. The reporter had read the "dossier" or been fed its contents verbatim when he met with Steele in September 2016 in a private upstairs room at a Washington restaurant that had been booked by Simpson.[53] What Isikoff did not disclose to his readers was that his "source" was working for Democrats who had paid for the material. Nor did he disclose that the same source had previously given the identical information to the FBI that triggered the investigation Isikoff was reporting.[54] His article referred to his source as a "well-placed Western intelligence source."[55] In fact, Steele was no longer an intelligence agent but a private contractor for hire. That was a significant distinction that created an illusion of importance where little existed. Readers were misled or fooled. Greater accuracy would have cast doubt on the credibility of the reporting. But most of all, Isikoff did not or could not confirm any of the "dossier's" allegations against Page that he had held secret meetings with Russians. More than two years later, when the "collusion" balloon burst, Isikoff admitted that he and other journalists "should have approached this, in retrospect, with more skepticism."[56] No kidding. This is the conceit and folly of hindsight. There can be no substitute for scrupulous reporting or excuse for the failure to perform rigorous fact checking, especially when using anonymous sources.

When Isikoff's story was published, it went viral. Page's shock turned to anger. He had done none of the things of which he was so recklessly accused. He had conspired with no one, including Russians. Within two days, the furious Page fired off a letter to FBI director James Comey advising that the accusations against him were "completely false."[57] He explained to the director that he had assisted the FBI and CIA for many years by virtue of his work as "an American consultant with Russian expertise."[58] Indeed, he had acted as a valuable source for the agencies. Instead of responding to Page or having the FBI interview him, Comey prepared and executed a warrant application to have him wiretapped and to secretly access all of his electronic communications. Deputy Attorney General Sally Yates also signed off on it. At the time, Page had no idea that the FBI was spying on him. Agents would continue to do so for roughly a year. For the next two years, his life would become a living hell as the media villainized him. He was spied on by the FBI and excoriated by Democrats.

When Page was finally allowed to defend himself more than a year later in a hearing before the House Intelligence Committee on November 2, 2017, he explained that his only contact with Russian officials during his July visit to Moscow had been a few brief encounters in the very hall where he had delivered his speech. He described them as "greetings and brief conversation."[59] They had all been benign and inconsequential. Page said he had shaken hands with

Arkady Dvorkovich, the deputy prime minister of Russia, after the speaking event. Their exchange had lasted less than ten seconds.[60] However, Representative Schiff grilled Page at length about that encounter, determined to demonstrate that it had somehow constituted an illicit "meeting" to steal the presidential election from Clinton.[61] In typical fashion, Schiff turned the congressional hearing into a theater of the absurd.

Page denounced Steele's uncorroborated "dossier," calling it "totally preposterous." Secret meetings that were detailed in the document had never happened, he told Schiff. He insisted he had not discussed the Trump campaign with Russians, had never "colluded" with them, and had not been involved in the hacking of Democratic emails, as Steele had asserted.

Schiff was undeterred, convinced that he had cornered the hobgoblin of a grand conspiracy. Citing the "dossier's" October 18, 2016, memo (page 30), he accused Page of having been offered a "19 percent stake in the Russian energy company Rosneft in return for lifting sanctions on Russia." Pause for a moment to consider how idiotic that was. Would Moscow offer the equivalent of $11 billion to a volunteer "unpaid, junior advisor" who had never met the presidential candidate in exchange for the prospect of lifting Western sanctions, which could only occur if that candidate won the race that polls showed he would lose? That, of course, did not stop Schiff from advancing the theory in a lengthy harangue. Page sat there flabbergasted that anyone, much less a US congressman, would even consider such drivel. It was classic Schiff.

Of course, the ridiculous Rosneft bribe never happened. It was just one of more than a dozen false accusations that Steele had conjured up for his phony "dossier." It appears to have been Russian disinformation that he was more than delighted to accept. If Steele thought for one moment that it was true, he has to rank as the most gullible and inept ex–British spy ever to work for MI6. More likely, he knew it was untrue but was eager to exploit it to smear the man he admittedly "despised"—Donald Trump. Some of it Steele may have simply invented himself. He was being paid handsomely by the Clinton machine to manufacture derogatory information about Trump. To that end, he was an obsequious manipulator. The more gossip he could dish for the Clinton benediction, the more he would profit financially. Page was merely an expendable pawn in the spy game and an innocent victim of Steele's treachery.

THE FBI LIES TO THE FISA COURT TO JUSTIFY ITS CRUSADE AGAINST TRUMP

It is a crime to lie to a judge. Depending on the case and circumstances, it could constitute numerous felonies, including perjury, false and misleading statements, obstruction of justice, fraud, conspiracy to defraud, deprivation of rights under color of law, electronic surveillance under color of law, and contempt of court.[62] Those crimes were all described at length in chapter 7 ("Government Abuse of Surveillance") of *The Russia Hoax* and explained in the context of how senior officials at the FBI and Department of Justice appear to have committed those crimes by spying on Carter Page.[63] Four judges signed off on four successive warrants approving the surveillance. But under the law, if the government deceived those judges, it means that lawful warrants were obtained by unlawful means. To put it simply, the judges committed no crimes, but any person who lied to them did. The evidence is compelling that this is what happened.

The FISC was deceived in six material ways:

- Judges were *not* told that Clinton's campaign and the DNC had paid for the information used in the warrant application;
- Judges were *not* told that the FBI's source, Steele, had lied;
- Judges were *not* told that Steele had a known bias against Trump;
- Judges were *not* told that the FBI's evidence was unverified;

- Judges were *not* told of exculpatory evidence suggesting the innocence of Page; and
- Judges were *not* told that the wife of a senior DOJ official had cultivated some of the Clinton-funded opposition research used by the FBI.

Those deceptions remained largely hidden until July 21, 2018, when the Justice Department, under pressure, released 412 pages of heavily redacted documents that had been used by the FBI and DOJ to gain warrants to spy on Page both before and after the 2016 presidential election.[64] The first application was signed by FBI director James Comey and Deputy Attorney General Sally Yates. They vouched for the veracity of their representations to the court, the authenticity of the documents cited, and the credibility of their sources. Comey and Yates swore under penalty of perjury that their information was "true and correct" and that they were adhering strictly to the requirements of the FISA law.[65] Believing that the FBI director and the deputy attorney general were being honest and forthright, a FISC judge issued the first warrant on October 21, 2016, to wiretap Page and gain access to his electronic communications. They included texts and emails that Page exchanged with members of the Trump team during the campaign and afterward. When he sought the warrant, Comey's FBI was still desperate to confirm the allegations in Steele's "dossier." But under FISA and FBI regulations, confirmation or corroboration was required *before* the warrant was sought, not after. That the FBI did not do, rendering its warrant application defective and its actions lawless, if not criminal.

In 2003, new rules were instituted by the FBI called the "Woods Procedures," named for the top FBI official who created them.[66] They were written to ensure that false evidence could not be used to gain a warrant to surveil. Verification was the necessary predicate. No application could be sought unless a source was credible and the FBI independently corroborated the information he or she provided. Those meticulously detailed procedures were established under the leadership of none other than Robert Mueller, who was the director of the FBI at the time. He had been forced to appear before the FISC in 2002 because the FBI had been caught frequently employing false information to conduct surveillance.[67] Inaccurate applications had reportedly been filed in more than seventy-five FISA cases, which represents an astonishing level of corruption and abuse of power. In essence, the Bureau had been lying to the judges and cheating on the rules. Mueller promised comprehensive reform with a new set of procedures to be scrupulously followed. They were not, at least in the case of Carter Page.

The "Woods Procedures" are memorialized in the FBI's stringent set of internal rules called the Domestic Investigations and Operations Guide (DIOG), which every FBI agent and official must rigorously follow. The relevant passage reads as follows:

> The accuracy of information contained within FISA applications is of utmost importance. Only documented and verified information may be used to support FBI applications to the court.[68]

The operative phrase is "Only documented and *verified* information may be used." What does it mean? The DIOG offers comprehensive instructions to insure that all applications presented to the court are "thoroughly vetted and confirmed."[69] It is not enough to simply rely on a "source" who passes along information. The information *itself* must be vetted.[70] This is where the FBI and DOJ were derelict. Before seeking a warrant, they were warned that Steele was a compromised and dubious source who was politically biased against Trump and his campaign. Indeed, he was fired by the FBI for leaking and lying about it.[71] This made him unreliable and his information inherently suspect. Moreover, Steele was not even a "source" in the true sense. He witnessed nothing. Instead, he "purveyed" hearsay information from supposed sources that were anonymous.[72] He was nothing more than a conduit. An analogy can be made to computer science and the concept of "garbage in, garbage out." If a "source" conveys garbage information to the FBI, the Bureau is not allowed to simply repeat that garbage to the FISA court to obtain a warrant to surveil someone. If it does, the result is a "garbage" warrant that was unlawfully secured. The FBI has a duty to investigate and corroborate the information to determine that it is not "garbage" before submitting it to a judge.

Nonetheless, the politically motivated opposition research financed by the Clinton campaign and the DNC that the FBI relied on for the Page surveillance was *not verified* when Comey and Yates signed their application insisting that it had been verified. Nowhere in the warrant materials is there an indication that the FBI had vetted or confirmed anything.[73] Comey all but admitted to that when he testified publicly in June 2017.[74] When he testified privately, he was reportedly even more candid, confessing "that the FBI had not corroborated much of the Steele dossier before it was submitted as evidence."[75] Six months later, when Trump fired him, Comey conceded that it was still not verified. His deputy, Andrew McCabe, made the same concession to Congress when questioned.[76]

Although many sections were blacked out, it is abundantly clear that the first application was drawn almost exclusively from the "dossier." Absent Steele's document, there was little or no evidence to justify a warrant to spy. That was confirmed by McCabe when he testified "that no surveillance warrant would have been sought from the FISC without the Steel dossier."[77] The Senate Judiciary Committee, which had previously reviewed the FISA application, also concluded that the "dossier" had comprised the "bulk" of the FBI's request for a warrant.[78] Democrats such as Representative Adam Schiff insisted that that was completely wrong. He wasn't alone. On a tour to hawk his book and profit from his own misconduct, Comey claimed that there had been a "broader mosaic of facts" other than the "dossier" that was used for the warrant.[79] That turned out to be completely untrue. Once the FISA application was made public, there it was in black and white.

Page after page of the application repeated Steele's allegations against the Trump campaign and Carter Page. It was nearly identical to what the ex-spy had written in his specious document: that is, that Page had met secretly during his Moscow lecture with Sechin and Divyekin and that lifting sanctions had been discussed in exchange for helping Trump get elected. The FBI knew that Page had denied that those meetings ever took place, and the Bureau had developed no evidence to the contrary. The FBI leadership didn't care.

Worse, Comey *pretended* that the news report by Isikoff had been independent corroboration of the "dossier." The application he signed had cited the Yahoo! News story as validating information extracted from a source that was separate from Steele. It was not. Isikoff admitted that Steele had been his source, and the ex-spy confirmed the same. So the FISA application had only one source, not two. The FBI knew that because Isikoff's attribution made it obvious where he had gained his information: from Steele. Equally troubling is how the FBI was willing to rely significantly on an anonymously sourced news story as a justification for violating an American's Fourth Amendment rights.

In all four of the FISA applications to spy on Page, the FBI and DOJ represented to the FISC that Isikoff's story was a secondary and autonomous confirmation of the evidence derived from Steele's "dossier" and that the ex-spy had not been the reporter's direct source. Look closely at what the FBI told the court:

> The FBI does not believe that Source #1 [Steele] directly provided this information to the identified news organization that published the September 23rd News Article.[80]

At roughly the same time the FBI first made this false representation to the FISC in October 2016, Bureau agents learned that Steele had, in fact, been talking to the media and sharing his "dossier" with them in direct violation of their rules and agreement with him. He had then lied about it to the FBI. Under its strict policy, the agency was forced to fire him. It couldn't be swept under the rug. Too many people, especially journalists, knew that Steele was sedulously feeding his document to the media, and that was obvious from the reporting. Yet Comey and Yates did not alert the FISA court that Isikoff's story should no longer be considered independent corroboration of Steele. Over the course of three remaining warrant applications, the FBI maintained its deception to the judges. Comey, Yates, and others involved in the application had a legal obligation to correct the record. They did not. Instead, they perpetuated the falsehood to keep their surveillance of Page going.

The FBI's misconduct got worse. As it knew that Steele had been lying, he was no longer a reliable and credible source. That did not deter either the FBI or the DOJ, although it should have. Instead of fully telling the FISC the truth about Steele's leaks and lies, the government kept covering it up and advising the court that Steele was "credible" when he was not. The initial application assured the FISC that the FBI knew of no "derogatory information pertaining to [Steele]."[81] That was untrue. Later applications admitted that the ex-spy had leaked information to the media, but the FBI kept insisting that he was still "credible."[82] At no point was the FISC told that the Bureau's main—and only—source had blatantly lied. Importantly, Steele's credibility should not have been the focus at all. He had *assembled* the information in his "dossier" based on multiple hearsays from anonymous sources. Therefore, "it is *their reliability* the FBI should swear to, not Steele's. [author's italics]"[83]

Comey's FBI neglected to advise the court that Page had once assisted the Bureau and the CIA in its 2015 prosecution of Russians who had tried to recruit Page years before he joined the Trump campaign. The application omitted this critical information and the fact that Page had not been charged. He had instead served as a cooperating witness who was instrumental in helping the government succeed in achieving convictions that brought down a spy ring.[84]

All of that was important exculpatory information that the FBI and the DOJ had a legal obligation to disclose to the FISA judges.

They were duty bound to inform the court. Before petitioning the FISC, the FBI was required to exhaust all "normal investigative techniques."[85] That would include *talking* to Page, who had *asked* Comey for an interview. In retrospect, it seems evident that Comey wasn't concerned about Page as an American with constitutional rights. He was a useful tool who would become collateral damage.

HIDING THE "DOSSIER'S" SKETCHY ORIGINS AND CONNECTIONS WITH THE CLINTON CAMPAIGN

One of the most egregious defects in the Comey-Yates FISA application was how the FBI and the DOJ deliberately hid from the court the origins of their so-called evidence. As we now know, it was funded by the Clinton campaign and the Democratic National Committee (DNC). The FBI knew it, but the judges did not. Comey and Yates camouflaged the fact that the "dossier" was a shady political document and the malignant forces that commissioned it. Cryptic references were buried in two of the application's footnotes that intentionally disguised specific names and identities.[86] Steele's partisan funding by Clinton and the DNC should have been clearly and deliberately *highlighted* for the court. Assuming the judges even read the veiled details in footnote 8, for example, they would have had to be telepathic to comprehend that it was Clinton and Democrats who were financing an attack on her rival's campaign and that the whole thing had been conjured up by a former British spy who was either inventing anti-Trump and anti-Page material or relying on Russian hearsay or disinformation, or both.[87] The court was being "played."

There was no valid reason for obscuring that vital information. The FISC operates in secret, with no opposing side present. Why not name names? Why shroud identities—unless, of course, doing so would diminish the chances of the court's approving the wiretap? The political motivations and funding were crucial and indispensable facts. The judges deserved to be given a clear and honest version of how the evidence had been obtained. Had they been told the undissembled truth, the judges would have instantly recognized that it was nothing more than a "hit job" by a political opponent and that the FBI/DOJ was aiding and abetting the effort. The wiretap warrant would surely have been rejected.

Those machinations underscore just how sneaky and misleading government officials like Comey were in their misrepresentations to the court. Comey and Yates knew they could mask the truth and get away with it, given how often the FISC had approved their paper submissions. Clearly, the Clinton campaign and Democrats were the driving force behind the "dossier" that was used by Comey's FBI to spy on her political opponent and investigate the Trump campaign. Yet the director appears to have covered up the FBI's interactions with Steele. After investigating the matter, the Senate Judiciary Committee accused Comey of having provided it with an account that was "inconsistent with information contained in FISA applications."[88] No surprise there, considering Comey's well-documented history of evasions and obfuscations.[89]

The FBI's most unconscionable and self-incriminating act was its attempt to bribe Steele to substantiate his phony "dossier." In early October, Steele's FBI handler met with him in Rome and offered $50,000 if the ex-spy could somehow produce evidence that verified his hearsay allegations.[90] That was astonishing on two levels. First, it establishes persuasive evidence that the FBI had no confidence whatsoever in the flimsy document and knew that it could never be relied on to pursue its improbable conspiracy theory. Second, the FBI was so desperate to implicate the Trump campaign in Russian "collusion" that it was willing to *buy* proof of it. Though it is true that the government will compensate informants, Steele had already been on the FBI payroll for many months. He was also being paid by the Clinton campaign and the DNC for the same work. That tantalizing $50K "bonus" created the inevitable risk that Steele would compose additional fiction to pocket even more cash. The money was reportedly never paid because Steele couldn't possibly corroborate his fabrications and/or Russian disinformation.

At that point, the FBI should have fed the "dossier" into the nearest shredder. But it didn't. Why? Michael Doran, who served as deputy assistant secretary of defense and senior director of the National Security Council, penned a lengthy story for *National Review*, explaining it this way:

> Because without Carter Page who appeared in the Steele dossier—without the Marvel Comics villain, there existed no credible intelligence pointing to a criminal conspiracy between Trump and Putin. If the investigation was to be sufficiently broad to dig up dirt on Trump, it had to include the fanciful allegations against Page. These, however, were impossible to corroborate—because they were fictive.[91]

A mere nine days before the FBI and the DOJ applied for their surveillance warrant to spy, internal text messages obtained by Fox News show that the two agencies were bickering over the "potential bias of a source pivotal to the application," likely Steele.[92] The same texts suggest that John Brennan's CIA was involved. Though a Justice Department lawyer was reticent about the warrant, Lisa Page and Andrew McCabe at the FBI were determined to gain court approval to spy, regardless of whether they had probable cause to do so. Reporting by Fox's Catherine Herridge and Gregg Re revealed that McCabe and Page were circulating anti-Trump blog stories, including one by a Comey friend, that averred that the GOP presidential nominee was "among the major threats to the security of the country."[93] Their bias was laid bare in other texts disparaging a prominent Republican lawmaker, Representative Trey Gowdy (R-SC), who was delving into the FBI's incomprehensible decision to clear Clinton of mishandling classified documents in her email case.[94]

The dearth of plausible evidence did not dissuade the FBI from appropriating Steele's cooked-up novel as it moved forward. Evidence was concealed, and the judges on the FISC were deceived. The FISA warrant was secured. Spying on Carter Page began, and a backdoor entry into Trump campaign electronic communications and documents was opened. Nothing of value was ever found. In its applications to spy, the FBI had declared with confidence that "Carter Page is an agent of Russia."[95] He was no such thing. The FBI well knew it. Comey's true target in his counterintelligence operation was not Page, but Trump.

Comey and Yates signed the first two applications to surveil in October and January 2017. Comey and Acting Attorney General Dana Boente signed the third application in April. Deputy FBI director (and temporarily acting FBI director) Andrew McCabe and Deputy Attorney General Rod Rosenstein signed the last renewal in June 2017. Rosenstein's decision to sign the final FISA application was, as before, based on unverified information. This is especially troubling, given what he later said at a forum in Washington:

> A FISA application is actually a warrant, just like a search warrant. In order to get a FISA warrant, you need an affidavit signed by a career law enforcement officer who swears the information is true . . . And if it is wrong, that person is going to face consequences. You can face discipline and sometimes prosecution.[96]

Rosenstein's words seemed to frame a prophetic indictment of his own wrongful actions, as well as those of Comey, Yates, McCabe, and Boente. They all swore under oath that the information was true and accurate. They knew the opposite was true. They were acutely aware that the FBI had spent months trying in vain to verify the evidence that they vowed was verified. The front page of each application they signed is entitled "Verified Application." It affirmed that the information submitted to the FISC was "verified."[97] It was not.

A month after Rosenstein's public remarks about the honesty and diligence demanded in FISA applications, he was questioned about it by the House Judiciary Committee. Rosenstein seemed to suggest that he might not have even *read* the FISA application that he had signed. Grudgingly, he admitted that he did not always read what he was signing.[98] At one point, he refused to reveal to committee members whether he had read the Page application specifically. Then he offered that he had been "briefed" on it by "a team of career DOJ attorneys."[99] It is ironic that the man who so sternly lectured an audience about how imperative it was for every prosecutor to ensure that a spy warrant contained truthful information is the same man who may never have bothered to read the one he signed against Carter Page.

If there is any doubt about how the FBI and DOJ engaged in shameless deceptions, consider the inconsistent statements and memory lapses of Comey. While promoting his book in the spring of 2018, he claimed he still had no idea who had funded Steele's "dossier." But that was not what he told Congress. In private sessions before joint House committees, testimony of which was later released to the public, Comey admitted he had known *before* the first FISA warrant application that Democrats had paid for the "dossier."[100] But then, in a statement that defies all common sense and belief, he claimed he did not know that the "dossier" had been used for the application to spy on Carter Page. Think about what Comey was saying. He conceded that he had read the application he signed, but he would have us believe that he did not know where the information contained therein came from? That is incomprehensible. Hold on, there's more:

REPRESENTATIVE JOHN RATCLIFFE: Do you know whether the application that you signed states that the FBI has reviewed this verified application for accuracy?
COMEY: I don't remember that specifically. It sounds like the kind of thing that would be in there as a matter of course, but I don't remember. . . .

RATCLIFFE: Does that cause you any concern about the fact
that you signed a verified application for a warrant to surveil
Carter Page when the Steele dossier was only minimally corroborated or in its infancy in its corroboration?
COMEY: I don't know enough or remember enough 2 years
later to have a reaction. I don't know their testimony. I haven't
looked at the thing.[101]

As this exchange reveals, Comey is a master of prevarication. As
will be described further in the next chapter, he feigns amnesia whenever it serves his purpose or when the truth might incriminate him.
He went on to admit that the FBI was still attempting to verify the
"dossier" when he was fired by Trump in May 2017. That was a full
seven months *after* he swore to the FISC that his surveillance application, which was based on the "dossier," was "verified information."
About all that the FBI had corroborated was that Page had traveled to
Moscow to deliver a speech. By itself, that was meaningless.

The five top FBI and DOJ officials who signed off on the FISA
warrants to spy on Page cannot credibly blame others beneath them.
Testimony by a former top FBI counterintelligence lawyer shows that
several of them examined the faulty warrants "line-by-line." Trisha
Anderson reported directly to James Baker, who was FBI general
counsel during the time the warrants were sought. Here is the relevant part of her testimony before the House Judiciary and Oversight
Committees on August 31, 2018:

> The sensitivity level of this particular FISA resulted in lots of
> very high level attention both within the FBI and DOJ. The
> general counsel (Baker) . . . personally reviewed and made ed-
> its to the FISA, for example. The deputy director (McCabe)
> was involved in reviewing the FISA line-by-line. The Deputy
> Attorney General (Yates) over on the DOJ side of the street
> was similarly involved, as I understood, reviewing the FISA
> application line-by-line.[102]

Anderson said those senior officials gave the FISA applications "their own de novo independent review" and scrutinized them "carefully."[103] How is it possible that Yates, McCabe, Baker, and others never asked themselves or one another the fundamental question "Are we being honest with the court?" Fidelity to the truth would have meant apprising the court that they had not even identified Steele's sources, much less verified their allegations. They would have had to 'fess up that Clinton's campaign had funded the "dossier" and that Steele had leaked and lied after admitting that he was riven with bias. They would have had to inform the judges of the exculpatory evidence they knew. Instead, the FBI and the DOJ exploited an unproven and fictitious "dossier" as a pretext to spy on a Trump campaign associate, perpetrating a fraud on the intelligence court.

Evidence emerged that Comey's FBI had known in advance of the first FISA warrant that Steele's "dossier" lacked credibility because the Bureau had been warned of its falsity and the severe anti-Trump political bias of its author. In May 2019, hidden documents were unearthed in a transparency lawsuit that showed that Steele met with Deputy Assistant Secretary of State Kathleen Kavalec on October 11, 2016, some ten days before Comey signed off on his court application to surveil Carter Page. As first reported by John Solomon of The Hill, the former British spy "admitted that his research was political and facing an Election Day deadline."[104] A reference to his funding by the Hillary Clinton campaign was reportedly made. Steele gave the impression that he was desperate to sabotage Trump's candidacy before voters went to the polls, and he shared his "dossier" material with Kavalec. Kavalec, who had no formal training in investigative techniques, quickly determined that Steele was peddling false information. For example, his claim that the Russians were running their secret operation out of the Russian Consulate in Miami was easily disproven by her. "It is important to note that there is no Russian consulate in Miami," she wrote.[105]

If a State Department employee could figure that out in one day, so, too, could the FBI before exploiting the "dossier" as the principal basis of the Page surveillance warrant. Kavalec also emphasized in her notes that Steele had been disseminating his information to the media. That directly contradicted the statement made by the FBI and the DOJ to the FISA court that "Steele did not have unauthorized contacts with the press prior to October 2016," as pointed out by Senator Charles Grassley (R-IA).[106] Steele made other wild claims that were also readily debunked.[107]

Records show that Kavalec and others at the State Department had been instructed to forward such information to the FBI.[108] She did so on October 13, eight days before the FISA warrant was considered and granted. Her notes, contained in an email, were sent to a top FBI counterintelligence official, who delivered them to "the FBI team leading the Trump-Russia investigation, headed by then–fellow Special Agent Peter Strzok." [109] The Department of Justice was also alerted. In a statement, Representative Mark Meadows (R-NC), who spent two years trying to uncover the truth, reacted with disbelief and disgust:

> This once again shows that officials at the FBI and DOJ were well aware the dossier was a lie—from very early on in the process all the way to when they made the conscious decision to include it in a FISA application. The fact that Christopher Steele and his partisan research document were treated in any way seriously by our Intelligence Community leaders amounts to malpractice.[110]

If ever there were incontrovertible evidence that both government agencies had been warned ahead of time that their FISA warrant application was based on false information, this was it. However, that did not deter Comey and Deputy Attorney General Sally Yates from signing and submitting a "verified" application that they *knew* was not verified at all. Just the opposite; they knew it was deeply flawed and driven by political motivations. They also knew that Steele was not suitable as a source, yet they vouched for his "credibility" to the court as they sought to spy on Page. The Kavalec memos had never been turned over to Congress, despite repeated requests for such evidence. Even the DOJ inspector general may not have been aware of their existence during his yearlong investigation of FISA abuse.[111]

What is equally troubling is that the FBI, under current director Christopher Wray, endeavored to cover up the Kavalec memos. Much of the material was belatedly redacted and retroactively classified on April 25, 2019, two and a half years after the memos were written, with the attached notation at the top, "SECRET . . . Declassify on December 31, 2041." [112] That's right, we will not see Kavalec's full memos for the next twenty-two years unless Attorney General William Barr and/or Secretary of State Michael Pompeo intervenes. What exactly does Wray fear? That current and former officials at the FBI will be held accountable by exposure to the truth? That the Bureau's reputation will be sullied even more? It appears that Wray is adhering to the disgraceful tradition of secrecy and obfuscation that was perfected to an art form by his predecessor, Comey.

When powerful forces in government misuse their positions of trust to circumvent the legal process and target or punish people for personal or political reasons, democracy is threatened. Reverence for the rule of law is lost. Corruption prevails.

UNDERCOVER SPYING ON GEORGE PAPADOPOULOS

Lying and spying by the FBI were not limited to surveillance and the collection of electronic communications. The Bureau mobilized several undercover informants to insinuated themselves into the Trump campaign to gather incriminating evidence. It was exactly the kind of covert human spying that Attorney General Barr referenced in his controversial appearance before Congress in April 2019. The AG already knew about the operation.

The most prominent FBI and/or CIA informant was Stefan Halper, a professor at the University of Cambridge in Great Britain, who has long ties to both US and foreign intelligence agencies. The son-in-law of a former deputy director of the CIA, Halper worked with the CIA from the late 1970s and into the 1980s. In 2016, he was the head of an allegedly autonomous intelligence business that "consults" with other spy agencies—notably the British Secret Intelligence Service, known as MI6, where Steele operated as a spy.[113] These connections cannot be overlooked: John Brennan's CIA, Christopher Steele's ties to MI6, and Halper's extensive work for both the CIA and the FBI. As noted in the previous chapter, there is persuasive evidence that British intelligence was providing information on the Trump campaign to Brennan, the CIA, and eventually the FBI. Halper has a close relationship with Sir Richard Dearlove, the ex-chief of MI6. Another British spy chief, Robert Hannigan, flew to Washington, DC, to meet with Brennan as the "collusion" hoax was taking shape in the summer of 2016.[114] Though the work by Halper and Steele was later exposed, the intelligence emanating from UK agencies is still cloaked in secrecy.

By most accounts, Halper is a freelance spy-for-hire kind of guy. His professorial appearance and practiced comportment serve as the perfect artifice. He doesn't *look* like a spy; more like a portly, frumpy academic. His position as a foreign policy scholar provided a useful cover. Well educated, he graduated from Stanford University and earned two PhDs at Oxford and Cambridge. But his real skills seem to have been practiced in the impenetrable underworld of espionage.

Beginning in July 2016, Halper was the beneficiary of more than $400,000 in payments for "research" from a somewhat abstruse Pentagon think tank. The timing of the income matches the beginning of his deployment as an informant targeting the Trump campaign. It could have been coincidental and unrelated, except when you consider that intelligence agencies have been known to launder payments through intermediaries to hide their true purpose and render the tracking of money difficult. This became even more apparent when the Inspector General of the Department of Defense audited the monetary outlays to Halper and discovered that there was little documentation in the Pentagon files of the work he did or the justification for the $411,000 in payments.[115] That raises a logical question. Was Halper being compensated for work as spy or as a legitimate and conventional defense contractor? Perhaps it was both.

What exactly was Halper's paid "research"? It is difficult to trace because neither he nor the FBI has been willing to disclose the exact nature of his work. However, Trump campaign associates have spoken about how Halper contacted them for seemingly legitimate reasons but then pivoted to suspicious behavior and strange questions about Russia. Page, Clovis, and Papadopoulos have all confirmed that Halper engaged them out of the blue. He also surreptitiously attempted to gather information from Michael Flynn, a foreign policy adviser to the Trump campaign.[116] Page stated that Halper had "intensified" communications with him just before the FBI applied for its surveillance warrant to wiretap his phone lines and access his electronic communications in October 2016.[117] There was frenzied activity to verify the heretofore unverified "dossier."

In early September 2016, Papadopoulos received an email from Halper, whom he had never met or heard of. In his book, *Deep State Target: How I Got Caught in the Crosshairs of the Plot to Bring Down President Trump*, Papadopoulos wrote that he was invited to London by Halper and the promise of $3,000 compensation to discuss Mediterranean oil and gas fields—a subject about which Papadopoulos was well acquainted.[118] Luring him overseas was legally advantageous because US intelligence agencies' spying on an American citizen there is less constrained by law. Before Papadopoulos met with Halper, a woman who pretended to be Halper's assistant arranged to have drinks with Papadopoulos at a bar. She identified herself as Azra Turk, although that turned out to be an assumed name. Here is Papadopoulos's account:

> Azra Turk is a vision right out of central casting for a spy flick. She's a sexy bottle blonde in her thirties, and she isn't shy about showing her curves—as if anyone could miss them. She's a fantasy's fantasy.[119]

Within minutes, Turk began peppering him with what he described as "creepy" questions about Russia and demanding to know whether the Trump campaign was working with Moscow. Papadopoulos insisted he didn't know what she was talking about. He denied any cooperation between the campaign and the Kremlin. The more she pushed, the more suspicious he became. When she didn't receive the answer she wanted, Papadopoulos said she changed tactics to a sexually seductive approach. She kept advancing the nonexistent thesis that would later become known as Trump-Russia "collusion." He repudiated the silly notion and rejected what he called her "honey-pot act," leaving the bar, returning to his hotel room, and telephoning his girlfriend to describe what had just occurred. Later reporting would identify Turk as an FBI investigator.[120] Papadopoulos believes she was "a Turkish agent or working with the CIA."[121]

Whether the woman who called herself Azra Turk was operating undercover as a spy for the CIA, the FBI, or both, she was undoubtedly searching for confirmation of the Russia "collusion" hoax fomented by the Clinton campaign, the DNC, Comey's FBI, and Brennan's CIA and propelled by the manipulations of Christopher Steele and Glenn Simpson. The witch hunt was on, and US intelligence was "all in" on the big lie that was aimed directly at Trump. They were shooting arrows blindly, hoping to hit the bull's-eye on the GOP candidate's back. It didn't seem to matter to them that Papadopoulos was a bit player in the campaign.

Turk had reportedly been sent by the FBI to oversee its other undercover informant, Halper, as he, too, attempted to elicit from Papadopoulos incriminating information that he did not possess. Two days after the "bizarre" meeting with Turk, Papadopoulos met with Halper at the Sofitel Hotel in London, where the informant placed his smartphone on the table next to Papadopoulos as if he were recording their conversation. He then promptly posed several leading questions about Trump-Russia "collusion":

It's great that Russia is helping you and the campaign, right, George?

George, you and your campaign are involved in hacking and working with Russia, right?

It seems like you are a middleman for Trump and Russia, right?

I know you know about the emails.[122]

Papadopoulos was both shocked and angry, so much so that he responded with a couple of profane denials and then said:

I have no idea what the hell you are talking about. What you are talking about is treason. And I have nothing to do with Russia, so stop bothering me about it.[123]

At that point, Halper picked up his phone and dropped it into his pocket. If the Cambridge professor was acting as a spy for the FBI and/or CIA and had been tasked with collecting valuable intelligence from Papadopoulos, he failed miserably. His obvious and clumsy attempts were exceeded only by the fact that his target had nothing incriminating to say. Trump's young foreign policy volunteer wasn't the fulcrum in some grand conspiracy between the candidate and Putin to steal the 2016 presidential election. It was a ludicrous supposition. However, it does underscore how desperate US intelligence agents were to implicate Trump in some illicit scheme and how determined they were to damage Clinton's rival for the White House. Attorney General Barr was particularly disturbed by their "dangling a confidential informant in front of a peripheral player in the Trump campaign."[124] It emitted the ugly odor of a rogue agency misusing its immense power.

Halper and Turk weren't the only ones who appear to have been spying on Papadopoulos. His book recounts others who "kept tabs" on him, including two military attachés at the US Embassy in London named Terrence Dudley and Gregory Baker, who he surmised were "intelligence operatives affiliated with the CIA or military intelligence divisions."[125] They even asked to be given jobs with the Trump campaign, sending Papadopoulos messages "up until the inauguration." Meanwhile, the British government was getting in on the game. A junior foreign minister and his colleague appeared out of nowhere, asking about Trump and Russia.[126] So did Sergei Millian, a US citizen with a Belorusan name who also has ties to the FBI. Charles Tawil, an American Israeli businessman, introduced himself and promptly began asking the same questions. Why all of the sudden and intensive interest in a peripheral figure in the campaign and his knowledge of any Trump-Russia connection?

Papadopoulos is convinced that he was set up—deliberately fed false information by US intelligence agents and their Western allies as an excuse to plant a scandal that would then justify an investigation of Trump and his campaign.[127] The false information came to Papadopoulos courtesy of a truly nefarious character by the name of Joseph Mifsud, a Maltese university professor, who was introduced to Papadopoulos by a man he would subsequently learn was "a former FBI employee."[128] Like Halper, Mifsud is an academic, with a PhD from Queen's University Belfast in Belfast, Ireland. Also like Halper, is it possible that Mifsud may have been a spy or "plant" for the FBI? It is difficult to know, since he seems to have made himself scarce. But in early March 2016, Papadopoulos attended what he thought was an innocuous three-day academic conference in Rome at an international venue called Link Campus University. Only later did he discover that it was a "training school for Western-allied spies, including CIA, FBI, and MI6," dubbed Spook University. Mifsud was also there.

Once the introduction was made to Papadopoulos, Mifsud kept in contact over the next few weeks by email. Over breakfast at the Andaz London Liverpool Street hotel in London on April 26, 2016, Mifsud did something strange and unexpected:

He leans across the table in a conspiratorial manner. The Russians have "dirt" on Hillary Clinton, he tells me. "Emails of Clinton," he says. "They have thousands of emails."[129]

Papadopoulos insisted that he was not only skeptical but wanted "no part in it."[130] He was determined to avoid anything to do with hacking or security breaches. It was a red flag. He maintains to this day that he never informed anyone on the Trump campaign of the rumor Mifsud had passed along. There is not a scrap of evidence that proves otherwise. No individual in the Trump campaign learned of it, and no emails show that the gossip was disseminated. Papadopoulos knew nothing about the Clinton emails themselves. So he kept his speculations to himself, except once—*maybe*. It's a big "maybe."

Roughly two weeks later, on May 6, Papadopoulos was invited by a woman who *claimed* to be a "senior adviser" to Alexander Downer, an Australian high commissioner with the rank of ambassador, to meet for drinks with her boss. They convened at the Kensington Wine Rooms in London. In a move that was identical to the one pulled by Halper, Downer hauled out his cell phone as if he might record the ensuing conversation. He was "so aggressive, so hostile, it's actually a bit intimidating," wrote Papadopoulos.[131]

And then something happens. Or more accurately, Downer later claims something happens. In his version of events, he asks me a question about Russia and Trump. I then tell him that the Russians have a surprise or some damaging material related to Hillary Clinton. I have no memory of this. None. Zero. Nada.

In my version of events, Downer brusquely leaves me and Erika [Thompson, his senior assistant] at the table, and we go our separate ways. I remember feeling completely disappointed by the meeting and pissed off about being treated so rudely. Downer's version, however, is the one that matters.[132]

Papadopoulos seemed convinced that Downer was spying on him—that the Australian diplomat was a vital component of the scheme to plant phony evidence on the Trump adviser to justify the FBI's counterintelligence investigation of the candidate's campaign:

> Did this man with extensive intelligence ties already know what Joseph Mifsud—a man who taught at the spook-training ground that is Link Campus Rome—had told me? Was he trying to bait me into saying something? Something that could spark an investigation? I believe so.[133]

When the *New York Times* first reported the story of that encounter a year and a half later, it ran a front-page headline stating that Downer had tipped off the FBI, which was what had triggered the opening of the "collusion" investigation.[134] Not coincidentally, that story was published at a time when the FBI was under intense pressure to justify why exactly it had commenced an investigation of Trump to begin with. If the probe had been opened based on an unverified "dossier" about Trump that had been paid for by the Clinton campaign and the Democrats, the investigation was highly suspect, if not lawless. The Bureau would be in serious trouble. Therefore, it is quite likely that the FBI deliberately fed the *Times* an invented or grossly exaggerated story about Papadopoulos to divert attention from the FBI's primary reliance on the fabricated "dossier" as the impetus for the investigation of Trump. The article added an extra tantalizing tidbit to suck in its readers and other members of the Trump-hating media: it implied that Papadopoulos had been drunk in a bar when he recklessly spilled allegedly incriminating information. In his book, Papadopoulos vigorously denied this:

> I don't know where the "paper of record" got this information, but it's completely wrong. I had one drink. A gin and tonic. Whoever leaked this false account also spun that Downer and I met in a random, chance encounter at a trendy bar. But that's

false, too. And Downer . . . obviously has extensive ties to intelligence operatives. So this meeting was anything but random. Intelligence operatives engineered it.[135]

Papadopoulos says Downer spent much of their conversation "praising" Hillary Clinton and "bashing Trump." Downer had a history with the Clintons, having organized a $25 million gift from Australia to their foundation.[136] Therefore, the ambassador had a motive to help the FBI manufacture a fraudulent investigation of Trump that would prove helpful to Clinton in her quest to become president. Two months after his conversation with Papadopoulos, Downer contacted the US Embassy in London after he read news reports that Clinton's emails had been hacked. He *assumed* that that was what Papadopoulos had been talking about. Or so the story goes.

But let's assume that what Downer told the FBI was true—that Papadopoulos had passed along conjecture about Russia having damaging material on Clinton. So what? It would still not come close to being sufficient evidence to launch a formal counterintelligence investigation of the Trump campaign. FBI regulations demand "specific articulable *facts*," not unverified hearsay allegations.[137] There must also be a nexus between the target of the probe (Trump and/or his campaign) and a potential crime. Where is the crime? None of this existed.

The Mueller Report tried to theorize a potential crime by inventing a conversation that had never occurred. Near the end of volume I, it stated that the FBI had approached Papadopoulos for an interview in January 2017 because he had suggested to Downer "that Russia had indicated that it could assist the [Trump] campaign through the anonymous release of information damaging to candidate Clinton."[138] Hold on a moment. Where in the world did that come from? There is no evidence of a proffer of campaign assistance anywhere in the record, by Mifsud or Papadopoulos or Downer. Mueller threw that statement out there as if no one would notice. He provided not a scintilla of support for it, nor did he source it to any person or any document. It appears to have been plucked from the special counsel's considerable imagination. Other such concoctions can be found elsewhere in the report, as chapter 5 will show.

As for Mifsud, he appears to be an enigma. The Mueller Report stated that the FBI interviewed him on February 10, 2017, and he "denied that he had advance knowledge that Russia was in possession of emails damaging to candidate Clinton." [139] He claimed that Papadopoulos must have misunderstood their conversation. That, of course, contradicts Mueller's own unfounded assertion that Russians offered campaign assistance. Months later, Mifsud told the British publication *The Telegraph* that "he had no knowledge of any emails containing 'dirt' on Mrs. Clinton." [140] Whether he is telling the truth is unknown, although the FBI never accused him of lying.

All of this is curious, if not suspicious, because the Mueller Report determined that Mifsud had lied several times to investigators during his interview. [141] Doing so is an obvious crime. Yet, he was never charged, even though the special counsel brought numerous criminal charges against Trump associates for lying, including Papadopoulos. Was Mifsud given a free pass by Mueller because he was not connected to the president but was, instead, the "catalyst" of the "collusion" hoax against Trump? [142] Or was Mifsud insulated from charges because he fed Papadopoulos false information at the behest of US intelligence agents in an effort to frame Trump with the phony Russia conspiracy allegation? When Mueller testified before Congress on July 24, 2019, he refused to answer any and all questions about the mysterious Mifsud. Not surprisingly, Democrats and the media were disinterested in knowing the truth behind the man who, according to the *New York Times*, was the genesis of the FBI's initial investigation of Trump.

Regardless, let's examine the rumor that Papadopoulos is *said* to have heard. An unidentified person in Moscow allegedly told a professor he had "dirt" on an American candidate and maybe some of her emails. The professor then told Papadopoulos, who then purportedly told an Australian diplomat, who then told the FBI. Any of the individuals along the chain of chatter could have lied or exaggerated or mistakenly conveyed the exchange of words. By the time the purported information was repeated numerous times and eventually reached the FBI, it was what lawyers call "quadruple hearsay." It contained no indicia of reliability and would never be accepted as trustworthy evidence in a court of law. It would be inadmissible in every courtroom in the United States. Hence, it should never have been used by the FBI as justification for opening a counterintelligence or criminal probe of any US citizen, much less a candidate for the presidency. Rumors, innuendo, supposition, and gossip are not a legal basis for an investigation. Hard *facts* are required. The alleged information must first be verified. Nowhere in the *Times* article did the reporters raise the essential question of why the Bureau would initiate an investigation with such scant and legally insufficient evidence.

Before opening its case, the FBI was required to identify a possible crime. Where was the crime in listening to someone claim that Moscow had "dirt" on Clinton? Where was the crime in repeating that bit of gossip? Even if someone in the Trump campaign had acted on the information in some way, it was still not a crime to do so. Conspiracy requires an agreement to do something illegal. There was no agreement to do anything at all in the aforementioned recitation of events. There was never any evidence that the Trump campaign was involved in hacking Clinton emails, as the Mueller Report later confirmed.

Papadopoulos was never charged with conspiracy or any "collusion"-related crime because there was nothing unlawful about his contacts or conversations involving Russia or Russians. There was never a *factual predicate* nor a *reasonable basis* that warranted the FBI's investigation in July 2016. Moreover, there was no reliable intelligence information to warrant a counterintelligence probe, which appears to have been a cover for a criminal investigation. "Counterintelligence investigations are not conducted for the purpose of building prosecutable court cases," according to former federal prosecutor Andrew C. McCarthy.[143] Yet that was what the FBI was doing.

The *Times* reporters may have been used, wittingly or unwittingly, to deflect attention from the dubious "dossier" and to provide an additional, albeit erroneous, justification for the Trump investigation. In other words, it appears that the whole story about Papadopoulos's conversation in a London bar was a convenient false front designed to conceal that it was Steele's "dossier" that the FBI had relied upon to initiate its investigation of Trump. If that was the goal, it was an implausible and clumsy attempt at misdirection. Under no legitimate circumstances could a rumor based on multiple hearsays be a proper and legal reason for opening a counterintelligence investigation, especially when the original source was not even identified and none of the allegations was substantiated. Yet the *Times* and, presumably, its readers believed it. It was then circulated and repeated so often and for so long that the chimera became accepted as gospel truth.

Even if *both* the uncorroborated "dossier" *and* the Papadopoulos "bar talk" were the basis for launching the Trump-Russia investigation, they were equally deficient under FBI regulations. Nevertheless, the Bureau was determined to ignore their own DIOG guidelines with impunity. Whether the FBI had one reason or two, it matters little. The entire probe was bereft of credible facts or verified evidence justifying its existence. The premise of the investigation was false on both counts.

The chronology of events indicates that it was Steele's specious memos that moved the FBI into taking devious action against the Trump campaign, not Papadopoulos's "bar talk." The Bureau learned of the "dossier" *first* on July 5, 2016. Michael Gaeta, the agent who saw the document, was so alarmed by what he had read (and, perhaps, unsuspecting) that he allegedly told Steele, "I have to report this to headquarters." [144] Washington was alerted, and a preliminary investigation was set into motion. It was not until two weeks later, on July 23, that the FBI was informed of Downer's purported conversation with Papadopoulos. However, it is clear that the FBI's interest in Papadopoulos soon waned after the investigation was formally opened on July 31, 2016. [145] Halper's spying provided the exculpatory evidence that Papadopoulos had known nothing about the hacking of Clinton emails or Russian "collusion." Instead, Steele's "dossier," funded by the Clinton campaign and Democrats, took center stage at the FBI as the agency sought the FISA warrants to spy on Page. Comey's Bureau turned all of its time, attention, and resources toward finding some evidence that would corroborate the document and implicate Trump. It was evidence that never existed. The FBI knew that. It knew the "dossier" was worthless junk.

The FBI spent the better part of a year attempting to verify the "unverifiable" memos produced by Steele. "FBI agents painstakingly researched every claim Steele made about Trump's possible collusion with Russia, and assembled their findings into a spreadsheet-like document," reported John Solomon of The Hill.[146] The spreadsheet was nearly empty of proof with "upward of 90 percent of the dossier's claims to be either wrong, nonverifiable or open-source intelligence found with a Google search."[147] This appears to be accurate based on all we know and confirmed by the conclusions in The Mueller Report. That means that when the FBI launched its investigation on July 31, 2016, it had no credible evidence to justify the probe. When the FBI and DOJ obtained a warrant to spy on Page in October 2016, its "verified" information presented to the FISA court was not only deficient but untruthful. It also means that when Robert Mueller was appointed special counsel in May 2017, there was an insufficient basis under federal regulations to do so. This will be explored in more detail in chapter 5.

The law safeguards US citizens, including political candidates for public office, from overzealous FBI officials. Under the agency's own operating guidelines, authorized by statutory law, agents are given a stern warning:

> These Guidelines do not authorize investigating or collecting or maintaining information on United States persons solely for the purpose of monitoring activities protected by the First Amendment or the lawful exercise of other rights secured by the Constitution or laws of the United States.[148]

As the Republican nominee for president, Trump was lawfully exercising a fundamental right guaranteed under Article II of the Constitution: running for office. Nearly every aspect of his campaign was also protected by the First Amendment's free speech clause. The evidence is persuasive that the FBI abridged those rights in violation of the law.

There is also compelling evidence that Papadopoulos was unfairly and illegally targeted because of his association with Trump. It is unlawful for a government agent to misuse his or her position of power to investigate someone for either personal or political reasons. Is that what happened? Consider the chilling account given by Papadopoulos of what happened on July 27, 2017, the day he was arrested:

> They handcuffed me and shackled my ankles. I spotted the two agents who had interviewed me months earlier in Chicago. When I asked them what was going on, I got no answer. When I repeated my question, another agent sneered, "This is what happens when you work for Trump." [149]

If that truly occurred, every American should be frightened at the unbridled and pernicious power of his or her own government.

To those who might be tempted to conclude that Papadopoulos is a crooked character who pleaded guilty to making a false statement to the FBI, you should consider reading the rest of his alarming chronicle of how he was treated by interrogators at the FBI and lawyers on Special Counsel Robert Mueller's team. In one session, he was subjected to seven hours of intimidation, harassment, and threats as prosecutors pressured him to lie to make their case. He told them, "I don't understand . . . it's as if you're trying to implant a memory in my mind of something that never happened." [150] Indeed they were. That's how unprincipled and unscrupulous prosecutors work. It should be a crime. That unconscionable treatment of George Papadopoulos will be explored in greater depth in a later chapter.

Papadopoulos may have been the classic "patsy" who was chosen to hold the bag. It could have been anyone in Trump's orbit, but it appears he was picked because he was, by his own admission, young, inexperienced, and gullible. He could be easily manipulated and framed; he wouldn't know any better. He was the perfect dupe. Blaming him was the ideal cover as the FBI exploited the phony "dossier" to frame Trump.

For all the spying that Halper and others did, they secured no incriminating evidence. Halper met with Papadopoulos, Page, and Clovis. He tried his level best to obtain something of value for his employer, the FBI. All he got was evidence of innocence. But that did not deter the Bureau one iota. It seemed to inspire Comey, McCabe, Strzok, Page, and Bruce Ohr at the DOJ. They accelerated their efforts to damage Trump.

BARR DEMANDS ANSWERS

Attorney General William Barr was right: spying on the Trump campaign most certainly occurred. The FBI's undercover informant, Halper, began his work well before the Trump investigation was initiated in late July 2016. Before that date, he was meeting with Carter Page, snooping for something that could be used against Clinton's opponent. Journalist Glenn Greenwald concluded that this "suggests that CIA operatives, apparently working with at least some factions within the FBI, were trying to gather information about the Trump campaign earlier than had been previously reported." [151] It wasn't just Comey's FBI that had targeted Trump, but likely John Brennan's CIA. This is a reasonable theory, given Halper's decades-long work for the CIA. As Greenwald explained, "Halper was responsible for a long-forgotten spying scandal involving the 1980 election" in which the CIA spied on President Jimmy Carter's administration, passing classified information to Reagan's campaign team. [152] Fast forward thirty-six years and Halper was at it again. Old spy habits die hard. Or not at all.

In an interview with CBS News, Barr made it clear that he was concerned about all of the lying and spying that had occurred before he took the helm at the Department of Justice:

These counterintelligence activities that were directed at the Trump Campaign, were not done in the normal course and not through the normal procedures as far as I can tell.

. . . It has to be carefully looked at because the use of foreign intelligence capabilities and counterintelligence capabilities against an American political campaign to me is unprecedented and it's a serious red line that's been crossed.[153]

On May 13, 2019, Barr appointed US attorney for the District of Connecticut John Durham to get to the bottom of it all.[154] Mueller's report had determined that there was no criminal conspiracy between the Trump campaign and Russia. In so doing, the special counsel has exposed the "dossier" as fictive and the "collusion" narrative as a hoax. How in the world did the "witch hunt" ever take flight?

The attorney general described the Durham investigation as focused on "a small group at the top," not the FBI or the DOJ as a whole. The actions of people such as Comey, McCabe, and Strzok would surely come under the microscope. In a letter to the chairman of the House Judiciary Committee, the DOJ outlined the parameters of Durham's probe, noting that it would include the "U.S. and foreign intelligence services as well as non-governmental organizations and individuals."[155] That meant Clapper and Brennan would be scrutinized and British and Australian intelligence would be examined. The actions of Simpson, Steele, the Clinton campaign, the DNC, Fusion GPS, and many others would almost certainly be looked at closely and carefully.

Barr fully realized that the media, which in their reporting are supposed to hold government accountable, would never do their job faithfully. "The fact that today people just seem to brush aside the idea that it is okay to, you know, to engage in these activities against a political campaign is stunning to me especially when the media doesn't seem to think that it's worth looking into," he said. "They're supposed to be the watchdogs of, you know, our civil liberties." [156] Sadly, they are not.

In their arrogance and ambition, the indomitable media seem more determined than ever to obscure facts and shade the truth in their dogged desire to pick winners and losers. Such matters, they reason, should never be left to mere mortal Americans.

CHAPTER 4

_____

THE ATTEMPTED COUP

QUESTION: Do you still believe the president could be a Russian asset?
MCCABE: I think it's possible.

—Former acting FBI director Andrew McCabe,
CNN, February 19, 2019

So many lies by now disgraced Acting FBI Director Andrew McCabe. He was fired for lying, and now his story gets even more deranged. He and Rod Rosenstein . . . look like they were planning a very illegal act, and got caught.

—President Donald J. Trump, tweet, February 18, 2019

A ndrew McCabe and Rod Rosenstein are living proof of the "Peter Principle." It is a concept that people in a hierarchy tend to rise to the level of their own incompetence.[1] Both men well exceeded their respective levels. Even worse, their ineptitude was surpassed only by their malice. That was a dangerous combination for the nation.

McCabe rose to become second in command at the FBI. When his patron saint, James Comey, was fired by President Trump in May 2017, McCabe briefly became acting director of the Bureau. But he was in command long enough to further encumber a presidency and cause lasting damage to the FBI. Rosenstein was deputy attorney general at the Department of Justice. After Attorney General Jeff Sessions wrongfully recused himself from any matters involving Russia, Rosenstein was elevated to acting AG, presiding over the FBI's counterintelligence investigation into the 2016 presidential election.

McCabe and Rosenstein, acting independently and at times in concert, abused their positions of power. Meeting privately in the aftermath of Comey's termination, they plotted a course of action to destroy the new president. They behaved rashly and out of vengeance. McCabe, furious that his boss had been fired, launched a new investigation of Trump without merit or cause. Rosenstein, in a fit of anger and wishing to deflect blame, appointed a special counsel to pursue Trump without justification. The two men secretly discussed deposing the president under the Twenty-fifth Amendment and recording him without his knowledge. It is of little consequence that they eventually abandoned their scheme as unworkable. Their willingness to embrace ethically reprehensible, if not lawless, conduct underscores how their stewardship was poisoned by enmity and arrogance.

THE MALEVOLENT MCCABE

It should be deeply troubling to all Americans that someone like Andrew McCabe rose to the top of the Federal Bureau of Investigation.

He was eventually fired from the FBI for dishonesty. But that is only a fraction of the sordid story. Before he was sacked, McCabe sought to punish the president of the United States because he disliked Trump's foreign policy statements and objected when the president dared to declare publicly that he had done nothing wrong. In addition, the president had the audacity to exercise his constitutional power to terminate McCabe's mentor Comey. To McCabe, these were unforgivable sins. In totality, they made Trump a traitor and only confirmed the acting director's belief that the president was in league with the Russians and trying to cover it up by obstructing justice. McCabe, a person possessed of chronic misjudgment, an inflated sense of self-importance, and utter disregard for both the law and executive authority, should never have been employed at the nation's premier law enforcement agency. Unfortunately, he is an all-too-typical example of the sort of elitist bureaucrat who rules our federal institutions and controls the narrative.

When McCabe voiced his opinion in February 2019 that the president might still be "a Russian asset," he was promoting his book *The Threat: How the FBI Protects America in the Age of Terror and Trump*, a title aimed not so subtly at Trump. The implication was that the constitutionally elected president was a threat to democracy because McCabe, an unelected and inferior government officer, thought he was. When asked why he believed Trump had been collaborating with Russia, he offered this explanation:

> From the very beginning, the president is referring to the investigation and our efforts as a witch hunt, as a hoax.
>
> In addition to that, he approaches Director Comey and asks him to drop the case against Mike Flynn. And after Director Comey fails to drop that case, he is in fact fired.[2]

From that statement, we gain three valuable insights into McCabe. First, he deemed the president's protestation of innocence and attack on the FBI's unjustified investigation to be an admission of guilt and, hence, evidence of treason.

Second, he had no qualms about misrepresenting the known facts about the president's conversation with Comey about Flynn. Nowhere in the memo that memorialized the president's alleged remark did Comey claim that it was a demand to drop the Bureau's investigation.

But even if "hoping" that the fired national security advisor would be cleared is construed as a request to drop an FBI investigation, the president would have been authorized to do so.

Third, he linked Comey's firing to his unwillingness to drop the Flynn case, even though there was *no* evidence that the FBI director was dismissed because of Flynn. Not even Comey has made that assertion. He was fired for insubordination in the Clinton email case and his refusal to repeat publicly what he was telling the president privately—that Trump was not under investigation in the Russia "collusion" case.

It's clear that McCabe's actions were driven by the liberal narrative, not the available facts. If the facts were not what he wanted, he twisted them to fit or manufactured outright lies to make his case.

In his book and during television interviews, McCabe offered another reason why Trump must be a Russian "asset": the president had been skeptical of a US intelligence briefing he had received on North Korea's ballistic missile capability, allegedly stating that President Putin had provided contrary information.[3] McCabe viewed that as a red flag of treason.

McCabe's presumptuousness was remarkable and disconcerting. Since when is the FBI entitled to substitute its judgment for that of the president in foreign policy? The Constitution grants the president the primary power—both expressed and implied—over foreign affairs. Other powers over foreign affairs were delegated to Congress. Americans and their political parties are free to disagree with decisions made or opinions expressed by a president, but they are not permitted to wrest control away from him or her through a bureaucratic agency. Change can always be effectuated every four years at the ballot box. Yet McCabe thought he knew better than Trump and sought to undermine the president by abusing his position of power at the FBI by targeting Trump. Whether the president was right or wrong in his rejection of intelligence information is immaterial and irrelevant. To put it bluntly, it was none of the FBI's business. Some presidents, such as George W. Bush, have relied, to the nation's detriment, on faulty intelligence and wished, in retrospect, that they had been less trusting.[4] Others, such as Barack Obama, have chosen to overlook negative intelligence in order to achieve a foreign policy objective such as the Iran nuclear accord.[5]

McCabe harbored a fundamental misunderstanding of the law and the Constitution. This is a peculiar characteristic in a trained lawyer serving as the deputy director of the FBI—until you consider all of the other misjudgments he has made. A pattern emerges. Like so many senior officials at the Bureau under Comey's misbegotten reign, McCabe considered his investigative operations to be autonomous and sacrosanct, as if he were answerable to no one but himself. That kind of pretension in law enforcement is exceedingly dangerous because it inevitably leads to abuses of power. And so it did.

In his book, McCabe laid bare his ignorance of constitutional law when he asserted, without evidence, that "Presidents don't weigh in" on federal investigations and prosecutions because "it is inappropriate."[6] He called it "a breach of propriety and of historical norms."[7] In truth, it is neither. Presidents do weigh in, and they have done so for more than two centuries. McCabe was stunningly wrong. Past presidents have involved themselves—sometimes quite directly—in federal cases because they have a *duty* under the "take care" clause of the Constitution to see that "the laws be faithfully executed."[8] We will examine this in more detail in chapter 5.

The important point is this: McCabe's misguided interpretation of presidential authority was what animated several of his own mistaken decisions at the FBI involving Trump and the Russia investigation. For example, McCabe decided that the president had, without doubt, committed obstruction of justice when he had removed FBI director James Comey.[9] Of course, neither the law nor the facts supported this. A president is empowered to fire an executive department head for any reason or no reason at all, as Comey admitted in a letter to his staff.[10] Moreover, when the director was fired, the FBI's investigation was, according to Comey, a counterintelligence matter, not a criminal case. It was impossible for Trump to have obstructed a counterintelligence probe since the information collected was for his own benefit. A person cannot, by definition, obstruct himself.

McCabe also alleged in his book that "on at least two occasions, the president asked the director to drop the inquiry regarding Mike Flynn," which was more evidence of obstruction.[11] Yet that was not what actually occurred, as confirmed by McCabe himself. Two days after Comey was fired, McCabe appeared before the Senate Intelligence Committee and conceded, "There has been no effort to impede our investigations to date."[12] His testimony clearly contradicted the revisionist story he was marketing to unknowing readers. Which was true? Comey unwittingly provided the answer. Six days before he was fired, he informed the Senate Judiciary Committee that no one had told him to stop anything for a political reason. "It's not happened in my experience," he said.[13] Days after Comey's termination, Rosenstein also told Congress, "There never has been, and never will be, any political interference in any matter under my supervision in the United States Department of Justice."[14]

Not only did those key people involved in the Russia case affirm that the president had never interfered or obstructed their investigations, but there was no other evidence that Trump was working for the Russians that would have justified the FBI's decision to launch its new investigation. Both Comey and FBI lawyer Lisa Page testified before House investigators that by the time the director was fired and Special Counsel Robert Mueller was appointed, there was no hard evidence of "collusion." The investigation had been running for ten months. Comey admitted, "In fact, when I was fired as director, I still didn't know whether there was anything to it."[15] FBI agent Peter Strzok, who had led the Bureau's investigation for all those months, texted Page about his reluctance to pursue the case under the new special counsel because there was no evidence of "collusion." On May 18, 2017, he wrote, "You and I both know the odds are nothing." He then added, "I hesitate in part because of my gut sense and concern there's no big there there."[16] Nevertheless, Strzok joined Mueller's team as a lead investigator.

Strzok, Page, McCabe, and Comey all knew the allegations against Trump were bereft of any proof and without merit. Nevertheless, McCabe opened a new or renewed investigation of Trump in May 2017 without sufficient evidence and in direct violation of FBI and DOJ regulations. Plainly, that decision was motivated by political bias and personal animus, not sustainable facts or credible evidence.

McCabe's anamorphic view of the law was so myopic that he actually believed that "there should be no direct contact between the president and the FBI director," except on matters of national security.[17] This is patently absurd, but it underscores how McCabe regarded many of Trump's actions through a distorted lens of presidential power.

It wasn't just constitutional ignorance that motivated McCabe, however, but a personal hatred for the president that is immediately evident in his book. McCabe's disregard for Trump's authority was compounded by his personal disdain for the man. Throughout his book, he revisited his real-time feelings whenever the president took some action that impacted the FBI. He blamed Trump for a "dank, gray shadow of uncertainty and bleak anxiety" in Washington.[18] Worse, he accused the president of "trying to destroy" justice in America.[19] Hyperbole aside, it is impossible to read McCabe's denunciations without recognizing that they must have dramatically influenced his decisions to target Trump. Passages in his book drip with contempt for the president. He seethed when Trump dared to exercise his right to free speech by "calling the Russia investigation a witch hunt."[20] In McCabe's cloistered world of law enforcement supremacy, falsely accused people are never entitled to proclaim their innocence. Such public expressions, according to McCabe, are witness tampering and obstruction of justice.[21] They are not, of course. Even a president has First Amendment rights.

McCabe also nursed deep resentment toward the president over the public comments that Trump had made while campaigning in 2016. They came about when the *Wall Street Journal* broke the story of how a close ally of Hillary Clinton had arranged exorbitant financial contributions to McCabe's wife when she had run for a state senate seat in Virginia.[22] McCabe had supervised the primary Clinton email investigation and been intimately involved in clearing her of criminal charges. The story immediately raised legitimate suspicions that the $675,000 in donations—which was an astronomical amount for a local race—might have had something to do with the favorable treatment Clinton had received, especially since Comey's explanation for declining to press criminal charges had made little sense. In his book, McCabe raged against Trump for raising that subject at a campaign rally, stating "His hatred was palpable. The crowd's angry response was chilling."[23] Only *after* the story was published did McCabe recuse himself from the Clinton probe when it was briefly reopened and closed. By that time, it was too late. Clinton had been cleared months earlier by Comey, on July 5, 2016. McCabe had endorsed that decision and taken an active role in absolving her.

The episode involving his wife embarrassed and angered McCabe, as his book makes apparent. He was blind to how his own conflict of interest did not end when his wife lost the election, and he was offended by the accusations of bias that dogged him. He blamed Trump for raising the issue, instead of himself for failing to step away from the investigation in the first place. His antipathy toward the president only grew as he closely aligned himself with Comey and against Trump in various disputes that ensued after the inauguration and as the Russia investigation evolved. When the president fired Comey, McCabe was incensed. He lamented that Trump had caused "insidious damage" to the FBI, without recognizing that he, together with Comey, Strzok, Page, and other top officials at the Bureau had done more to ruin the credibility of the FBI than any president could.[24]

Infuriated, McCabe went after Trump the moment he took the helm of the FBI from Comey. Unlike the previous counterintelligence probe of Russian interference and potential involvement with the Trump campaign, McCabe's new two-part investigation focused squarely on Trump himself.

One part pursued Trump for obstructing justice in his firing of Comey. The other was a renewed effort to gather proof that Trump was a covert Russian agent.

The accusations were ludicrous on their face. In addition, from a legal standpoint, McCabe's new probe constituted an egregious abuse of power. He had no probable cause, no credible evidence, and no reasonable suspicions. His new probe defied the law, ignored or perverted facts, and debased the integrity of a heretofore respected law enforcement agency. He investigated Trump simply because he could. He was now in charge at the FBI. He would do what he wanted. He was oblivious to the president's constitutionally conferred powers and determined to act against him. He would also utilize his newfound power to pressure Deputy Attorney General Rod Rosenstein to appoint a special counsel to go after Trump. But that's not all. In a breathtaking move, he and Rosenstein discussed a plan to force the removal of the president by marshalling forces that would evict Trump under the Twenty-fifth Amendment. If it worked, it would be the equivalent of a silent coup.

TRUMP WAS THE TARGET FROM THE BEGINNING

As FBI director, Comey repeatedly reassured Trump that he was not a suspect in the FBI's counterintelligence investigation of Russia that had been initiated in July 2016. The great weight of the evidence indicates that that was a lie. Trump *was* a suspect and had been so since the beginning. Three reasons stand out as proof. First, Comey consistently refused to tell the public what he was telling the president in private. If the director was telling Trump the truth, why not make that same truth available to the American people (unless, of course, it wasn't the truth)? Second, when he testified in March 2017, Comey implied to Congress that Trump might be a suspect. Third, the Mueller Report recited numerous examples of how the FBI had treated Trump as a suspect before handing over the case to the special counsel. Thus it is apparent that Comey was deceiving the president in his many private assurances. McCabe, as deputy director, was privy to all of that. Why the duplicity? Former federal prosecutor Andrew C. McCarthy surmised, "They were hoping to surveil him [Trump] incidentally, and they were trying to make a case on him." [25]

McCabe all but admitted that Comey was lying to the president. The FBI was certainly investigating whether the campaign had somehow coordinated with Russians before the election. As head of the campaign, Trump was obviously within the scope of the inquiry. Yet the FBI was pretending otherwise. In a meeting with McCabe, FBI general counsel James Baker complained that Comey was being, in McCabe's words, "jesuitical" in saying that "the president was not under investigation." [26] That is a polite description. A more accurate description is that the FBI was lying to Trump. Comey was the messenger, but McCabe was a party to that lie.

When Comey refused the president's persistent requests to inform the public that he was not under investigation, Trump viewed it as the last straw. Comey already deserved to be fired for insubordination and usurping the authority of the attorney general in the Clinton email case. When the president finally sacked the FBI director on May 9, 2017, McCabe was automatically elevated to replace him. The new acting director dropped all pretense of fairness. He immediately convened his team at FBI headquarters. [27] They were the identical people who had twisted the law into a pretzel to absolve Clinton of any criminality.

Within two hours of Comey's removal, FBI agent Peter Strzok texted his lover, Bureau lawyer Lisa Page, "We need to open the case we've been waiting on now while Andy is acting." Page replied, "We need to lock in [redacted]. In a formal chargeable way . . . soon." [28] It appears that everyone involved enthusiastically agreed to McCabe's idea of initiating the new and more expansive investigations of

Trump. Their true purpose was not to gather evidence for the *benefit* of the president so that he might counter foreign threats; that is how all counterintelligence cases are handled. Instead, they were attempting to assemble evidence *against* Trump. It was a secret criminal investigation of the president disguised as a continuation of the previous counterintelligence case. Make no mistake—Trump was the implicit target. Yet according to Page's July 2018 testimony before the House Judiciary and Oversight Committees, the FBI had had no real evidence of Russian "collusion" at the time Comey was fired.[29] There was no "articulable factual basis for the investigation," as FBI regulations demanded before opening such an investigation.[30] That should have stopped them. It did not. They proceeded to violate Trump's due process rights, as Representative John Ratcliffe (R-TX) observed during the hearing.[31]

THERE WAS NO JUSTIFIABLE BASIS FOR OPENING A CRIMINAL INVESTIGATION

In his book, McCabe explained that he had initiated his new investigation of Trump because he feared Trump would fire him, as he had Comey:

> If I was going to be removed, I wanted the Russia investigation to be on the surest possible footing. I wanted to draw an indelible line around it, to protect it so that whoever came after me could not just ignore it and make it go away.[32]

He later told CBS News that he had feared the case would "vanish in the night," especially if he were to be fired.[33]

This reasoning is incomprehensible coming from a veteran FBI official. It demonstrates a profound misunderstanding of how the FBI is structured and operates. Investigations do not suddenly disappear because a director, an acting director, or any individual agent happens to depart the Bureau. Once a case is opened, it is continued uninterrupted by others among the 35,000 agents and support professionals who are employed there. Moreover, it is a severe violation of regulations to open a case because an agent or director fears that something *might* happen in the future. Conjecture about a prospective event, such as being fired, that may or may not occur is not an identified basis for initiating a probe according to the FBI guidelines. McCabe surely knew he was misusing his authority, but he needed to invent an excuse that sounded good.

Simultaneous with the opening of a new investigation of Trump, McCabe met frequently with Rosenstein over the next eight days. The deputy AG was stung by the criticism he'd received for recommending Comey's termination. He was "glassy-eyed, emotional, upset . . . and shocked," wrote McCabe.[34] In the first meeting, Rosenstein confessed that he was mulling over the appointment of a special counsel to investigate both Russia and Trump. That was music to McCabe's ears. Over the course of four successive meetings, he pushed Rosenstein relentlessly to name a special counsel and eventually prevailed. Both men knew that such a move would badly damage Trump because independent probes, regardless of their merits, tended to disable a presidency with the cloud of suspicion.

At that point, revenge seemed a top priority. McCabe was furious at Trump for firing his friend and colleague Comey. For his part, Rosenstein felt he was a hapless scapegoat and blamed Trump for it, conveniently overlooking the fact that as deputy attorney general he had volunteered to write the memo recommending Comey's removal and had spent weeks and even months in advance discussing how to do it with his boss, Jeff Sessions.[35] Rosenstein was attempting to rewrite history to exonerate himself. The plan to name a special counsel was the perfect solution for McCabe and Rosenstein. It would serve as an ideal act of retribution against Trump and help salve their own wounds. It is a sad commentary on the lack of character in government officials who exploit their positions of power by resorting to petty vindictiveness.

MCCABE AND ROSENSTEIN PLOT A COUP

Having launched a renewed FBI investigation of Trump and having succeeded in persuading Rosenstein to name a special counsel,
McCabe then involved himself in one of the most diabolical plots
in US political history: discussion and deliberation over whether to
evict the duly elected president of the United States from office and
undo the 2016 election results. You would expect a lengthy excursus
of these events to be featured as a central part of McCabe's tell-all
book—at the very least, a full chapter. Amazingly, McCabe made
no mention of it. Not a word was written about the plot to depose
the president under the Twenty-fifth Amendment. Yet when it came
time for McCabe to peddle his book on a nationwide tour and gin up
interest so that he could profit handsomely from its sales, it was all he
could seem to talk about.

It began with an ingratiating interview with Scott Pelley on CBS's
60 Minutes in which McCabe offered a detailed account of how
Rosenstein had wanted to secretly record the president to gather incriminating evidence against him:

> The Deputy Attorney General offered to wear a wire into the
> White House. He said, "I never get searched when I go into
> the White House, I could easily wear a recording device, they
> wouldn't know it was there." Now, he was not joking, he was
> absolutely serious, and in fact he brought it up in the next
> meeting we had. I never actually considered taking him up on
> the offer, I did discuss it with my general counsel and my lead-
> ership team at the FBI after he brought it up the first time. . . .
> I think the general counsel had a heart attack.[36]

McCabe didn't stop there. He launched into an unnerving description of how Rosenstein had schemed to use whatever evidence
he might gather furtively as a basis for convincing cabinet officials to

remove Trump under the Twenty-fifth Amendment to the Constitution for want of mental capacity:

> Discussion of the 25th Amendment was simply, Rod raised the issue and discussed it with me in the context of thinking about how many other cabinet officials might support such an effort. I didn't have much to contribute, to be perfectly honest, in that conversation. So, I listened to what he had to say.
>
> I mean, he was discussing other cabinet members and whether or not people would support such an idea, whether or not cabinet members . . . shared his belief that the president was—was really concerning, was concerning Rod at that time.[37]

Pelley then asked if Rosenstein had talked specifically about whether there was a sufficient majority of the cabinet who would vote to remove the president. McCabe answered, "That's correct. Counting votes or possible votes."

The alarming dialogue between McCabe and Rosenstein about surreptitiously wearing a wire and recruiting cabinet members to depose Trump, if pursued, would have been a lawless misuse of power. Disliking a president's decision making, even if you believe it to be unjustified or irrational, is not a basis for evicting him from office. McCabe and Rosenstein are both lawyers, and one would assume they had known this. If it had been entirely Rosenstein's idea, McCabe would have had a duty to immediately notify the inspector general at the DOJ, Congress, or both. Kevin R. Brock, a former assistant director of intelligence for the FBI and a special agent at the Bureau for twenty-four years, was unsparing in his criticism of McCabe:

> The mere presence of an FBI acting director in such a meeting, let alone his active participation, is a monumental misuse

of position, betrayal of the trust of the American people, and humiliating embarrassment to the dedicated rank and file of the FBI.[38]

McCabe kept notes of the entire episode with Rosenstein and handed them over to the special counsel. The Mueller Report made no mention of it. That should come as no surprise since Mueller had served a lengthy tenure as director of the Bureau and had every reason to protect the reputation of the institution by covering it up. Congress lawfully demanded the documents pursuant to a subpoena, but both the FBI and the Justice Department were defiant in refusing to produce them. Rosenstein's role as Mueller's supervisor surely helped to suppress the incriminating records. The deputy attorney general was obstructing Congress while supervising an investigation into whether the president had obstructed justice. His rank hypocrisy was lost on no one.

In an attempt to tamp down the shocking revelation that Rosenstein had attempted to orchestrate the equivalent of a "soft coup," one colleague of his portrayed him as merely "joking" or being "sarcastic."[39] Right. Rosenstein has a history of claiming that his words were misinterpreted.[40] To hear him tell it, he is the most misunderstood man in Washington. More likely, he is peddling a deceptive excuse. But does this explanation make any sense? Secretly recording the president and mobilizing forces to remove him from the highest office in the land is no laughing matter. It is doubtful that the deputy attorney general was in a jovial or jesting mood, since McCabe depicted him as disoriented, erratic, and distraught over his pivotal role in the Comey firing and the backlash he had suffered. An act of reprisal against the president seemed more plausible for the "frustrated" Rosenstein, who was described by others as "conflicted, regretful and emotional."[41]

ROSENSTEIN DENIES EVERYTHING

Reeling from that televised account by McCabe, Rosenstein dismissed the story as "inaccurate and factually incorrect," issuing a meticulously worded statement:

> The deputy attorney general never authorized any recording that Mr. McCabe references. As the deputy attorney general previously has stated, based on his personal dealings with the president there is no basis to invoke the 25th Amendment, nor was the DAG in a position to consider invoking the 25th Amendment.[42]

A careful reading of this statement indicates two conspicuous nondenial denials. First, notice that Rosenstein claimed that he did not "authorize" wearing a wire. That is not the same as discussing the *idea* of wearing a wire or taking steps to do it. He might well have accomplished it without any authorization whatsoever. So that statement is a cagey diversion. Second, whether there was a "basis" to invoke the Twenty-fifth Amendment was not the issue. What's relevant was whether Rosenstein considered taking action against the president and actively sought to do so, regardless of a proper or legal basis. More to the point, he did not deny seeking an illegitimate basis for removing Trump. So that was another canny misdirection, all of which suggests that Rosenstein pursued the plot to expel Trump from office. There was little chance of obtaining a majority vote in the cabinet, but that was not why the plan was abandoned. The entire enterprise was deemed "too risky," according to the private testimony of FBI general counsel James Baker, who became privy to the plot.[43]

Given that McCabe was eventually fired from the FBI for lying and was later referred to federal prosecutors for potential criminal charges, his credibility is inherently suspect.[44] Except when you consider that there was at least one percipient witness to the McCabe-Rosenstein conspiracy. FBI lawyer Lisa Page accompanied McCabe to one or more of the meetings in which the subject was broached. Both Page and McCabe then promptly informed Baker of Rosenstein's desire to evict Trump from office. In closed-door testimony to Congress, Baker revealed that both McCabe and Page had believed that Rosenstein was "serious" about dislodging the president from office.

What's more, Baker learned that "the deputy attorney general had already discussed this with two members in the president's cabinet and that they were . . . onboard with this concept already."[45] Specifically, he understood that those two cabinet officials were "ready to support . . . an action under the Twenty-fifth Amendment."[46] Who were the cabinet secretaries? The *New York Times*, which first broke the story, identified Attorney General Jeff Sessions and Secretary of Homeland Security John Kelly.[47] "One participant asked whether Mr. Rosenstein was serious, and he replied animatedly that he was," reported the *Times*.[48] Other associates confirmed that. Where did the *Times* get its story? It is tempting to conclude that McCabe's fingerprints appeared all over it.[49] His motivation to expose Rosenstein's plot was strong because six months earlier the deputy attorney general had endorsed the decision to fire McCabe for his leaks to the media. What better way for him to trash Rosenstein than to leak once again to the media? If not McCabe, Baker was the probable leaker. He, too, had a motive to deflect blame away from the FBI and onto Rosenstein at DOJ. When rats get trapped in a cage, they begin to turn on one another.

Neither McCabe nor Rosenstein had authority to even consider what would be, under the Constitution, a cabinet-level decision that also requires the consent of the vice president. The acting head of the FBI and the deputy attorney general at the DOJ have no involvement under the provisions of the Twenty-fifth Amendment. It is a measure of their immense hubris that they would connive behind the scenes to instigate and organize the overthrow of a sitting president under false pretenses. They appear to have insinuated themselves into a decision that did not even remotely concern them. Even more significantly, someone's *opinion* that a president has committed maladministration is not a basis for invoking the amendment.

The Twenty-fifth Amendment provides for the president's removal and replacement in the event that he is "unable to discharge the powers and duties of his office."[50] Although incapacity was not defined, the legislative history summarized by the Congressional Research Service makes it evident that the framers of the amendment envisioned a chief executive who was stricken by a debilitating stroke, heart attack, or bodily injury or had been the victim of a failed assassination attempt and was physically or mentally disabled as a direct consequence. The authors of the amendment made it clear that it "was not intended to facilitate the removal of an unpopular or failed President" or for any other political purpose.[51] Senator Birch Bayh (D-IN), the architect of the amendment, stated that the word "unable" meant "an impairment of the President's faculties" such that he is "unable either to make or communicate his decisions as to his own competency to execute the powers and duties of his office."[52] On the House side, the principal framer, Representative Richard Poff (R-VA), cited "some physical ailment or sudden accident" rendering a president "unconscious or paralyzed and therefore unable to make or to communicate the decision to relinquish the powers of his Office."[53] Whatever the reasons that motivated Rosenstein's maneuvers, the Twenty-fifth Amendment had no application to the circumstances.

Rosenstein miscalculated the political consequences of his memo recommending Comey's termination. He assumed that both Democrats and Republicans would leap to their feet and give him a standing ovation for finally accomplishing what both parties had called for incessantly. When the opposite occurred, he became overwrought and "angry."[54] His venom was directed at Trump. Though he reportedly regretted what he had done, engineering the president's removal would have been, by any standard, an extraordinary act of retaliation.

McCabe and Rosenstein had every reason to act outside the law to rid themselves of the president. The FBI had used a fraudulent "dossier" funded by the Clinton campaign and Democrats to launch its original investigation of Trump in July 2016. Comey, McCabe, and others at the Justice Department presented the FISA court that same unverified Russian disinformation to spy on a Trump associate, misrepresenting to the judges that it had been verified. As noted in the previous chapter, they had concealed vital evidence and deceived the court. The "collusion" narrative had been leaked to the media in an attempt to influence the election. As one columnist for The Federalist put it, "They had vastly overreached, assuming that an inevitable Clinton win would cover up their sins." [55] When voters decided otherwise, their only option was to act outside the law to nullify and undo the election results by vanquishing Trump.

In the *60 Minutes* interview, McCabe portrayed himself as the mild-mannered Clark Kent turned Superman of the FBI—interested only in truth, justice, and the American way. Pelley played the adoring Lois Lane, lofting one softball question after another, never challenging his interviewee as he claimed credit for the appointment of a special counsel and feigned innocence as Rosenstein connived behind the scenes to pitch Trump from the Oval Office. Pelley also managed to misrepresent key facts while omitting others.[56] The interview was a combination of performance art and choreographed charade. The truth is that McCabe's abysmal judgment and rogue actions against Trump, together with his self-serving book and media tour, only served to further embarrass and disgrace the already beleaguered FBI.

In a scathing critique of McCabe's media leak and ensuing lies about it, the Justice Department scolded him for conduct inimical to "the highest standards of honesty, integrity, and accountability" that are demanded of employees at the FBI.[57] Michael Horowitz, the inspector general who had been appointed by President Barack Obama, found that McCabe had lacked candor (i.e., lied) or misled the FBI and federal investigators four times, and on three of these occasions had done so while he was under oath. The IG concluded that McCabe's leak had been entirely self-serving, "designed to advance his personal interests at the expense of department leadership." [58] The findings were sent to the US attorney in Washington for potential criminal charges over false statements.[59]

McCabe and his mentor Comey had much in common. Both men were fired for wrongdoing, and both of them pilfered government documents on their way out the door.[60] They lacked candor and scruples. They almost certainly abused their high positions of power for personal or political reasons. Once they were fired, they continued their campaign to castigate Trump at every turn through their books, tweets, and public denunciations. The bias they have exhibited since leaving the Bureau is persuasive evidence that they were acting on that same bias while serving at the FBI.

ROSENSTEIN BEGS TRUMP'S FORGIVENESS

It was duplicitous for Rosenstein to preside over the special counsel investigation of Trump at the same time that he was conspiring to overthrow him. How could he conceivably oversee Mueller's probe in an objective manner while harboring such extreme bias against the president? The deputy attorney general was hopelessly conflicted, but he dug in his heels and refused to step aside despite repeated calls to do so. The evidence suggests that he was determined to exploit his bias. No one at the Justice Department had the authority to force him off of the McCarthy-style investigation he initiated. Mueller seemed quite satisfied with a compromised sycophant as his direct supervisor. It gave him greater autonomy to do as he pleased. Rosenstein also resisted all efforts by Congress to have him testify about his conspiracy to depose the president. Rosenstein was an island unto himself—unreachable and untouchable. Only the president could get rid of him, but that was politically untenable. Any effort to fire someone connected to the special counsel, even for just cause, would have been met with mass hysteria among Democrats and the leftist media. Trump would have been crucified with phony portrayals that he was obstructing the investigation and, hence, justice.

Following the *Times* story in the fall of 2018 detailing Rosenstein's clandestine plan, Republicans in the House of Representatives demanded that he appear before the Judiciary and Oversight Committees to explain himself. The deputy attorney general adamantly refused to testify publicly. He then engaged in protracted negotiations to answer questions behind closed doors.[61] Every time a tentative agreement was struck, Rosenstein imposed new conditions at the last minute. He rebuffed the notion that he should be held to account for his words or actions. In so doing, he was obstructing Congress in much the same way he had done by defying lawful subpoenas to produce documents.[62] He was also buying valuable time.

It was fairly obvious that Rosenstein was trying to run out the clock until the 2018 midterm elections on November 8. He was banking on polling data showing that the GOP would lose control of the House. Once Democrats assumed the reins of power, all interest in Rosenstein's coup attempt would be consigned to the trash heap of political cover-ups. That was precisely what happened. In the end, when the House flipped control, Democrats saw to it that the entire matter was buried. Like Houdini, Rosenstein proved to be a proficient escape artist.

Congress's leverage over the deputy attorney general was diminished when the "emotional" and contrite Rosenstein retreated to the White House in late September, prepared to submit his resignation.[63] He reportedly pleaded his case to the president's aides that he had never sought to evict Trump from office, notwithstanding all of the evidence to the contrary. The gambit worked. With the Mueller investigation winding down, a "detente" was reached with the president.[64] If nothing else, perhaps a general commitment was secured from Rosenstein that he would treat the president fairly upon the conclusion of the probe. That, of course, had been his duty all along.

Why did Rosenstein resist answering questions under oath so vigorously? In the absence of a transcribed or recorded interview, he would be free to give deceptive answers without putting himself into legal jeopardy. If he gave a false or misleading statement, he could always deny it later. With no record, he might simply claim that he had never made a given statement. Another reason is that he could not readily explain or justify many of his actions. People who have nothing to hide generally don't hide. They are anxious to tell the truth. But Rosenstein knew that there were several credible witnesses to his illicit scheme to remove the president. Denying it under oath would have exposed him to perjury or false statement charges. It is easy to lie to the media through anonymous leaks. But under direct questioning before Congress, it is a crime not to tell the truth.

When news of Rosenstein's alleged plot became public, he rushed to the White House, expecting to be fired.[65] A meeting with the president, however, was delayed. On October 8, 2018, Rosenstein was invited aboard Air Force One to discuss the matter with Trump face-to-face.[66] Here is what the president told me about their meeting:

> **JARRETT:** Two witnesses who were in Rosenstein's office in the days after Comey was fired confirmed that he was serious about wearing a wire to secretly record you and recruiting cabinet members to remove you from office. You talked to him about it on Air Force One on the way to Florida. What did he say?
>
> **PRESIDENT TRUMP:** He said it didn't happen. He said he never said it. What he told other people is that he was joking. But to me he claimed he never said it.
>
> **JARRETT:** Did you believe him?
>
> **PRESIDENT:** I didn't really know what to believe.
>
> **JARRETT:** Why did you keep him on as deputy attorney general?
>
> **PRESIDENT:** Because I thought it would be bad to fire people in the middle of an investigation. And it turned out to be the right thing to do.[67]

So which was it? A joke or a fiction? Logically, it cannot be both. During which account was Rosenstein telling the truth? In one of

them, he must have been lying. It is no wonder he resisted all efforts
to be questioned under oath by Congress. Lying to the president is not
a prosecutable crime, but lying to Congress during an investigative
hearing is.

McCabe's book made it clear that Rosenstein was emotional, upset, and even angry at Trump when he made the reckless decision to
appoint a special counsel to investigate the president. He also knew
that what he was doing was fundamentally wrong. Given Sessions's
recusal, the deputy attorney general was now acting attorney general
for all matters involving Russia. He resolved to retaliate without notice to the White House.

Mueller had met with Trump on May 16 to discuss replacing
Comey as FBI director, an interview arranged by Rosenstein. The
president had been uninterested in naming Mueller to the same position he had previously held and immediately dismissed the idea. In
more than one conversation, Trump told me that he had felt Mueller had a serious conflict of interest because the two men had been
involved in a previous business dispute over membership fees when
Mueller had resigned from one of Trump's golf clubs. "I didn't think
he'd be fair, so I rejected him," said Trump.[68] Little did the president
know that Rosenstein was secretly discussing with Mueller the possibility of his serving as special counsel to investigate Trump. Such
duplicity was appalling.

The next day, Sessions was in the Oval Office with Trump, conducting more interviews for Comey's replacement, when he took a
call from Rosenstein advising that he had named Mueller as special
counsel. As the Mueller Report recounted, the president was understandably shocked and angry. He said, "Everyone tells me if you get
one of these independent counsels it ruins your presidency. It takes
years and years and I won't be able to do anything."[69]

Trump's assessment was correct, and that was surely what the acting AG had in mind when he appointed Mueller. It was the ultimate
reprisal against Trump by the scheming Rosenstein. The embarrassed

Sessions said he would prepare his resignation letter, although the president declined to accept it.

Attending these events was Sessions's chief of staff, Joseph "Jody" Hunt, who took fastidious notes. When he returned to the DOJ, he marched into Rosenstein's office. Trump's lawyer John Dowd told me he had reviewed Hunt's notes before they were provided to the special counsel and that the notes included the following confrontation:

> Rosenstein was literally cowering... hiding behind and somewhat below his desk. "Am I gunna get fired?" blubbered Rosenstein. Hunt replied, "What you did was despicable and unprofessional!" Hunt then stormed out of Rosenstein's office.[70]

Dowd's account is consistent with Rosenstein's erratic and unhinged behavior as depicted in McCabe's book. It is also congruous with other actions taken by Rosenstein that reflect a vindictive nature.[71]

In January 2018, he allegedly threatened to subpoena emails, phone records, and other documents from lawmakers and staff on the House Intelligence Committee in what was described as "retaliation" for the committee's inquiries into the Russia probe.[72] A similar episode happened in May, when Rosenstein launched a "very personal and very hostile attack" on the same committee staff and its chairman, Representative Devin Nunes (R-CA), according to a congressional email documenting the meeting and corroborated by two additional sources.[73] In both instances, a Justice Department spokesperson claimed Rosenstein's remarks had been mischaracterized.

Either Rosenstein is persistently misunderstood and his words are forever being misinterpreted, or he is a petulant bully, who tries to cover up his misconduct by peddling the same deceptive excuse again and again. The answer should be obvious. Rosenstein has a history of abusing his position of power. When caught, he claims that everyone else is mistaken and confused. That was also his excuse when reports surfaced that he had schemed to secretly record his conversations with the president and depose him under the Twenty-fifth Amendment. It was the same pattern of conduct that underlay his vindictive decision to name a special counsel to investigate Trump.

CONFLICTS OF INTEREST AND AN UNNECESSARY SPECIAL COUNSEL

Rosenstein's misconduct and malfeasance were set into motion when Attorney General Jeff Sessions wrongfully removed himself from the FBI's Russia case. As explained in *The Russia Hoax*, neither the facts nor the law required it.[74] Ethical rules demand a recusal in criminal cases and prosecutions.[75] But at the time, the FBI's investigation into Russia was a counterintelligence matter, not a criminal case. That was confirmed by the Bureau's director, James Comey, when he testified before Congress.[76] A recusal under federal regulations has no application to counterintelligence probes. Sessions's decision to remove himself was illogical and mistaken. There was no legitimate reason for it. There must first be an identified crime for a recusal to arise. When Sessions announced his decision on March 2, 2017, the FBI had no evidence of any crimes.

Even worse, Sessions had not advised the president that that was his intention when he was sworn in. The attorney general later informed Congress that he had taken preliminary steps to recuse himself on his first full day in office.[77] He deceived his boss by concealing that. Failing to disclose such a matter was a serious betrayal. The president was entitled to know the truth, but Sessions actively hid it from him. Trump was understandably disgusted and angry. Sessions's deception also deprived him of the president's confidence and trust, which are essential to the job of attorney general. That ethical impropriety rendered Sessions unfit to serve. But for his deceit and wrongful recusal, it is unlikely that a special counsel would have been appointed.

Sessions's incomprehensible actions left Rosenstein in charge of the Russia investigation as acting attorney general. He had been on the job a scant two weeks when Comey was fired. A week later, he named Robert Mueller as special counsel. Hence Rosenstein's knowledge of the FBI's nearly ten-month-long counterintelligence investigation into Russian activities had to have been at best superficial. Yet that did not give him pause, as it should have. He promptly misconstrued—and thereby misused—the special counsel regulations authorizing the appointment of Mueller. He did so in not one but four ways.

First, the Code of Federal Regulations (28 C.F.R. § 600.1) specifically states that a special counsel is to be appointed in cases where an investigation or prosecution "would present a *conflict of interest* for the Department of Justice. [author's italics]"[78] If there is no identifiable conflict of interest, there can be no special counsel. But where exactly was the conflict? Who had it? Attorney General Jeff Sessions had already recused himself, albeit mistakenly, from any matters involving Russia. No other known conflicts of interest existed for Rosenstein or others at the Justice Department that would have prevented the department's regular prosecutors from handling the case. Indeed, the authorization order signed by Rosenstein discloses no such conflict of interest because none existed.[79] He has never offered an explanation for that fundamental defect in the appointment.

Proof that a conflict was nonexistent is the fact that Mueller enlisted top prosecutors from within the DOJ to serve as his primary staff. If there had been a conflict, all of the lawyers on the team should have been culled from private law firms or elsewhere. Additionally, the special counsel investigated and/or charged at least fourteen matters that he then referred to the Justice Department itself for prosecution or disposition.[80] Transferring cases to the presumably conflicted department would have been highly improper. All of this confirms that the necessary predicate of a conflict of interest was never present at the outset. That made the appointment of Mueller illegitimate at the outset.

The same code section alternatively provides for a special counsel in the event of "other extraordinary circumstances." However, none existed at the time or was cited in the appointment order. Comey's firing did not alter or end the FBI's counterintelligence investigation, which still had uncovered no evidence in support of the "collusion" allegations.

Second, the operative regulation states that the attorney general or acting attorney general must "determine that *criminal investigation* of a person or matter is warranted. [author's italics]"[81] An "articulable criminal act" must be identified.[82] This, too, is a necessary predicate for a special counsel. Evidentiary support for a criminal investigation must antecede the appointment, not vice versa.[83] But on May 17, the FBI had developed no such evidence. That was confirmed by the subsequent testimony of both Comey and FBI lawyer Lisa Page, as noted earlier. Nowhere in Rosenstein's authorization order did he bother to identify a suspected crime because the FBI had no evidence of one. It is not a crime to talk to, or have a relationship with, a Russian. There would have to have been some evidence of espionage or a conspiracy to commit an illegal act. The FBI had no such evidence in its possession. The evidentiary premise of a crime was conspicuously missing. Thus Mueller began an investigation in search of a crime, reversing the legal process mandated under the regulations.

Former US attorney general Michael Mukasey, who was also chief judge of the US District Court for New York, noted that prior investigations utilizing independent counsels "identified specific crimes" and "the public knew what was being investigated."[84] The whole point of the special counsel regulation was to avoid limitless investigations by unaccountable prosecutors and to apprise the public of what potential crimes were being pursued. The public, after all, has a right to know if its tax dollars are being used or misused for an unbounded search for unidentified crimes. But with Mueller's appointment, Americans were given no idea what the special counsel was up to because Rosenstein's assignment order violated the regulation that demanded the identity of a specific crime to be investigated.

As the Mueller Report grudgingly admitted, "collusion is not a specific offense . . . nor is it a term of art in federal criminal law." [85] It was, however, a potent word that journalists and pundits could toss around ad infinitum to defile the president as a purported Russian agent. As Mukasey observed, "[it] may sound sinister, but it doesn't define or suggest the existence of a crime." [86]

Third, the appointment of a special counsel applies only to *criminal* investigations, not to counterintelligence probes. [87] The FBI's case at that point was, according to the testimony of Comey, a counterintelligence matter. The distinction is an important one. Counterintelligence collects information about foreign threats to be given to the president so that he may fulfill his responsibility to protect the United States' national security. No suspicion or evidence of a crime is needed. Contrast that to a criminal investigation, which develops evidence for potential prosecution. It must be premised on a crime. In the absence of any articulable facts supporting a crime, the FBI should have quietly continued its counterintelligence probe. If in the process it had found possible crimes, it could have referred them to the DOJ for a determination as to whether charges should be brought. There was no need whatsoever for a special counsel.

Finally, the importance of *specifically stating* a suspected crime as a necessary predicate for appointing Mueller is reinforced by the condition set forth in 28 C.F.R. § 600.4, which states, "The Special Counsel will be provided with a *specific factual statement* of the matter to be investigated." [88] The scope of the probe is to be strictly circumscribed in order to limit the special counsel's work to a defined subject. But Rosenstein did not comply with that requirement. The order announced a sweeping mandate with ill-defined parameters. Mueller was granted broad and nearly unlimited license to investigate matters that were not criminal at all.

Though it is true that 28 C.F.R. § 600.4 allows a special counsel to investigate and prosecute obstruction, it is expressly confined to obstruction "with intent to interfere with the special counsel's probe." [89]

But potentially obstructive acts, such as Trump's supposed comments to Comey about Flynn and the subsequent firing of the director, fell well outside this stated jurisdiction inasmuch as they *predated* Mueller's appointment. To put it another way, Mueller had no authority under the order to pursue any obstruction that might have occurred *before* he was appointed. He could pursue only obstruction impeding the investigation itself *after* it began.

Once again, Rosenstein failed to adhere to the law. He charged the special counsel with investigating something without authority. Rosenstein and Mueller knew they were trampling all over the special counsel regulations. In later court documents, the special counsel admitted that his original authorization order had been *deliberately vague* because anything specific would have been confidential and might have jeopardized his investigations.[90] This is not, and has never been, a bona fide excuse for disobeying federal regulations derived from statutory law. It was an astonishing confession that the acting attorney general had carefully crafted his appointment order in a way that willfully violated the regulations requiring a specific statement of facts.

As the Mueller Report confirmed, the special counsel had conducted an intensive examination of whether the president's firing of Comey had been a "corrupt" act intended to obstruct any investigation into Trump-Russia "collusion." The president had relied, at least in part, on Rosenstein's written recommendation that Comey be terminated for the reasons detailed in the deputy AG's memorandum of May 9, 2017.[91] He had also had extensive conversations with the president and then attorney general Jeff Sessions about discharging Comey for a justifiable cause.[92] That made Rosenstein a central witness in any obstruction case.

Under those circumstances, Rosenstein was undeniably the wrong person to decide whether a special counsel should be named, and it was an indefensible mistake for him to have proceeded with the appointment given his status as a key witness in the case. The deputy attorney general was seriously compromised by his own role in firing

Comey. Sessions had recused himself because of the *appearance* of impropriety arising from a *perceived* conflict of interest. By comparison, Rosenstein was afflicted with a *genuine and indisputable* conflict of interest that necessitated his recusal.[93] His failure to recognize the obvious was incomprehensible. His desire to make the appointment notwithstanding is strong evidence that he nurtured a hidden agenda.

As previously noted, Rosenstein was an emotional train wreck in the aftermath of Comey's firing. As reported by the *Times*, colleagues and friends described him as simultaneously angry and remorseful, shaken and unsteady, overwhelmed and conflicted.[94] That was affirmed by McCabe, who had several meetings with him.[95] "He alternately defended his involvement, expressed remorse at the tumult it unleashed, said the White House had manipulated him, fumed how the news media had portrayed the events and said the full story would vindicate him," reported the *Times*.[96] Those accounts by people close to Rosenstein underscore his personal and professional involvement in the Comey firing. Rosenstein should not have been the one at the Justice Department to make the decision as to whether a special counsel should be appointed. Someone else should have taken charge while the deputy attorney general stepped aside.

Rosenstein's recusal from the entire case was mandatory. Section 28 C.F.R. § 45.2 requires disqualification if a person is "substantially involved in the conduct that is the subject of the investigation."[97] Rosenstein was substantially involved as a vital witness to the reasons behind Comey's firing. That bore directly on whether the president had obstructed justice. Rosenstein had a "personal and professional" interest in the outcome—another disqualifying circumstance cited in the regulation. His refusal to recuse himself also violated the Rules of Professional Conduct that govern lawyers.[98]

Once Mueller was appointed, he began investigating whether Trump had obstructed justice. The most important person who could attest to the president's motivations was the very man who was now supervising the investigation into Trump's intent. The provisions of 28 C.F.R. § 600.7 placed Rosenstein in charge and made him Mueller's immediate superior.[99] As acting attorney general on the case, Rosenstein was overseeing every aspect of it and directing both the scope and the decision making. He wielded the ultimate power to determine whether a prosecution would be brought but was also the principal witness who would influence that decision.

Was Rosenstein interviewed by Mueller or his investigators? He must have been. The *Wall Street Journal* reported that the deputy AG was among the first witnesses questioned by the special counsel's team.[100] Cryptic footnotes in the Mueller Report seem to confirm this.[101] Although Mueller answered to Rosenstein, it was Rosenstein who had to answer the special counsel's questions. That was about as warped and improper as a case can get. The boss was questioned by his subordinate; then the boss played an instrumental role in deciding whether charges should be brought based, in whole or in part, on his own testimony. It does not take a genius to realize that the scenario was fundamentally wrong and unethical. Rosenstein, as a chief witness, should not have been permitted to contemporaneously judge whether a prosecution was merited. That is the equivalent of a prosecutor putting his boss or himself on the witness stand to testify in the very case he is prosecuting. A lawyer cannot be an impartial investigator, witness, prosecutor, and judge all rolled into one. Rosenstein played all of those roles in a single case. His refusal to recuse himself was an atrocious lapse in judgment. It rendered the entire special counsel investigation inherently suspect amid persistent questions of self-interest, political bias, and a lack of objectivity.

The Mueller Report left little doubt that Rosenstein was interviewed by the special counsel. Private conversations between the deputy attorney general and Trump are recited almost verbatim in the section on obstruction of justice.[102] Since Trump wisely chose not to be interviewed, those conversations were likely recounted by Rosenstein to the special counsel.[103] Mueller seems to have accepted his boss's version of events as gospel, without questioning whether Rosenstein had a veiled motive to shade the truth in a way that was self-serving and adverse to the president. That was lunacy: the deputy

AG advised Trump how to get rid of Comey, then hired Mueller to investigate whether that decision constituted obstruction of justice by the president. Rosenstein's refusal to recuse himself and his determination to supervise the case were unconscionable. No responsible and ethical lawyer would do such a thing.

Throughout the probe, Rosenstein was evasive and cagy about his conflict of interest. He allegedly consulted with an ethics adviser at the Justice Department and "followed that individual's advice."[104] If so, that person should be cashiered for incompetence. At one point, Rosenstein said he would leave it up to Mueller to decide about his recusal.[105] This is nonsensical in the extreme. Ethical rules and federal regulations impose the duty to recuse on the conflicted person, not his subordinate.[106] If the subordinate is also involved in the case, as was Mueller, any advice on recusal would inevitably create his *own* conflict of interest because he would have an interest or bias in protecting his star witness as his boss.

There is another essential aspect to this: What if Rosenstein did not tell the truth when he was interviewed by the special counsel? What if other witnesses or documents offered proof that he lied? As supervisor of the case who would make the ultimate prosecutorial decisions, he would be free to disregard it and cover it up. Would he really decide to charge himself for false statements or perjury? Not a chance. That is exactly why a prosecutor cannot wear another hat as a witness in the same case: there is no penalty for dishonesty; it is a license to lie.

Rosenstein knew he should recuse himself because McCabe told him so to his face during an angry confrontation right in front of Mueller, according to the *Washington Post*.[107] It happened just days after the special counsel was appointed. In a room with Mueller present, the two men hurled verbal assaults, each accusing the other of compromising behavior that had left them riddled with conflicts of interest requiring their recusals. Rosenstein assailed McCabe over his wife's connections with Clinton allies, as well as "public and private statements expressing deep loyalty to Comey and unhappiness over his firing." [108] In response, McCabe impugned Rosenstein's objectivity since he had "abetted" Comey's termination and was now a fact witness in the case he was overseeing. Both men were correct.[109] Both of them should have disqualified themselves and put as much distance as possible between their jobs and the special counsel investigation. They did not. To his discredit, Mueller made no move to force their departure, despite the "tense standoff" and the validity of their respective arguments.[110]

Rosenstein had not just one but two untenable conflicts of interest. In addition to presiding over an obstruction case that involved himself directly as a witness, he was tangled up in the "collusion" case, as well. He signed his name on the fourth FISA warrant application to surveil Trump campaign associate Carter Page.[111] He vouched for its veracity and authenticity, even though the statements he submitted to the court had not been verified as required by federal rules and regulations. Did he even read the document he signed and "certified" as truthful? Under questioning by Representative Louie Gohmert (R-TX) at a House hearing in June 2018, Rosenstein deflected answering.[112] Though he admitted that he had signed the third Page warrant renewal, he declared that he didn't need to read all the FISA applications he signed. "Understanding" them was sufficient, he said.[113] Had he read the renewal, he would have said so. His evasion is evidence that he did not. That exchange was a harbinger of the alibi Rosenstein would likely use for deceiving the FISA court. That is, he couldn't have defrauded the judge because he hadn't actually read the misrepresentations to which he had affixed his signature. It's a standard ploy by wily prosecutors to shift the blame elsewhere.

Rosenstein's suspected plot to secretly record Trump and then use the recording as evidence to remove him under the Twenty-fifth Amendment should also have disqualified him from any involvement in the special counsel probe. It was not possible for someone so noticeably antagonistic to the president to be an unbiased and neutral party overseeing that investigation. Even if one accepts Rosenstein's dubious explanation that his discussion about deposing Trump was all in jest, the *appearance* of a conflict of interest was more than enough to merit his recusal. Rosenstein had agreed that the Sessions recusal had been appropriate.[114] Applying that same standard, how could the deputy AG not disqualify himself? It made no sense.

So many of Rosenstein's decisions failed to pass the smell test. As the number two person in charge of the Justice Department, he had scarcely been on the job when he counseled the president to fire the FBI director. When a political firestorm erupted and blame was heaped on him, he grew angry and emotionally distraught. That prompted his hasty decision to appoint a special counsel when none was merited under federal regulations. The result created a debilitating cloud over the presidency and embroiled the nation in a scandal that consumed public discourse for more than two years.

But Rod Rosenstein did not act alone. The man he helped fire was also instrumental in the wrongful appointment of a special counsel.

COMEY'S MACHIAVELLIAN MOVES AND DECEPTIONS

James Comey was not the kind of person to take the indignity of his termination without consequence. Extraordinarily tall at six feet, eight inches, he is a man who tends to dominate whatever room he occupies, for better or worse. His imperious nature is consistent with his physical stature. His sense of self-righteousness can be suffocating. At the FBI and the DOJ, he flanked himself with a cadre of loyal lieutenants, but others loathed him as a smug and egocentric overlord, questionable in both character and scruples. Joseph diGenova, who spent decades as a US attorney, independent counsel, and special counsel to the House of Representatives, offered this assessment:

> In the annals of U.S. law enforcement, no individual has reached such depths of disgrace as James "Cardinal" Comey. The "Cardinal" was a sobriquet that FBI agents used to denigrate their leader. As Deputy Attorney General, he was called

"drama queen." Known for his pomposity and self-regard, Comey cut a swath of arrogance unmatched in the histories of either of those illustrious organizations. He was the cult of personality. He surrounded himself with sycophants, people dedicated to his promotion and their advancement. The Bureau and the public became the losers.[115]

The director and his FBI confederates reviled Trump. They despised the man himself more than his politics. Proof of that was laid bare in the hateful and profane text messages shared between Peter Strzok and his lover, Lisa Page, as well as in McCabe's spiteful book and Comey's own publication wrapped in the fulsome title *A Higher Loyalty: Truth, Lies, and Leadership*.[116] It is a nauseatingly preachy book by an author who preens behind false rectitude in an effort to paper over his own lies and lack of leadership. Truth is to Comey what military music is to music. Apologies to John Philip Sousa, but you get the idea.

Comey viewed himself as the savior of the nation who would oust Trump from the seat of power by any means necessary, legal or illegal. The august director surely knew better than all those deplorable voters who were crippled by levels of intelligence and wisdom inferior to his. So in the summer of 2016, Comey and his lieutenants launched an investigation of Trump without a whiff of credible evidence to legally justify the probe. Comey cleverly disguised it as a counterintelligence investigation, but its aim was to target Trump for treasonous acts he had not committed. On the very day Comey contorted the law to clear Hillary Clinton of crimes, FBI agents met covertly with ex–British spy Christopher Steele and enlisted him to provide the phony evidence that Trump was a latent Russian agent. It was utterly preposterous, of course, but that didn't matter. If Trump could be tarnished with the specter of treason, that would suffice.

Armed with the unverified Steele "dossier," secretly funded by the Clinton campaign and Democrats, Comey signed the first FISA warrant to spy on a Trump associate in hopes of gathering some incriminating evidence in support of his imaginary case of Russian "collusion." [117] When Trump unexpectedly won the election, Comey doubled down and met privately with the president-elect to debrief him on the "salacious" document so that it could then be leaked to the media to damage Trump before he was even sworn into office. The plan worked, and the "collusion" narrative took flight with the complicit media merrily on board. While the calculating director privately assured the new president that he was not a suspect, he suggested to Congress that indeed he was. [118] Comey was conducting a secret counterintelligence operation against the president of the United States. Trump could sense it. By May 2017, the FBI director's manipulations had caught up with him, and he was summarily canned. But Comey was far from finished.

Comey should have taken a page out of Deuteronomy and retitled his book "Vengeance Is Mine." [119] Stewing over his termination as director of the FBI, the notion of how to exact retribution against Trump came to him, he claims, in the middle of the night just a few days after he was banished from the J. Edgar Hoover F.B.I. Building. [120] Before he was fired, Comey had poached copies of several presidential memos he had composed that recited the content of conversations he had had with the president. As detailed in *The Russia Hoax*, those memos were the property of the US government, not Comey's personal property. [121] He had written them in the course and scope of his employment for the government. Bureau regulations required him to "surrender all materials that contain FBI information . . . upon separation from the FBI." [122] Comey did not do that. He stashed copies of the memos in his home.

In essence, Comey stole government records and converted them to his own use. He has tried to justify the theft by claiming that the memos were "like a diary" and therefore belonged to him. [123] This is a ridiculously disingenuous characterization. He was literally on the job when he wrote them, used taxpayer-funded government resources to do so, and recited almost verbatim the words that were exchanged in an official meeting with the president of the United States. Comey was speaking with Trump in his capacity as FBI director, not private citizen. Those conversations with the president were both privileged and confidential. Comey was in the Oval Office as a function of his job. Indeed, he shared the documents with senior officials at the Bureau and they discussed their contents, proving that they were work-related. Under no reasonable interpretation can the memos be described as a private "diary." They were strictly work documents that were not his to plunder. The originals were stored in a secure place inside the FBI building. He admitted that he had made a conscious decision about their classification status, which only proves that they were important work documents. Comey's rationalization bears no semblance to the facts and the law.

In his book, Comey asserted that since he was now a "private citizen," he could do as he pleased to exploit the government-owned property that he'd pilfered in order to harm Trump.[124] After an allegedly sleepless night, he came up with the idea that if he could stealthily leak the memos to the media, it would surely precipitate the appointment of a special counsel and the inevitable demise of the president who had unceremoniously fired him. Comey wanted to keep his fingerprints off the leak. On Tuesday, May 16, he contacted his "good friend" Daniel Richman, a professor at Columbia Law School, and arranged to deliver one or more of the memos so that Richman could communicate their contents to the *New York Times* for publication.[125] Comey directed the operation, and the newspaper dutifully printed a story on May 16.[126] A day later, Mueller, Comey's longtime friend and colleague, was appointed special counsel. It was an insidious plan that worked flawlessly.

Comey told readers that he had given his friend just *one* memo, but that was untrue.[127] Richman later confirmed that he had received *four* out of the seven memos Comey had authored and kept.[128] Two other people were also given the memos, something Comey conveniently omitted from his book. For a man who once bragged to Trump, "I don't do sneaky things, I don't leak, I don't do weasel moves," it was the quintessential leak and the archetype of a weasel move.[129] Comey insisted that the single memo he had given to his friend was "unclassified." However, the *Wall Street Journal* reported that at least one, if not two, of the four memos in Richman's possession was indeed classified.[130] The inspector general at the Department of Justice launched an investigation of Comey.[131]

It was asinine for Comey to insist in his book and during interviews for his endless promotional tour that his leak was "not a leak." [132] No serious person believed that. The Freedom Forum Institute, which is dedicated to First Amendment protections, defines a media leak as "a government insider (including a former employee) sharing secret information about the government with a journalist" without authorization.[133] Comey's actions fit the definition like a snug glove. His contention that he did not "leak" defies credulity. However, it does demonstrate that he is insufferable and without conscience.

Under 18 U.S.C. § 641 it can be a felony punishable by up to ten years in prison for a person to "convert" a government record to his own use or "convey" it to another person without authority.[134] This is what Comey appears to have done. He most certainly violated the FBI's standard nondisclosure contract, which prohibited him from leaking information.[135] That was a binding, enforceable, and actionable contract that did not lapse after he was fired. As FBI director, he was legally obligated to adhere to its terms, which state that he could be sued and face criminal prosecution for breaching the agreement. Moreover, if any of the memos he gave to Richman and others contained classified information, he could be prosecuted for the same crimes Hillary Clinton surely committed under 18 U.S.C. § 793.[136]

Why did Comey leak the presidential memos? The obvious answer is vengeance. He seemed to have written the documents as a form of "insurance policy," realizing that he might be fired sometime soon. He kept copies in his home in the event he was shown the door before he could access his FBI files at work. The February 13, 2018, memo was what Comey tried to sell as a "smoking gun" of presidential maladministration. In reality, it wasn't even a popgun; it was nothing at all. In both the memo and his testimony before the Senate Select Committee on Intelligence on June 8, 2017, the FBI director recited a brief conversation in the Oval Office shortly after National Security Advisor Michael Flynn was forced to resign. According to Comey's version, the president said of Flynn, "He is a good guy and has been through a lot. I hope you can see your way clear to letting this go, to letting Flynn go." [137] Comey told the committee he had *interpreted* the president's words as a request that "we drop any investigation of Flynn in connection with false statements about his conversations with the Russian ambassador in December." [138]

I asked Trump about Comey's account of the alleged Oval Office conversation and whether he had ever spoken the words about Flynn that were attributed to him.

> **JARRETT**: Did you ever say that?
> **PRESIDENT TRUMP**: No. He totally made it up. It's made up. His memos, in my opinion, were just made up stories. He was a bad guy for our country. He was a sick puppy and a bad guy. [139]

Trump may have been correct in his assessment of Comey. But let's assume, arguendo, that the president did express his "hope" or wish that Comey and/or the FBI "would let it go." So what? Those words do not constitute the crime of obstruction of justice, as the Mueller Report would later conclude because the special counsel conceded that "the President has broad discretion to direct criminal investigations." [140] But wait . . . what investigation? Comey and McCabe had sent two agents to interview Flynn about his conversations with Russian ambassador Sergey Kislyak under the pretext of the Logan Act. [141] That exercise was nothing more than a clever ruse. There was no legal justification for the FBI to have visited the White House to question Flynn. The act prohibits private citizens from interfering in diplomatic disputes between the US and foreign governments. For more than two hundred years since its passage in 1799, no one has ever been prosecuted, much less convicted, under this fallow law because it is universally regarded as unconstitutional.[142] Regardless, the Flynn-Kislyak conversation did not even fall under the legal language of the act. Flynn was serving as a designated representative of the incoming government during the transition period, not as a private citizen, as the act requires. Flynn should never have been duped into being questioned by the FBI.

Remarkably, Comey later admitted that he had intentionally broken protocol in requesting the Flynn interview. He actually *boasted* about it during an MSNBC audience forum, saying "It was something I probably wouldn't have done or maybe gotten away with in a more . . . organized administration."[143] The FBI director knew he was flaunting established standards in bypassing the White House counsel when he said, "I thought: It's early enough, let's just send a couple guys over."[144] To make matters worse, McCabe got on the telephone to Flynn and encouraged him to avoid having a lawyer present so as to seduce him into thinking that it was just a casual conversation—no big deal.[145] It *was* a big deal, because a year later, Mueller would resurrect the interview from the dusty FBI archives to charge Flynn with making a false statement. It was a deliberate effort to pressure him into saying something incriminating against Trump in the special counsel's "collusion" probe. That was especially confounding—not to mention wrong—since Comey had informed the Senate Judiciary Committee that the agents "saw nothing that led them to believe [Flynn] was lying."[146] That assessment was corroborated by FBI written reports (called "302s") that had been filed by the agents and eventually released by the special counsel.[147]

But in February 2017, when Trump allegedly commented to Comey that he "hoped" he could let it go, there were no real investigation of Flynn and, according to the agents who interviewed the national security advisor, no false statements appeared to have been made by him. The FBI knew that Flynn had not violated the

Logan Act, and, further, Bureau agents believed that Flynn had not lied to them. Acting Attorney General Sally Yates "was not happy" that Comey and McCabe had pulled a fast one on Flynn and the new administration. Why, then, during the Oval Office meeting didn't the FBI director tell the president the truth? If he had been honest with Trump, he would have advised the president that "there's nothing to 'let go,' sir . . . because there's no investigation of any wrongdoing." Comey should have apologized for sending agents to bamboozle Flynn with a fictitious investigation. Naturally, he did not do this. Instead, he penned a memo that he might later need to either extort Trump or, if fired, destroy him.

There are other reasons why Comey's memo was not evidence of obstruction. If the director sincerely thought Trump was obstructing justice in the Flynn matter, why is there no mention of that in his memo? The commission of a suspected crime would surely merit a declarative statement. It's not there in the memo because Comey *knew* that the president's words did not constitute obstruction. In fact, he said so when he testified in front of the Senate Judiciary Committee three months after his discussion with Trump and just days *before* he was fired. He was asked if anyone had encouraged him to stop an investigation, to which he replied, "That would be a very big deal . . . it's not happened in my experience." Only *after* he was dismissed did he reverse himself and testify on June 8 that he had "understood" the president to be instructing him to drop the Flynn matter. So which was the truth and which was a lie? It is likely that Comey's first account, when he was still the director, was the more accurate one. Aggrieved or disgruntled ex-employees tend to massage the truth and revise history as a way to avenge their termination. If there was any doubt, it was removed by Comey's most senior aide, Andrew McCabe, who testified two days *after* his boss was fired, "There has been no effort to impede our investigation to date." [148]

If Comey earnestly thought Trump was obstructing justice, he was duty bound to immediately alert the attorney general, the deputy attorney general, or Congress. People who serve in law enforcement have a special responsibility to report knowledge of a felony to a person of higher authority, not simply bury it somewhere in a file or take it home to be resurrected for future blackmail or retribution. Comey admitted that he had not informed anyone at the Justice Department that the president might have tried to interfere in the Flynn matter.[149] It is worth emphasizing that in his May 2017 testimony, he assured Congress that no one had attempted to intercede in his investigations.

After he was fired, Comey testified that he had "understood" that the president wanted him to drop the Flynn investigation.[150] But how Comey interpreted the president's words is largely irrelevant. As the obstruction statute makes plain, the specific intent of the person speaking the words, not how the listener may construe them, proves obstruction. Comey, who was schooled in the law, knew that Trump's remarks were not obstructive. During his June 2017 appearance before the Senate Intelligence Committee, Senator James Risch (R-ID) posed the key question:

> **SENATOR RISCH**: You don't know of anyone ever being charged for "hoping" something, is that a fair statement?
> **COMEY**: I don't as I sit here.[151]

Of course he didn't. Wishful thinking and hopeful observations are not the same thing as an order. Comey also knew that Trump was constitutionally authorized under both the "vesting" and the "take care" clauses to end any criminal pursuit and for any reason. That was also conceded by Comey when he told the Senate Intelligence Committee:

> As a legal matter, the president is head of the executive branch and could direct, in theory . . . anyone being investigated or not. I think he has the legal authority. All of us ultimately report in the executive branch to the president.[152]

The fired FBI director knew that a lawful act performed under constitutional powers cannot, by definition, be an obstructive act. Even a corrupt motive cannot magically transform a constitutional act into a crime.[153] This will be explained in detail in the chapter on Mueller's report. The only reason Trump was accused of obstruction is because Comey, Democrats, and the liberal media shaped it into one of their favored anti-Trump narratives.

After investigating Comey's leak of the presidential memos, the Justice Department's Inspector General, Michael Horowitz, referred his case to the DOJ for potential prosecution. As reported by *The Hill*, the I-G determined that Comey "lacked candor" and concluded that there was sufficient evidence to support possible criminal charges for mishandling government documents.[154] Thereafter, federal prosecutors evaluated the case and reportedly decided against filing charges.[155] Why would Comey not be prosecuted? Not all corrupt acts are crimes that can be proven beyond a reasonable doubt in a court of law. As of this writing, other actions by Comey are under review.

COMEY IN THE MIRROR

For much of his professional career, Comey has carefully crafted his image as a selfless government servant with unimpeachable integrity. Like many of his illusions, the truth can be found in a paper trail of financial disclosure forms. In 2005, after leaving the Justice Department, where he worked closely with then FBI director Robert Mueller, Comey entered the private sector as senior vice president and general counsel of Lockheed Martin, the largest defense and surveillance contractor in the world. According to the book *Compromised: How Money and Politics Drive FBI Corruption*, Comey's net worth "skyrocketed over 4,000 percent."[156] In one year alone, he received $6.1 million in income, excluding highly lucrative stock options. How did that happen? The book's author, Seamus Bruner of the Government Accountability Institute, explained:

During Comey's time at Lockheed, Mueller's FBI granted Lockheed contracts exceeding $1 billion for various IT and surveillance programs.[157]

Lockheed was grateful to Comey and enriched him accordingly. Financially, the career move was a colossal windfall. But it also tapped (or compromised) Comey's insider expertise. While in government, he had been a strong advocate for increased government surveillance. Then he joined one of the biggest recipients of surveillance contracts, Lockheed. Comey was not above cashing in on his public service. It appears that his friend Mueller was instrumental in helping Comey make millions of dollars off of the same surveillance programs he had championed.

A review of Comey's record demonstrates that he was also a shrewd architect of manipulation. When caught, he becomes a master of prevarication. His malevolence is exceeded only by his skill at deception and flagrant misstatements. He assured the president, "I don't do sneaky things, I don't leak, I don't do weasel moves," and he also testified before Congress that he had *never* "been an anonymous source in news reports in matters relating to the Trump investigation."[158] Then he proceeded to leak presidential memos to the media while insisting they weren't a "leak." He told lawmakers in a private session that his FBI agents had believed that Flynn was being truthful. He later, in a televised interview, denied ever having made that statement.[159]

He gave false testimony to Congress about the number of Clinton emails found on her aide's laptop; the FBI had to correct the record because documentary evidence proved that Comey's count was not even close to being accurate.[160] In another hearing, he testified that he had never been given "any kind of memorandum from the Attorney General outlining his recusal."[161] That was false. Documents provided by the DOJ showed that he had received a detailed email informing him of Sessions's "recusal and its parameters."[162] Finally, he testified that he had not decided to clear Clinton until after her interview.[163] Yet he had written her exoneration statement months in advance.[164] It is a mathematical challenge to keep track of all Comey's false and misleading statements.

Comey is also prone to pleading ignorance. In his testimony before the House Oversight Committee, he claimed he hadn't known that Christopher Steele worked for Fusion GPS, hadn't known that Steele had been fired from the FBI as a source, and hadn't even known that Bruce Ohr at the DOJ had served as a conduit transmitting information between Steele and the FBI.[165] It gets worse, if possible. Comey told Bret Baier on Fox News that he didn't know if Steele's "dossier" had been funded by the Clinton campaign and the Democratic National Committee.[166] He then suggested that it had been funded by *Republicans*. Those are the kinds of statements that would make the average person's head explode.

The "dossier" was never funded by the GOP. That was an old canard that had been disabused long before by Glenn Simpson, the founder of Fusion GPS, when he had testified before Congress that early opposition research, paid for by a Republican group, had ended well before he had hired Steele to compose the "dossier."[167] That was confirmed publicly by the group itself.[168] How could Comey not know that? How could he not know that the Clinton campaign and the DNC had commissioned the anti-Trump "dossier"? How could Comey know almost nothing about Steele, Ohr, and Fusion GPS? He had signed not one but *three* FISA warrant applications to surveil Carter Page, based almost entirely on the "dossier" composed by Steele. Are we to believe that the director of the FBI, who prided himself on being meticulous and conscientious, did not bother to learn the fundamental facts and evidence that *he presented* to the FISC in support of a warrant to spy on an associate of a candidate for president of the United States? He never even asked the paramount question "Where did this come from—and who paid for it?"

Whenever Comey is confronted with uncomfortable truths exposing his own wrongful acts, he feigns ineptitude and a failed memory. Unless he is the most incompetent FBI director of all time, it is unimaginable that he did not know all of those salient facts. Yet he pretended to be clueless because there were no viable explanations for his inexplicable actions. He knew exactly where the "dossier" came from and who had paid for it. He had used it as the primary basis for the warrants, used it as part of the nonpublic version of the intelligence community assessment, and used it to debrief President-elect Trump so that it could be leaked to the media in January 2017. To contend otherwise is the equivalent of believing in the tooth fairy and the Easter bunny.

Comey must have been cognizant of the genesis of the "dossier" and how it had been misused by the FBI. Over time, the hoax began to unravel. Although he wanted to peddle his book and relished the bright lights of television celebrity, he did not want to respond candidly to penetrating questions about his central role in The Big Lie. He made a calculated decision to obfuscate and dissemble. During tough interviews, he bobbed and weaved like a prize fighter.

He also made a career out of trashing Trump in opinion columns, labeling the president delusional, amoral, deeply unethical, and a liar.[169] He tweeted cornball photographs of himself standing tall amid gigantic trees with the metaphysical query "So many questions."[170] Another photo showed him gazing pensively at the ocean accompanied by the caption "Geologic time offers useful perspective."[171] It was all so vapid and banal that Twitter had a field day mocking him, as did columnists. Kevin Brock, a former assistant director of intelligence for the FBI, writing for The Hill, derided Comey as the "Pontiff-of-the-Potomac working his beads."[172] William McGurn of the *Wall Street Journal* quipped, "The glowing picture of Parson Comey as a paragon of virtue is a self-portrait."[173]

Only an audacious and arrogant man accuses others of lying when he is guilty of doing the the same. In December 2018, after testifying behind closed doors, Comey stood outside the hearing room in front of the clicking cameras to accuse President Trump of "lying about the FBI, attacking the FBI, and attacking the rule of law in this country."[174] That from the man who had been fired for abusing his authority at the FBI and usurping the power of the attorney general in the infamous Clinton email scandal. His lack of self-awareness was breathtaking. The truth is that Comey's absence of probity and his defiance of Bureau rules and principles of law were his downfall. His unchecked ambition and desire to thrust himself into the public limelight only exacerbated his mistakes of judgment and deed. Comey, not Trump, is the one who subverted the rule of law and ruined the good name and reputation of the FBI.

To this day, Comey refuses to accept responsibility for his misdeeds. Instead, he embraces the mantle of purity, while shifting blame to others or attempting to cover it up. When faced with vexatious evidence he says, as he did more than two hundred times in his December congressional testimony, "I don't know" or "I don't remember." His amnesia is, quite literally, unbelievable. He can't recall who had drafted the document that launched the Trump-Russia investigation in July 2016. He claimed he had never known that Clinton, Fusion GPS, and the DNC had all been responsible for the anti-Trump "dossier." To hear him tell it, he had known hardly anything at all about the involvement of Steele and his phony "dossier." Yet he was all too willing to exploit the "dossier" as a pretext to spy on a Trump campaign associate. As FBI director, he affixed his signature to the "verified" warrant applications, even though he and Bureau agents had never verified their contents.

Comey loves to sermonize about lies and liars. This is perversely ironic coming from a man who, more than anyone else, is responsible for the most notorious hoax in modern US history. There was no credible evidence and hence no legal basis to justify the Trump-Russia investigation when Comey opened it in July 2016. Lisa Page, the FBI's lead lawyer in the Russia case, admitted in her closed-door deposition that the FBI had discovered no evidence of a crime, which is a legal prerequisite for invoking a special counsel.[175] But that did not stop Comey from filching presidential memos and delivering them to a friend, who then leaked them to the media to carry out his nefarious scheme to trigger a special counsel.

Comey is not the heroic or noble figure that he imagines he is. He twisted the facts and contorted the law to clear Clinton. He launched an investigation of Trump without legally sufficient evidence. He tasked undercover informants to infiltrate Trump's campaign. He deceived the FISA court by withholding exculpatory evidence. He misappropriated government documents and furtively fed them to the media to precipitate an illegitimate special counsel investigation. And he has repeatedly given deceptive or misleading statements to Congress, the media, and the American people. As FBI director, Comey was a cancerous tumor. He betrayed the public trust. As a private citizen, he continues to spin his web of deception while exalting his status as a martyr. He is a monument to vanity.

Thomas Jefferson once wrote, "He who permits himself to tell a lie once, finds it much easier to do it a second and third time, until at length it becomes habitual; he tells lies without attending to it, and truths without the world's believing him."

James Comey may be the only one who believes the stories he's selling.

———

CHAPTER 5

THE FOLLY OF MUELLER'S MAGNUM OPUS

Rosenstein is the guy who picked Bob Mueller. He could have picked anybody, and he picked a killer. He picked a trained assassin to be the special counsel.

—FORMER NEW JERSEY GOVERNOR CHRIS CHRISTIE
ON FOX NEWS, JANUARY 31, 2019

The Mueller Report should have come with a stern warning on its cover: "Do not swallow whole. Guaranteed side effects include confusion, frustration, dizziness, stomach upset, and nausea." These symptoms become more severe with every page. As you try to digest all 448 of them, you will feel the pain of a colonoscopy without Propofol. The report is excruciating because it is so hopelessly disorganized, self-contradictory, redundant, and often incomprehensible. Some sections are so chaotic and schizophrenic, you'll get whiplash. The more you read, the more it impairs your ability to think. It is fatuous to try to make sense of all the senselessness. Too much of it is slapdash and tangled up in extraneous or mundane facts. Other parts, especially those that feign legal analysis, are so mind-numbingly convoluted that their meaning is impenetrable. The entire document would have benefited from a copy editor's red pen. Or more lucid lawyers who were not blinded by their bias.

If you manage to wade through the detritus, one manifest truth emerges from the wreckage: Special Counsel Robert Mueller did not complete the assignment he was given. As a prosecutor, he failed to do his job. If he were a student, his deficient work would have earned him either an "F" or, at best, an "incomplete." If he were employed by a private business, he would be summarily fired.

Only in government can someone squander an exorbitant amount of time and money, fail miserably at the designated task, and still retain a modicum of esteem.

Mueller spent nearly two years conducting an exhaustive investigation. He employed 19 lawyers and roughly 40 FBI agents, forensic accountants, and intelligence analysts. His team issued in excess of 2,800 subpoenas and executed roughly 500 search warrants. Approximately 230 orders for communications records were obtained, and 50 pen registers were secured, allowing intrusive access to telephone and other electronic information. An astounding 500 witnesses were interviewed, and 13 foreign governments were involved. Millions of pages of documents were examined. Throughout the endeavor, the special counsel was armed with vast power and unlimited resources, spending more than $30 million. The nation was held in suspense and a presidency hung in the balance while waiting for Mueller to wrap up the job by making two critical decisions.

Yet Mueller finished only half of his assignment. He made a determination on "collusion" but not on obstruction of justice. He booted it. It was a shameful abdication of the trust and obligations vested in him. The duty and parameters of his undertaking were defined by the very regulations that authorized his appointment: to reach "prosecution or declination *decisions*. [author's italics]"[1] To put it plainly, he was hired to make decisions. He did manage to conclude that no "members of the Trump Campaign conspired or coordinated with the Russian government in its election interference activities."[2] It never happened. Neither the president nor anyone on his behalf engaged in criminal acts with Russia, notwithstanding the constant two-year barrage of accusations to the contrary by the media and Democrats. No one was prosecuted for that by Mueller. It was the so-called "collusion" aspect of the special counsel's investigation—the Russia hoax.

However, on the question of whether President Trump had obstructed justice by attempting to interfere in the FBI's or Mueller's investigation, the special counsel said that he had "determined not to make a traditional prosecutorial judgment." He had decided *not* to decide one of only two questions he had specifically been hired to resolve. He was like a truck driver who hauls goods only halfway to the agreed-upon destination. Or a chef who doles out half the dinner for which you paid handsomely. You would feel cheated. And so should every American.

By way of an excuse, Mueller bemoaned that the evidence "presents *difficult issues* that prevent us from conclusively determining that no criminal conduct occurred. [author's italics]"[3] What? Did we read that correctly? He seemed to be confessing that his assignment was simply too onerous for him to complete, in much the way a pupil might tell his teacher, "Gosh, the exam question is just too hard for me to answer." Both the student and Mueller deserve a failing grade. When Attorney General William Barr testified before the Senate, he said reprovingly that it is the essential role of a prosecutor to make a charging decision.[4] It's his *job*, for goodness' sake. The special counsel declined to meet his responsibility. That confounded Barr, especially since Mueller offered no coherent reason. When asked to make sense of what he did (or didn't) do, the attorney general replied, "I'm not really sure of his reasoning . . . we didn't really get a clear understanding of the reasoning."[5] Barr wondered why an appointed prosecutor would spend twenty-two months investigating, "if at the end of the day you weren't going to reach a decision."[6] Why indeed. It was beyond baffling.

One hypothesis is that Mueller did so for a malign reason, not the labyrinthine one he recited in his tortured explanation that can be found in the "Introduction" section of volume II of his report.[7] Perhaps the real reason the special counsel chose to deflect such an important decision is that he *knew*, under the law, that the president had not obstructed justice. A studious examination of his analysis, in consideration of both constitutional principles and statutory law, makes this seem certain—that he could not bring himself to publicly confirm that there was a dearth of evidence to justify an obstruction offense. He could not or would not validate the president's repeated reproval that the accusations against him were nothing more than a perfidious "witch hunt." Is it possible that Mueller couldn't muster the fortitude to be fair? There is little doubt that if the special counsel had had sufficient evidence to outline an obstruction offense against Trump, he would have said so and specified that evidence.

It appears increasingly evident that Mueller and his prosecutors were driven by personal and/or political animus toward Trump. It is also conceivable that his team of partisan lawyers was divided, with some willing to contort the law to pronounce Trump culpable of obstruction, while others were reluctant to do so. Regardless, Mueller's refusal to decide the issue immediately precipitated declarations by opponents of the president that "the obstruction of justice material laid out by Mueller is damning enough."[8] That was likely the special counsel's intent all along after recognizing that the evidence he had collected did not support a case for obstruction. In his solitary news conference that followed his report, he all but hung an "impeachment" sign on the lectern.

Mueller spent 183 pages smearing Trump by *implying* that under certain circumstances (which did not exist), the facts might sustain an obstruction case—if only the law could be used innovatively. He deliberately left a trail of faux obstruction crumbs so that Trump critics, unschooled in the law, might gobble them up and accuse the president of a crime that was no crime at all. Others, famished for impeachment, would pursue the president under the constitutional rubric of "high crimes and misdemeanors."[9]

MUELLER'S PRESUMPTION OF GUILT

Mueller's determination to *imply* presidential obstruction was a remarkable achievement in creative writing. He set forth in luxurious
detail what the attorney general described as "evidence on both sides
of the question." [10] But this is not the job of any prosecutor anywhere.
Mueller was not retained to compose a masterpiece worthy of Mar-
cel Proust. He was hired to investigate potential crimes arising from
Russian interference in a presidential election and make a reasoned
decision on whether charges were merited. Instead, he produced a
"magnum opus" that presented competing sides of the legal argument
on obstruction, without rendering any decision whatsoever. [11] That
was his folly. And his artifice.

Mueller's dereliction of duty was explained by the liberal former
Harvard Law School professor Alan Dershowitz:

> The job of a prosecutor is to make decisions. To charge or not to
> charge. It is not to write law review essays that lay out "on one
> hand, on the other hand" arguments. In law, as in life, there
> are close cases, about which reasonable people can disagree.
> But the job of the prosecutor is to decide those close cases.
> [Mueller] failed to come to a clear decision about obstruction
> of justice. That was his job and he should have done it. [12]

By ducking the decision and kicking it to the attorney general,
Mueller could have it both ways. Barr would be forced to do what the
special counsel had been tasked to do, all the while absorbing criticism for making the ultimate decision that obstruction was not supported by the evidence or the law. Meanwhile, the special counsel's
opus would give ammunition to Trump's enemies, who were resolute
in their pursuit of impeachment to drive him from office. It was a
cunning maneuver.

Sure enough, the attorney general dutifully studied Mueller's
report and the jumble of evidence cited therein. Despite its length,

excessive verbiage, and conflicting findings, he was quickly able to resolve that the president's words and actions could never form the legal basis required to justify an obstruction offense. For every negative there was a countervailing positive. For every seemingly incriminating piece of evidence cited by Mueller, there was a corresponding exculpatory explanation. The special counsel surely knew that Trump's enemies outside the Justice Department would focus only on the incriminating elements, while ignoring the exculpatory components. However, as an experienced and scrupulous lawyer, Barr looked at both.

In concluding that obstruction was not warranted, the attorney general did not operate in a vacuum. He was joined in his opinion by the deputy attorney general and top lawyers at the Justice Department, including those in the highly regarded Office of Legal Counsel.[13] They studied the evidence and the law. They consulted the same DOJ lawyers who had been guiding Mueller on the subject of obstruction throughout his long investigation. They reached a firm consensus that, under the law, President Trump's actions had been constitutionally authorized and that he had not acted with "corrupt intent" to obstruct "a pending or contemplated proceeding."[14]

As a legal matter, President Trump was absolved of any crime related to "collusion," such as conspiracy to defraud the government. He was also cleared of obstruction when Barr revealed that "the evidence developed during the Special Counsel's investigation is not sufficient to establish that the President committed an obstruction of justice offense."[15] That should have marked the end of the "witch hunt." Of course, it did not. That may have been precisely what Mueller desired when he wrote his opus. Many of the Democratic presidential candidates running to unseat Trump in the 2020 election predictably demanded Trump's immediate impeachment by citing Mueller's report.[16] Their motives were obvious and can be dismissed as rank opportunism or an electoral "litmus test."[17] But other Democrats in Congress called for the president's prompt impeachment, as well.[18]

If that was Mueller's plan, it worked flawlessly. His opus was never going to be a "confidential report," as stated in the special counsel regulations. The length and composition make that self-evident. It appears to have been designed for public and congressional consumption. It could serve as an inciting document that would agitate and inflame. If the law couldn't drive Trump from office, embroidered storytelling might do the trick. Mueller disregarded the guiding principles for prosecutors and the express language of the special counsel regulations that "prosecutors are to speak publicly through indictments or confidentially in declination memoranda." [19] He ignored established constitutional law and indulged in inconclusive observations that seem to have been cleverly framed to arouse and induce action against the president by his harshest critics, including both Democrats and the media.

Mueller's actions were not only noxious but patently unfair to Trump. The special counsel publicly besmirched the president with tales of suspicious behavior instead of stated evidence that rose to the level of criminality. This is what prosecutors are never permitted to do. The rules of the Justice Department forbid its lawyers to annunciate negative narratives about any person absent an indictment. How can a person properly defend himself without trial? This is why prosecutors such as Mueller are prohibited from trying their cases in the court of public opinion. If they have probable cause to levy charges, they should do so. If not, they must refrain from openly disparaging someone whom our justice system *presumes* is innocent.

In that regard, Mueller shrewdly and improperly turned the law on its head. Consider the most inflammatory statement that he leveled at the president. It was guaranteed to ignite the impeachment fire:

> While this report does not conclude that the President committed a crime, it also does not exonerate him. [20]

To reinforce the point, he stated it thee times in his report. [21]

Prosecutors are not, and have never been, in the business of exonerating people. That's not their job. An experienced federal prosecutor, Mueller certainly knew this. It seems he had no intention of treating Trump equitably or applying the law in conformance with our criminal justice system. In a single sentence, he managed to reverse the legal duty that prosecutors have rigidly followed in America for centuries. Their legal obligation is not to exonerate someone or prove an individual's innocence. Nor is any accused person required to prove his or her own innocence. Everyone is entitled to the *presumption* of innocence. It is the bedrock on which justice is built. Prosecutors must prove guilt beyond a reasonable doubt. To bring charges they must have, at minimum, probable cause to believe that a crime was committed.

The special counsel took this inviolate principle and slyly "inverted" it.[22] He argued that he could *not* prove that the president had *not* committed a crime. In the annals of American jurisprudence, that was a new one. Think about what it really means. It is a double negative. Mueller was contending that he can't prove that something didn't happen. What if this were the standard for all criminal investigations? Apply it to yourself. Let's say you deposit your paycheck at the bank on Monday, the same day it's robbed. A prosecutor then announces that he cannot prove you didn't rob the bank, so you are neither criminally accused nor "exonerated." The burden of proof has now been shifted to you to disprove the negative. How would you feel? You've been maligned with the taint of criminality and no longer enjoy the presumption of innocence.

This is the equivalent of what Mueller did to Trump. The special counsel created the *impression* that Trump *might* have engaged in wrongdoing because he could not (or would not) prove otherwise. The consequential injustice and harm that inevitably followed is what happens when we reverse the burden of proof and abandon the presumption of innocence standard that is revered in a democracy as a fundamental right. Yet that was what Mueller did: he improvised a new extralegal standard that applies only to Trump: *presumption of guilt.* Under this novel "guilty until proven innocent" paradigm, it is up to the accused to show that the allegations are false. When Mueller testified before two committees of Congress on July 24, 2019, he was asked how he came up with this unprecedented standard and whether it had ever been used before in any case. The special counsel stammered and then replied, "I cannot, but this is a unique situation."[23] This was an evasion. Trump deserved the same presumption of innocence as any American. There were no examples of when it had ever been used before because none exist.

Attorney General Barr recognized that Mueller had queerly mangled the legal process, describing his statement as "actually a very strange statement."[24] Barr told Congress that he had been forced to correct Mueller's intolerable mistake. "I used the proper standard," said Barr. "We are not in the business of proving someone did *not* violate the law—I found that whole passage very bizarre."[25] The attorney general was being polite. It wasn't just bizarre, it was legally deranged.

If Mueller had adhered to the special counsel regulations, his report would have been brief—not 448 pages long. On "collusion," it should have been easy to explain his declination decision, stating that no evidence of a conspiracy had been uncovered. Perhaps a short and succinct recitation of the major facts would have been in order. As to the question of obstruction, Mueller decided not to decide. Hence, he should have delivered to Barr a concise explanation of why he had chosen to abdicate his responsibility and simply handed over the evidence files for the attorney general to consider. Nothing more was appropriate, especially not a nearly two-hundred-page obstruction chimera.

Prosecutors do not compose reports for public consumption. They make a binary choice: charge someone or decline to do so *without* commentary. There is no middle ground. The reason for this should be obvious. It is intrinsically unfair to sully an uncharged individual who has no opportunity to counter the adverse effects that derogatory information can have on that person's name and reputation. Evidence cannot be challenged in a court of law because no indictment was rendered. Instead, the prosecutor's one-sided point of view of noncriminal wrongdoing is presented untested in the court of public opinion without a counterbalance from the party who was never charged for lack of evidence or proof. An innocent person can be forever stained by the imputation of suspicious acts. There is little recourse once the stigma sticks.

That did not seem to bother Mueller in the least. He produced a prodigious exegesis of every perceived or imagined sin Trump *might* have committed—maybe, sort of, if you squint your eyes at just the right angle. As the president's lawyer Emmet Flood observed, he "produced a prosecutorial curiosity—part 'truth commission' report and part law school exam paper."[26] That is a generous portraiture.

Our system of justice is designed to protect the innocent. That is why there are laws that prevent disclosure of grand jury testimony and even more expansive rules at the Justice Department that prohibit prosecutors to disclose negative information about uncharged individuals. Mueller was well aware of that. In the introduction to volume II on obstruction, he recited the duty of prosecutors to be fair by refraining from comment. In the case of a sitting president, wrote Mueller, "The stigma and opprobrium could imperil the President's ability to govern."[27] In that he was correct.

Ironically, the special counsel then proceeded to ignore his own warning. He produced his own "dossier" on Trump that was filled with suspicions of wrongdoing. He refused to recommend a future indictment of the president in a court of law but was more than willing to indict him in the court of public opinion. His report was a nonindictment indictment. It was calumny masquerading as a report. The special counsel cleverly accomplished it under the guise of providing a "confidential" report to the attorney general, as regulations required. But Mueller knew his report would be made public because political pressures would demand it, and Barr had already acquiesced by pledging transparency during his confirmation hearing.

Rules of professional responsibility and conduct forbid prosecutors to make "extrajudicial comments that have a substantial likelihood of heightening public condemnation of the accused."[28] Mueller's report was an accusatory document, albeit not a charging one, that was replete with such extrajudicial comments. Moreover, leaks to the media that claimed Trump was being investigated for obstruction of justice could only have come from the investigators themselves, the special counsel team.

Mueller also demolished other ethical rules. He recklessly abridged the attorney-client privilege by seizing protected documents and then insisted that counsel be questioned (interrogated), thereby invading confidential communications.[29] Worse, he relentlessly pursued a "collusion" investigation that he must have known for months, if not a year, was unsupported by the evidence. Hundreds of innocent people were harassed under threat of prosecution, and many were left financially in debt, having been forced to retain lawyers to assist in their defense. One lawyer, offended by those tactics, penned the following description of what Mueller had done in The Federalist:

> Consider that 500 Americans were interrogated by federal agents during the probe—one per business day. Imagine federal agents looking through your private email, pictures, and other electronic data. Imagine FBI agents swarming your house with guns drawn. Mueller executed approximately 500 search warrants against our fellow Americans, all to no end.[30]

MUELLER'S BIAS WRIT LARGE

Though Mueller's report is notable for his failure to decide the obstruction issue, it is conspicuous for another, equally egregious, failure. He did *not* investigate how the Clinton campaign and the Democratic National Committee (DNC) had "colluded" with others, especially Russia, to interfere in the 2016 presidential election.

In the order authorizing Mueller's appointment, the special counsel was instructed to examine not only "the Russian government's efforts to interfere in the 2016 presidential election" but "any matters that may arise directly from the investigation."[31] It would have been impossible for Mueller to have studied Russian interference without discovering how the FBI's "collusion" investigation had originated at the hands of the Clinton campaign and Democrats through financial payments and other surreptitious operations. Yet nowhere in the 448 pages is there a word about how the Clinton campaign and the DNC commissioned Russian disinformation that was fed to Christopher Steele, who, in turn, delivered it to the FBI and more than a dozen members of the media.

There is nothing in the report about the composition of the anti-Trump "dossier" and its alleged Russian sources, how it appears to have been used as a pretext for initiating the FBI's investigation of the Trump campaign, the extent to which the FISA court was deceived by unverified evidence presented to it, the suspected improper surveillance of a Trump adviser, and the source of government leaks that fueled the "collusion" narrative that permeated the public discourse for more than two years. Only a passing reference is made of the notorious "dossier." The role of Fusion GPS, Glenn Simpson, Nellie Ohr, and others? Apparently written with invisible ink. When confronted about these omissions during his congressional testimony, Mueller either refused to answer or said it was not in his "purview."[32] It should have been. Astonishingly, the special counsel did not seem to know what Fusion GPS is.[33]

The arrant misconduct and biased decision making by senior FBI officials, including Mueller's friend James Comey, were also omitted from the Mueller Report. So, too, were the suspect actions of intelligence agencies that appear to have gathered information on the Republican candidate and his associates. Was law enforcement weaponized for partisan advantage in an "attempt to neutralize and denigrate a political opponent," as Senator Charles Grassley wondered aloud in a Senate floor speech?[34] Was the intelligence community, under the supervision of James Clapper and John Brennan, intimately and strategically involved? It is inconceivable that Mueller could have conducted his comprehensive probe without discovering that kind of incendiary evidence. It was hiding in plain sight. He either turned a blind eye to what he saw or chose to remain silent to protect the agency he had once led, as well as his friend Comey. In either case, it was wrong to do so.

Mueller's anti-Trump bias was writ large throughout his opus. Observations were presented as evidence. Numerous "facts" were inconclusive or untested. Key findings presented by Mueller were later discovered to be false or obscured. For example, Konstantin Kilimnik is identified as working for Russian intelligence. In fact, documents show he was an important source for the US State Department "who informed on Ukrainian and Russian matters."[35] Joseph Mifsud is portrayed as Russian-connected or maybe a Russian agent. In fact, he is neither, but the reader must "wade through the fine print of Mueller's report" to discover this.[36]

In another instance, the transcript of a phone call from the president's lawyer, John Dowd, to the attorney for former national security adviser Michael Flynn was edited by the special counsel to make it appear sinister when, in truth, the full transcript shows it was not.[37] Facts and evidence seem to have been massaged in a way most detrimental to Trump. There was no attempt at fairness or objectivity.

This should come as no surprise, given Mueller's disqualifying conflicts of interest and the team of partisans that he assembled. As explained in detail in *The Russia Hoax*, Mueller should never have been selected as special counsel. Once the job was offered to him, he was obligated to decline it.

MUELLER SHOULD HAVE BEEN DISQUALIFIED FROM THE START

The regulations governing Mueller's appointment prohibited him from serving as special counsel if "he has a personal or political relationship with any person or organization substantially involved in the conduct that is the subject of the investigation or any person or organization which he knows has a specific and substantial interest that would be directly affected by the outcome."[38] The regulation then clarifies that a personal relationship includes "friendships" and is a cause for disqualification if the friendships pose "a close and substantial connection of the type normally viewed as likely to induce partiality."[39]

The *first* conflict was Mueller's close personal and professional association with fired FBI director James Comey. It is well documented and indisputable that they have been friends, allies, and partners for many years.[40] Various publications have profiled their bond as driven by a mentor-protégé relationship, which makes the likelihood of favoritism and partiality self-evident. This represented an acute conflict of interest. Even the *appearance* of a conflict merited recusal under the law, as well as the rules of professional responsibility to which lawyers must adhere.[41]

Comey was "substantially involved" in any obstruction investigation of Trump. Indeed, he was the only witness to the alleged Oval Office discussion about the firing of National Security Advisor Michael Flynn and was involved in conversations with the president that led to his termination. That was one part of Mueller's obstruction of justice investigation of Trump. The other part involved the firing of Comey himself and whether it constituted an obstructive act by the president to interfere with the FBI's counterintelligence investigation of "collusion" with Russia. In both areas of Mueller's probe, Comey was an instrumental witness.

By his own admission, Comey was the one who engineered the appointment of his friend as special counsel. When he was fired as FBI director, he improperly removed presidential memos from the FBI building, converted them to his own use, and then purposefully delivered them to a friend to leak to the media. Appearing before the Senate Intelligence Committee in June 2017, he had the audacity to boast about his actions, saying "I thought that might prompt the appointment of a special counsel."[42] It was a typical Comey maneuver—deviously effective.

Clearly, he had a keen interest in the outcome of Mueller's probe and the potential prosecution of the man who had fired him, Trump. Comey was hardly an innocent or disinterested bystander. More important, Mueller could not be viewed as a neutral party, either. His close friendship with the key witness raised the likelihood of prejudice or favoritism, which is anathema to the fair administration of justice. How could Americans have confidence in the result of Mueller's investigation if they suspected that the special counsel harbored a bias in favor of the witness testifying against the target? Which man would the special counsel believe? His good friend or the president who had fired his good friend? How could Mueller fairly and impartially assess Comey's credibility versus Trump's?

A *second* conflict of interest that should have forced Mueller's recusal was his previous service as director of the FBI for twelve years. He had a relationship with an "*organization* substantially involved in the conduct that is the subject of the investigation. [author's italics]"[43] That disqualified him under the regulation cited earlier. There is no doubt that Mueller felt a strong allegiance to the agency he had once led. Protecting its reputation and its people was a factor. Naturally, he was predisposed to trust the evidence the Bureau had already collected against Trump by the time he assumed the special counsel position. He would also have been predisposed to believe the version of events told by Bureau officials and agents as opposed to that told by Trump, and he might have been inclined to slant facts in favor of the very agency he had overseen in order to protect the interests of the FBI.[44] That meant that both the FBI and Mueller had a "specific and substantial interest" in an outcome that he alone would determine. Each had a stake in the result. This is not allowed.

Mueller had no choice but to disqualify himself. The binding regulations afforded him no discretion because the recusal was mandatory in its language. It does not say "may" or "can" or "might." It states unequivocally that the special counsel "shall" recuse himself in such instances. Mueller ignored it. It is known that Justice Department officials granted him "an ethics waiver to serve as special counsel."[45]

They were wrong to do so and equally wrong to cover up the details of how they had reached their decision.

The special counsel regulation contemplates the selection of someone "outside" the Department of Justice, which includes the FBI. It specifically uses that term in the appointment provision.[46] Although Mueller had retired from the Bureau, he remained the ultimate "insider." He had spent years as a top official at the Justice Department and then served as director of the FBI for thirteen years. He knew just about everyone involved in the case. That undermined the whole point of choosing someone who would be impervious to personal influences and institutional prejudices. Mueller was exactly the kind of person the regulations sought to avoid. Many other qualified lawyers in America would have been better suited for the job and could have brought a greater sense of impartiality and legitimacy to the process.

There is a *third* reason why Mueller should have disqualified himself: he was interviewed by Trump to possibly return to the FBI as Comey's replacement just one day before reversing course and accepting the job as special counsel to investigate the president.[47] That was extraordinary and ethically problematic. Is it even possible that a discussion about Comey, his aberrant conduct, and the reasons for his firing did *not* take place during the Trump-Mueller meeting in the Oval Office on May 16, 2017? It quite likely did, especially given Trump's proclivity to talk openly about Comey's misconduct. "With obstruction a focus of the investigation, that made Mueller a possible witness in his own investigation," wrote law professor Jonathan Turley.[48] A lawyer is strictly prohibited from being both a prosecutor and a witness in the same case.[49] Even if Mueller did not solicit or obtain evidence from the president for the ensuing obstruction investigation, his interaction created the appearance of a conflict of interest, which is also disqualifying.

Incredibly, the Mueller Report dismissed that conflict by suggesting that the Oval Office meeting had not been a job interview at all but a discussion whereby Mueller had offered the president insights and "a perspective on the institution of the FBI."[50] Steve Bannon, a former White House chief strategist, is cited as the source of this statement. There are several problems with this account. First, Bannon was not present at the meeting. Second, when I interviewed the president he pointedly told me he had not been interested in gaining a perspective on the FBI but only in finding a proper replacement for Comey and that Mueller said he wanted the job.[51] Third, Trump's personal secretary, Madeleine Westerhout, who arranged the meeting with Mueller and was privy to its purpose and content, confirmed to me that he had been there for a job interview, nothing else.[52] The explanation in the report is not only wrong but self-serving. But all of this is irrelevant. The meeting between the president and Mueller unquestionably occurred the day before he accepted the job to investigate Trump. Even if their discussion was broadly about the institution of the FBI and its future in the wake of Comey's firing, it makes it more likely that the director's misconduct and the reasons for his removal were discussed. When he testified before Congress, Mueller was directly asked whether the president discussed with him the firing of Comey. The special counsel replied, "Cannot remember."[53] The ensuing exchange took place:

> QUESTION: You don't remember? But if he did, you could've been a fact witness as to the President's comments and state of mind on firing James Comey.
> MUELLER: I suppose that's possible.[54]

It is difficult to fathom that Mueller could not remember a critical Oval Office conversation with the president that would have required his recusal. More likely, the special counsel knew his disqualifying conflict of interest had been exposed and sought to deflect it by claiming total memory lapse.

The appointment of a special counsel does not happen spontaneously. Surely Mueller knew he was under consideration when he met with Trump. We know from McCabe's book that Rosenstein had been pursuing the idea for a week. He must have been consulting Mueller about accepting the position. Regardless of the purpose of the Oval Office meeting, Mueller sat there talking with Trump but never advised the president that he might decide to investigate him. That duplicity was more than a sufficient basis to require that he disqualify himself from the special counsel position. It is hard to believe that Mueller had the temerity to accept the job of investigating the very person with whom he had just met the day before. Maybe that had been his plan all along. Perhaps his intent was to secretly gather evidence from Trump himself that he could then use against him in some way as special counsel, while pretending to be interviewing for the job of FBI director or offering some "perspective." Mueller's actions were deceptive and dishonest.

And what is one to make of the cozy relationship among the three men who were at the heart of the special counsel appointment? Comey admitted having leaked the presidential memos to trigger the naming of a special counsel. Was it a mere coincidence that Rosenstein just happened to pick Mueller, of all people? Or was it all coordinated? Was there a plan hatched whereby Mueller would become either the new FBI director or the special counsel in order to take down Trump? From either position, he could investigate the president. Why would he want to return to the job he had already held for twelve years, having retired from government service? Why the sudden motivation? According to Trump, it was Rosenstein who suggested Mueller to replace Comey, and the deputy attorney general was present during the Oval Office interview on May 16. When the president rejected Mueller, Rosenstein appointed him as special counsel the very next day. Comey and Mueller were close, and Rosenstein had worked with both men. The timing of their actions is suspect. They have never divulged the curious details of how it all came about. Congress should demand answers, and so should the Department of Justice.

During my June 2019 interview of President Trump at the White House, I asked him directly whether he and Mueller had discussed the reason Comey had been fired during their Oval Office conversation:

JARRETT: Was there any discussion at all about your having just fired Comey? Did you say to him, "This is why I fired Comey"?

PRESIDENT: [long pause] I have no comment.[55]

Trump then told me he could easily answer the question because he had a strong recollection of their discussion that day. However, he was reluctant to delve back into it.

"I just want to put this stupid thing to bed," he offered by way of explanation.

The reader can draw his or her own conclusion. But it was abundantly clear to me, sitting there in the Oval Office with the president and judging his reaction and demeanor, that the answer was "yes." Of course they had talked about why Comey had been fired. It would have been a natural, and perhaps necessary, subject. It is hard to imagine a scenario in which it would not have come up. Again, it cannot be overemphasized that all of this would have made Mueller a witness in his own case, which is strictly forbidden. The president provided more context:

PRESIDENT: Mueller wanted very badly to have the job as FBI director and to return to the FBI. I didn't want him. I rejected him. By the way, how much of a conflict is it when a guy comes in wanting a job, I say no, and the next day he's your special prosecutor? It's outrageous.[56]

Trump then spoke about another matter that he felt was a significant conflict of interest. Mueller had resigned from the Trump National Golf Club in Sterling, Virginia, and asked that he be refunded his membership fee.

PRESIDENT: We had a business conflict, too. We had a dispute over his golf fees. He wanted his money back. I said no.

It would set a bad precedent. Everyone else would want their money back if they left the club at some point. I wouldn't do it. He didn't like it. I never gave him back his $15,000, approximately, and I think he was mad about that.[57]

One can debate whether that "dispute" and any underlying ill will or anger over it constituted a true conflict of interest requiring disqualification. Mueller, in footnote 529 of volume II of his report, seemed to dismiss their squabble as insignificant, but that may have been another self-serving explanation.[58]

However, all of the other conflicts of interest, either individually or taken as a whole, meant that Mueller should never even have interviewed for the job as special counsel, much less accepted it. His failure to recognize that calls his awareness and prudence into serious question. "Bob Mueller should never have been allowed to do this case," said Trump.[59]

Investigating a president of the United States is a serious matter and therefore necessitated a special counsel who was beyond reproach in both conduct and appearance. Mueller's decision to accept the position in the face of the federal regulations and ethical rules discouraging it indicates that his judgment was compromised.

Then he made matters worse. If there had been legitimate reservations about his ability to approach the Trump-Russia investigation impartially, he removed all doubt by assembling a team of prosecutors composed almost exclusively of Democrats.

MUELLER'S TEAM OF PARTISANS

After ignoring his own conflicts of interest, Mueller went about the business of selecting as his prosecutors a group of people, many of whom were bias ridden. Their lack of impartiality polluted the report they produced. Although they found no crimes related to "collusion," their monograph on obstruction of justice was a decidedly political document aimed at Trump. Of course, that was consistent with the partisan composition of the special counsel team. It called into question their motives and delegitimized their observations. In that regard, Mueller sabotaged his own investigation.

It is significant that none of the nineteen lawyers hired by Mueller is registered as a Republican, according to investigations by Fox News and Daily Caller.[60] Public records show that the vast majority were Democrats and most had donated to their party. Many had given money to the Obama and Clinton presidential campaigns, as well. Though Mueller himself was once a Republican, the status of his "current party registration is not clear."[61] President Bill Clinton nominated him to be a US attorney, and it was President Barack Obama who asked Mueller to remain as director of the FBI.[62]

The "partisan in chief" on the special counsel team of prosecutors was Andrew Weissmann. Infamous in legal circles for wrongful prosecutions that had earned both reversals and rebukes, he attended what was supposed to be Clinton's election-night victory party on November 8, 2016, that eventually resembled a funeral service.[63] When Acting Attorney General Sally Yates, an Obama holdover, was fired by Trump shortly after he took office, Weissmann expressed his contempt for the president in a fawning email to Yates.[64]

Weissmann is a prosecutor who is known for what has been described as abusive tactics, using the law as a weapon in an unprincipled quest to convict. He has been accused of suppressing evidence and threatening witnesses.[65] Innocent people have been victimized by his maneuvers. Some of his biggest cases have been overturned by higher courts. Former federal prosecutor Sidney Powell observed that "the truth plays no role in Weissmann's quest" and "respect for the rule of law, simple decency and following the facts do not appear in [his] playbook."[66] Weissmann's tactics don't end there. Justice Department documents show that he quietly met with reporters for an "off-the-record" discussion about Paul Manafort in April 2017.[67] The next day, the story broke that Trump's former campaign manager was being investigated for financial improprieties. A month later, Weissmann joined the special counsel probe that eventually charged Manafort, although none of the crimes involved "collusion."

Mueller knew full well what he was doing when he brought Weissmann on board, since the two had worked closely together before at the FBI. John Dowd, Trump's lawyer, told me that whenever he had met on behalf of the president with the special counsel team, he had refused to be in the same room as Weissmann. "He was so unethical and untrustworthy, I said I wouldn't talk with him present," recalled Dowd.[68] According to Dowd, Weissmann was banished from those conferences. His notorious reputation did not seem to bother Mueller in the least. Just the opposite: he gave Weissmann complete authority to lead the effort to hire other members of the special counsel team, as evidenced by documents obtained through a Freedom of Information Act lawsuit.[69] It is no wonder the deck became loaded against Trump.

When I interviewed another of the president's lawyers, Rudy Giuliani, he offered an equally harsh assessment:

> Weissmann can see a crime if you're just walking down the street. He was out of control—unscrupulous and unethical. Implicating the president for "collusion" was all Weissmann's plan. He was steering the ship. Mueller was along for the ride. Bob lets other people take over, run the ship, and then he takes credit for it.[70]

Mueller also brought to his staff Jeannie Rhee. The two had been partners in private practice. The hiring of Rhee was especially brazen since she had defended the Clinton Foundation in a civil racketeering case and donated $5,400 to Clinton's presidential campaign. When Clinton had jeopardized national security by using a private server for classified documents, it was Rhee who had reportedly represented the former secretary of state in litigation that had sought to force her to produce the emails she was hiding.[71] Suddenly, Rhee was in a position to prosecute the person who had defeated the candidate she had supported and defended.

Given how obvious a conflict of interest that represented, it was shameful for the special counsel team to even consider hiring her. But there was more. Compounding Rhee's conflict was the fact that she had previously worked for Deputy FBI director Andrew McCabe, an important witness and source of information in the Trump investigation. She had also represented ex–Obama aide Ben Rhodes during the congressional investigation into the Benghazi attack, an inquiry that had focused primarily on Clinton's misconduct.[72] Rhee's multiple disqualifying conflicts of interest were stunning. Mueller seemed unbothered and undeterred by them, which may speak volumes about an underlying motivation to damage Trump by utilizing people who held the same anti-Trump prejudices. Amazingly, Mueller sheepishly admitted during his congressional testimony that he had no idea of Rhee's involvement with Clinton when he hired her.[73] It appears that Mueller either didn't care about conflicts of interest or had left it all up to Weissmann, who had his own glaring conflict.

The Clinton connections extended to others on Mueller's team as well. Aaron Zebley was chosen, even though he had served as the lawyer for Justin Cooper, who had worked for Clinton and set up her private email server, registering the domain in his own name instead of hers. Cooper had admitted to destroying Clinton's Black-Berry devices, "breaking them in half or hitting them with a hammer."[74] He had not had security clearance for any of the classified documents on Clinton's server, yet had had access to it. Zebley's involvement in the Clinton cases created yet another glaring conflict of interest and, at the very least, the appearance of impropriety in the Mueller probe. Mueller must have known all of this because he and Zebley and Rhee all worked together at the WilmerHale law firm that was actively involved in representing Clinton during her email scandal.[75]

Another of Mueller's prosecutors was Kyle Freeny, a donor to both the Obama and Clinton campaigns.[76] As a lawyer acting on behalf of the Obama administration, she had been excoriated by a federal judge in a 2014 case for "intentional, serious and material" misconduct.[77] After an apology from the DOJ, the judge had accepted the department's claims that her representations to the court had been unintentional. But is this the kind of lawyer who should be involved in an investigation of the president of the United States? If nothing else, the optics reflect poorly on Mueller's judgment. Perhaps a person capable of misconduct was exactly what he wanted on his team of prosecutors.

Two of the lawyers Mueller selected had known as early as the summer of 2016 that the FBI investigation of Trump and his campaign over "collusion" was rife with bias and that the "dossier" was likely a phony document. That was revealed in the testimony of Justice Department official Bruce Ohr.[78] After meeting with the "dossier" author, Christopher Steele, in late July 2016, Ohr said, he had immediately shared his improbable intelligence with Andrew Weissmann and Zainab Ahmad, warning them of Steele's anti-Trump bias, the Clinton campaign's funding of the document, and the document's unreliable allegations.[79] Incredibly, those two lawyers were hired by Mueller, despite their early involvement in the defective "dossier" and the spurious FBI investigation that had been inspired by it.

It should not be forgotten that FBI agent Peter Strzok and his paramour, FBI lawyer Lisa Page, were also picked by Mueller to join his investigation. Their animus toward Trump and admiration for Clinton were embarrassingly chronicled in their numerous text messages. Page departed the probe, and Strzok was removed by Mueller and McCabe when the inspector general at DOJ discovered the shocking communications.[80] Yet when Congress demanded to know why Strzok had been terminated from the probe, the true reason, together with the prejudicial texts, was concealed.[81] Mueller must have known that the Strzok-Page messages would raise bright red flags that warned of a partisan witch hunt against Trump. Under questioning before the House Judiciary and Oversight Committees, Strzok conceded that when he had met with Mueller the day of his dismissal, the special counsel had never bothered to ask him if his bias had un-

duly influenced the special counsel investigation.[82] Did he not care?
How much of the probe up to that point was contaminated by the
lead investigator's malice toward Trump? Deputy Attorney General
Rod Rosenstein, who had appointed Mueller, attempted to deflect
the controversy by insisting that none of those conflicts or blatant
antipathy toward the president constituted even the appearance of
impropriety.[83] It was an absurd claim to make.

Like a deck used in a crooked game of cards, it was stacked.
The special counsel could not have put together a group of Democrats more "terminally hostile to Trump, or at least certifiably pro-
Hillary."[84] Beyond Mueller, there were no Republicans on the team
and no prosecutors who could be described as apolitical. The team's
hyperpartisan composition could not have been accidental. It had to
have been done by design. The handpicked lawyers would be highly
motivated to damage Trump by pursuing a case "based on false pretenses and a fake dossier funded by Democrats."[85] They were armed
with unlimited resources and unfettered power. Nineteen politically
biased prosecutors eyed Trump through their investigatory scopes
and took direct aim. The president never stood a chance. He was a
sitting duck for Mueller's assembled team of legal sharpshooters.

THE LEFT'S HYSTERIA

The truth always has a nemesis. Representative Adam Schiff spent
more than two years pretending that he was privy to evidence that
he did not have. That illusion gave him a comfortable home on television networks that offered no pretense of their antipathy toward
Trump, especially CNN and MSNBC. The more Democrats and the
media worked in concert to advance their hallucination that Trump
had conspired with the Kremlin, the more audacious Schiff became
in his public denouncements of the president. He frequently insinuated that he had special access to damning information that few

others could procure. Even after the House Intelligence Committee issued its majority investigative report concluding that it had all been a hoax, Schiff announced, "I can certainly say with confidence that there is significant evidence of collusion between the campaign and Russia."[86] He produced no such evidence, because it did not exist. But that didn't stop him from perpetuating the "collusion" delusion. On CBS's *Face the Nation*, he ventured that Trump "may be the first president in quite some time to face the real prospect of jail time."[87] Schiff was so heavily invested in the scam and the celebrity it brought him that there was no reversing course. Maybe a part of him really believed his claims, and he was hoping that some magical revelation would appear out of nowhere to vindicate them. Like a guy with a counterfeit bill, he kept trying to pass it off to others.

Schiff kept good company. There was no shortage of fellow Democrats (and even an outlier Republican or two) who echoed false allegations, some with unhinged warnings. The most ridiculous came from Norman Eisen, the former White House ethics attorney, who stated that the criminal case against Trump was "devastating" and he had been "colluding in plain sight."[88] If so, few could see it, including the special counsel, who peered furiously into every obscure corner and crevice for some proof that the president had coordinated or conspired with Russia. People such as Eisen and Schiff were convinced that Trump's election had been illegitimate. They could not conceive how Americans had voted him into office absent some treasonous conspiracy hatched by him in the bowels of the Kremlin. Here are just a few of the members of Congress who pronounced Trump and his campaign guilty of "collusion" in advance of the Mueller Report:

REPRESENTATIVE JERROLD NADLER (D-NY): It's clear that the campaign colluded, and there's a lot of evidence of that.

REPRESENTATIVE ERIC SWALWELL (D-CA): In our investigation, we saw strong evidence of collusion.

SENATOR RICHARD BLUMENTHAL (D-CT): The evidence is pretty clear that there was collusion between the Trump campaign and the Russians.

SENATOR RON WYDEN (D-KS): There was clearly an intent to collude.

DNC CHAIRMAN TOM PEREZ: And over the course of the last year we have seen, I think, a mountain of evidence of collusion between the campaign and the Russians to basically affect our democracy.

REPRESENTATIVE MAXINE WATERS (D-CA): Here you have a president who I can tell you, I guarantee you, is in collusion with the Russians to undermine our democracy.

FORMER REPRESENTATIVE DAVID JOLLY (IND-FL): Donald Trump is done. He's done. There is no question.[89]

Even lawyers and professors steeped in the law made erroneous predictions of criminality by contorting conspiracy statutes in an attempt to conform them to facts and evidence that had no application. Neal Katyal, a law professor at Georgetown University and acting solicitor general under Obama, stated that Trump had ordered the commission of felonies and "has got to know his future looks like it's behind bars."[90] Four months later Katyal doubled down to say that the Mueller Report marked "the beginning of the end" for Trump.[91]

Jens David Ohlin, a law professor and vice dean at Cornell University Law School, said that emails produced by Donald Trump, Jr., confirming a meeting with a Russian lawyer at Trump Tower were "a shocking admission of a criminal conspiracy."[92] They were not, and the president's son was never charged with conspiracy or anything else. Paul Butler, a former federal prosecutor turned legal analyst at MSNBC, said that "what Donald Trump Jr. is alleged to have done [is] a federal crime" because he was "conspiring with the U.S.'s sworn enemy to take over and subvert our democracy."[93] Except, of course, that he had not "conspired" under the law, and no such crime had occurred.

The moment Steele's "dossier" was published by BuzzFeed in January 2017, the media embraced the "collusion" delusion without first verifying any of the allegations therein. It was their holy grail of evidence that Trump was a "pretender" occupying the Oval Office that rightly belonged to their favored candidate, Hillary Clinton. As anchors, reporters, and commentators spun their collective narrative, they argued that an unproven secret pact between Trump and Putin was the reason she had lost. Day after day, they accused the president of being a Russian asset whose crimes would soon be exposed:

> **RACHEL MADDOW, MSNBC:** We're about to find out if the new president of our country is going to do what Russia wants . . .

> **RACHEL MADDOW, MSNBC:** This is unsettling because if the worst is true, if the presidency is effectively a Russian op, right, if the American presidency right now is the product of collusion between the Russian intelligence services and an American campaign—I mean, that is so profoundly big.

> **NICOLLE WALLACE, MSNBC AND NBC:** This cloud about collusion with Russia will hang over him no matter where he stands.

> **MIKA BRZEZINSKI, MSNBC:** This is not funny. This is really bad. Just for the record, we're all really nervous.

> **MIKA BRZEZINSKI, MSNBC:** I think they're shocked that the noose is tightening.

> **LAWRENCE O'DONNELL, MSNBC:** The beginning of the end of the Trump presidency.

> **DONNY DEUTSCH, MSNBC:** You could feel the thread being pulled, you can feel the clothes starting to come off the emperor. I believe this is the beginning of the end.

> **DONNY DEUTSCH, MSNBC:** We have a treasonous president.

> **CARL BERNSTEIN, CNN:** We now have to figure out how to deal with a president of the United States who wittingly or unwittingly has been compromised.

ANDERSON COOPER, CNN: The president sees the walls closing in, and he's lashing out.

JOY REID, MSNBC: What if he refuses to open the White House door? What if the Secret—or if he fires any Secret Service agent who would allow in the federal marshal? What if Donald Trump simply decides, I don't have to follow the law? I refuse to be held under the law, no marshal can get into this White House?[94]

Some of those statements were inane, while others were uninformed. But they all shared a common denominator: Trump's guilt was predetermined. His imminent removal from office was preordained. The mainstream media *said* so; thus it must *be* so. Facts mattered little. The law was irrelevant. In the echo chamber occupied by the vast majority of journalists in the United States their repeated condemnation of Trump reinforced one another's convictions. "Collusion" would lead them to their political Promised Land, where the president was behind bars instead of the *Resolute* desk. As it turned out, the Mueller Report was not their salvation and liberation.

Both Democrats and the media lapsed into shock when the results of Mueller's investigation were announced to the public on March 24, 2019. There would be no deliverance from Trump. That became apparent when the full report was released on April 18. Mueller burst the bubble of the "collusion" hoax when he stated that "the investigation did *not* establish that members of the Trump Campaign conspired or coordinated with the Russian government in its election interference activities. [author's italics]" In plain language, Trump did not criminally "collude" with Russia.

Mueller's conclusion should have come as no surprise to anyone who was paying attention. The notion that Trump had coordinated or conspired with Moscow to steal the presidential election from Clinton was pure sophistry. Talk with anyone involved in his campaign, and they'll tell you that it was an undisciplined, unmethodical shoestring operation. The whole of it was an embarrassing model of entropy. It was a *dis*organization.

The truth is that the Trump campaign could not have organized a free lunch, much less an elaborate plot with a foreign power. His candidacy was built on ideas, personality, and a collective sense of indignation. Trump was the driving force and chief messenger. He tweeted out heated denunciations of the Washington elite to millions of embittered and angry voters. He traveled from one venue to the next in all the key electoral states, holding rallies that drew many thousands of voters. Those events were frequently televised live and were seen by a much larger audience of voters everywhere. He was the symbol of a new empowerment and radical change. He embodied government transformation.

This strategy was so fundamental—even pedestrian—that it actually worked. The candidate's name identification gave him an enormous advantage, to be sure. It also helped that his opponent was a lackluster campaigner and an uninspiring candidate. But Trump's penchant for brutal candor and well-placed derision of the DC establishment was what attracted so many people to his corner. He was the antipolitician at a time when Americans had come to loathe politicians. The "swamp," as Trump called it, was a chronically dysfunctional group of stale and self-interested careerists. The people they were supposed to be serving were fed up with the incompetent service. Voters felt they were often deceived. Promises were invariably left unfulfilled. The electorate deeply resented that treatment. When Trump vowed to drain the swamp, millions cheered him on. His timing was fortuitous. The moment had arrived for an unconventional choice. There was no one more unconventional than Donald Trump.

The media, which never understood how Trump could be elected, was more than willing to sign on to the "collusion" narrative. It justified their own mistakes and rationalized their misjudgments. Trump *had* to be an illegitimate president, they reasoned. That was the only explanation for why Clinton had lost. The race must have been stolen from her—and them. They never bothered to examine the facts, evidence, and law. They embraced "collusion" as if simply saying the word meant it was a crime. Any conversation or contact with a Russian was "collusion" and therefore criminal. Any chance encounter or casual handshake with a Russian was vilified. Journalists never bothered to cite a specific statute. They were satisfied, indeed anxious, to level the accusation as a legal conclusion. They insinuated a crime by using a word that connoted a crime.

In *The Russia Hoax*, I argued that " 'collusion' is not a crime, except in anti-trust law." [95] I was lambasted for making that legally correct statement. But it was eventually affirmed by the special counsel's report when it observed that "collusion is not a specific offense or theory of liability found in the U.S. Code, nor is it a term of art in federal criminal law." [96] After spending two years suggesting that "collusion" must necessarily be criminal, the *New York Times* confessed, "Outside of specific factual contexts—such as price fixing in antitrust law—the word 'collusion' has no legal meaning or significance." [97] The "newspaper of record" finally got it right.

At its core, "collusion" is merely an agreement to do something. Historically, the term is laden with criminal overtones. It is often associated with secrecy. But it is criminal only when two or more people agree to commit an illegal act. Over the stretch of two chapters in *The Russia Hoax*, I examined several federal statutes that might have some application to the Trump-Russia investigation, such as conspiracy to defraud and federal campaign finance laws. [98] That was what Mueller did, as well. To the chagrin of the media and Democrats, he arrived at the same conclusion that I had in my book: there was no criminal "collusion."

THE COLLUSION DELUSION

The special counsel's inability to find a scintilla of incriminating evidence against Trump was not for want of trying. Mueller consumed

199 pages in what he called "volume I" of his opus describing in excruciating minutia every interaction—however arbitrary—between Trump campaign associates and any known Russian. No trivial contact or insignificant communication was spared a full-throated analysis. If a courteous handshake was extended or a casual "Hello" was said, Mueller gave it an exhaustive examination. Even a conversation with a Russian diplomat, which is frequent and standard in Washington, was fully scrutinized for some hidden illicit purpose. The report made it evident that Mueller and his team were obsessed with establishing wrongdoing. They worked assiduously to find it. As Kimberley Strassel of the *Wall Street Journal* observed:

> What stands out is just how diligently and creatively the special counsel's legal minds worked to implicate someone in the Trump World on something Russia . . . and how—even with all its overweening power and aggressive tactics—it still struck out.
>
> While it doubtless wasn't Mr. Mueller's intention, the sheer quantity and banality of details highlights the degree to which these contacts were random, haphazard and peripheral. By the end of Volume I, the notion that the Trump campaign engaged in some grand plot with Russia is a joke.[99]

It was no joke to President Trump, who was encumbered with false accusations from the outset of his presidency. The attacks inhibited his ability to do his job. The demands of the special counsel consumed valuable time and resources at the White House that could have otherwise been spent more productively on matters of state and for the benefit of the American people. Untold time and effort were misspent by his having to respond to relentless allegations of "collusion" that never happened. The president's critics, especially the media, focused like a laser on several events that had occurred during his campaign. They assured their viewers and readers that any one of them rose to the level of criminal "collusion" with Russia. Mueller addressed six of them at length in his report.

First, despite chronic reporting that the Trump campaign had somehow been involved in the hacking of the Clinton campaign and Democratic Party email accounts, there was not a shred of evidence that this was true. None. The Russians are adept at cyberespionage. Why would they need someone on the Trump campaign to help them? Besides, there was no one on the team who had the skills to contribute anything of value. The Mueller Report concluded that Russian actors operated on their own. In 2016, they allegedly hacked into computer networks and stole thousands of documents that were then published by WikiLeaks beginning in July 2016. According to Mueller, the Trump campaign "showed interest" in learning the contents before they became public.[100] Of course they did. So did hundreds of curious journalists, myself included, who reached out to WikiLeaks or otherwise tried to locate the missing emails through what are known as "open-source" methods. That's not criminal, it's normal.

At a public event in July, Trump jokingly quipped that he hoped Russia would be "able to find the 30,000 emails that are missing" from Clinton's server.[101] Those were the emails she had deliberately deleted (including work-related documents), even though Congress had ordered that she turn over all of her communications. By the time of Trump's flippant remark, it was well known that Clinton had exposed her private email account to outside theft and that the FBI had stated publicly, "It is possible that hostile actors gained access."[102] Predictably, Trump critics such as Norman Eisen and many others declared that the GOP candidate had broken the law by making his comment.[103] He had not, and it's ridiculous to say so. Trump's sarcastic swipe at his presidential opponent did not violate campaign finance laws and did not come close to meeting the legal requirements of a conspiracy with Russia to hack computer systems. A mocking public comment is hardly an agreement to commit a crime. Mueller did not conclude otherwise.

Second, the special counsel meticulously detailed how Trump adviser George Papadopoulos had *allegedly* heard from Maltese professor

Joseph Mifsud that "the Russian government had 'dirt' on Hillary Clinton in the form of thousands of emails."[104] Papadopoulos had then *allegedly* repeated the rumor to Australian diplomat Alexander Downer, who had eventually notified the FBI. As explained in an earlier chapter, passing along hearsay gossip is not a crime, even if it has something to do with Russia. Moreover, Papadopoulos denied he had even repeated the gossip. Indeed, he had no memory of it, which tended to underscore its unimportance. In addition to determining that Mifsud lied, Mueller found no evidence that Papadopoulos had advised anyone in the Trump campaign that he had been the recipient of the rumor.[105] No one in the campaign knew anything about it. The special counsel rendered this conclusion:

> No documentary evidence, and nothing in the email accounts or other communications facilities reviewed by the Office, shows that Papadopoulos shared this information with the Campaign.[106]

Third, Mueller tried mightily to connect Trump's erstwhile campaign manager, Paul Manafort, to some amorphous conspiracy with Moscow by virtue of his business dealings in Ukraine. Enormous pressure was brought to bear by the special counsel. The president's lawyer, Rudy Giuliani, said he had been made aware of the tactics allegedly employed by Mueller's team:

> Manafort's attorney told me how the prosecutors would play this charade. They would yank him out of solitary confinement, order him what to say, and when he wouldn't say it because it was untrue, they'd dump him back into solitary and bring him back a week later. They did it a dozen times. They wanted Manafort to lie about the president, but he refused. He kept telling them he didn't know anything about Trump colluding with Russia. The prosecutors didn't care. They just wanted to get Trump even if it was all a lie.[107]

In the end, Manafort was charged with a series of financial crimes that well predated his brief service for the campaign. This will be discussed at length in a subsequent chapter. No indictment involving Russian "collusion" was ever brought against Manafort or, for that matter, anyone else.

The special counsel uncovered no evidence that Trump was acquainted with Manafort's foreign consulting activities before he was hired by the campaign. When the media raised questions about his lobbying work for Ukrainian oligarchs, Manafort was forced out of the campaign in August 2016.[108] In his report, Mueller regurgitated in agonizing detail every contact, no matter how frivolous or trite, that Manafort had had with persons of Russian-sounding names. It was not only tedious but pointless. At the conclusion of that section of the report, no grand conspiracy was revealed. The special counsel did mention that Manafort had given a longtime employee, Konstantin Kilimnik, "polling data" and expected him to share it with others in Ukraine.[109] Most of the data turned out to be public information.[110] Manafort had been trying to curry favor with his clients but had received nothing in return. Naturally, the media instantly claimed that that was the smoking gun that would prove criminal "collusion" once and for all.[111] The special counsel team, despite its anti-Trump bias, knew better. It is not a crime to share polling data with someone. Campaigns do it all the time.

Fourth, the special counsel dug deep into the life and times of Carter Page. No wrongful conduct was exhumed. As noted before, Page had traveled to Moscow to deliver a speech. That was about the extent of his supposedly collusive Russian activities—nonexistent. He had *not* held secret meetings with the Kremlin. He had *not* been the conduit for "a well-developed conspiracy."[112] He had *not* negotiated the "lifting of western sanctions" in exchange for a "19 percent stake in Rosneft," the Russian energy company.[113] Those accusations against him in the Steele "dossier" had been debunked, although Mueller declined to specifically call out the document for what it obviously was: a fraud. That was a curious omission, inasmuch as the

FBI had launched its investigation of Trump-Russia "collusion" based largely on the "dossier," then exploited its contents to spy on Page for a year. Was Mueller protecting the FBI's illegitimate reliance on a phony document that the Bureau had failed to verify? Quite likely.

Fifth, the special counsel devoted much attention to a subject that deserved, frankly, precious little. The Trump Organization had spent decades exploring and, in many instances, consummating development projects all over the world. As Trump had extended his brand through real estate deals and licensing arrangements, his company had grown more profitable and the Trump name had expanded its reach. The idea of constructing a tower in Moscow had been one of many potential projects discussed. After years of intermittent negotiations with prospective Russian partners, a letter of intent (LOI) had been signed in the fall of 2015.[114] It was not a binding contract to build anything. It was not a "deal" that had been consummated. It had simply been an agreement to engage in "further discussions," as Mueller's report recognized.[115] Government approval would be required, and Trump associates had moved forward to obtain it. Whatever the motives of people such as Michael Cohen and Felix Sater, associates who were pursuing the venture on Trump's behalf, no evidence emerged that the candidate himself had sought to leverage the contemplated project in his campaign for president. In early 2016, the discussions stalled. Over the ensuing months, Trump declined to travel to Russia and subsequently ended his organization's discussions about building a tower in Moscow. Although Democrats and the media inferred "collusion," the evidence gathered by the special counsel did not support it.

Sixth, the infamous Trump Tower meeting with a Russian lawyer by the name of Natalia Veselnitskaya was given a painstaking review by the special counsel. As I explained in considerable detail in *The Russia Hoax*, it is not a crime to talk to a Russian. Nor is it a crime for a candidate to receive negative information about a political rival that is volunteered by a foreign national, even if it benefits the campaign. When the *New York Times* published a story on July 8, 2017, about the meeting among the Russian lawyer, Donald Trump, Jr., and other campaign officials, Democrats and the media immediately seized upon it as evidence of "collusion" and all manner of illegality.[116] Clinton's former running mate, Virginia senator Tim Kaine, branded it "potentially treason," while a former White House ethics lawyer, Richard Painter, suggested that the president's son should be "in custody for questioning."[117] Those hyperbolic claims had no support in the law. The idea that it was treason is so preposterous that it does not merit an explanation here. Both men should go back to law school.

Here's what actually happened. Mueller's dissection of the law relative to the Trump Tower meeting replicates almost identically the analysis I provided to readers in my book, written a year before the special counsel produced his report. Two possible offenses were examined and discounted. One was "conspiracy to defraud the United States," which makes it a felony for two or more persons to enter into an agreement to interfere or obstruct a lawful function of the government.[118] An election would be a lawful government function. However, the US Supreme Court has said that such an agreement must be done by "deceit, craft or trickery, or at least by means that are dishonest."[119] Yet Mueller noted the absence of "surreptitious behavior or efforts at concealment."[120] More significantly, there was no evidence that the Russian lawyer had conveyed any information whatsoever to the Trump campaign about Clinton. Mueller found no agreement or conspiracy that had arisen out of the Trump Tower meeting. He further determined that there had been no attempt to "interfere with or obstruct a lawful function of government."[121] Other statutes that have their own conspiracy language were also abandoned by the special counsel for similar reasons.[122]

The other law that Mueller considered but rejected was the complicated set of campaign finance laws. After the Trump Tower meeting was disclosed, Nancy Pelosi, the Democrat leader in the House at the time, declared during a news conference that the president's son had violated those laws "plain and simple."[123] Pelosi was wrong, and so were others who joined the chorus of condemnation based on campaign laws they had surely never read. The Federal Election Commission makes it clear on its website that it is perfectly lawful for foreign nationals to "volunteer for a campaign or political party."[124] They are permitted to attend meetings, contribute ideas, and convey information. None of that is considered to be a donation or "other thing of value" under the campaign statutes, as some have alleged. It is true that the Federal Election Campaign Act prohibits foreign nationals from making a "donation or money or other thing of value."[125] But the campaign laws have never interpreted the giving of information as having value in this way. In fact, a close reading of the statute identifies specific examples of what "other thing of value" is supposed to mean. Providing information is not among them as a prohibited contribution.[126] Additional regulations state that "the value of services . . . is *not* a contribution."[127] Mueller noted:

> No judicial decision has treated the voluntary provision of uncompensated opposition research or similar information as a thing of value that could amount to a contribution under campaign-finance law.[128]

Finally, as I explained in my book—and as Mueller acknowledged in his report—prosecutors could never bring a criminal case against Trump Jr. because they would have to show that he *had known* that he was somehow breaking the law in collecting information from a foreign national. How many people *know* that it might be a crime? I dare say, very few. It is doubtful that many lawyers understand the legal requirement of what Mueller called an "elevated scienter element."[129] This is a unique aspect of campaign finance laws. They demand proof of what is called specific intent. That is, the person who accepts the donation must "knowingly and willfully commit a violation" of the campaign law.[130] For that reason, the Trump Tower meeting could not have been the subject of criminal prosecution because the special counsel team openly admitted that they couldn't prove that essential element.[131] This is why most campaign finance violations result in civil penalties, not criminal charges.

The Mueller Report negated several other events that Democrats and the media had assured us would prove how Trump was a clandestine agent of Russia who had swindled Americans out of a free and fair presidential election:

- **Did the Trump campaign alter the party platform at the Republican National Convention in a way that favored Russia in its stance with regard to Ukraine?** That canard was peddled incessantly for the better part of two years by Democrats and the media.[132] They offered it as irrefutable proof that Trump was in league with Putin. However, as columnist Byron York has consistently explained, "the final platform was tougher on Russia and Ukraine than the original platform draft; it was not 'gutted,' it was strengthened."[133] Mueller's report vindicated York's account.[134]

- **Did Trump's personal attorney, Michael Cohen, travel to Prague for secret talks with Russians, as the "dossier" alleged?** The Mueller Report says no.[135] That wrecked yet another story advance by the media that Trump had "colluded" with Russia.

- **Did campaign members and supporters share or "like" tweets from Russian-controlled Internet Research Agency (IRA) accounts on social media?**[136] They did. But, as Mueller confirmed, they had no idea that Russian operatives were behind the disguised accounts. It is not a crime to be fooled.

- **Did Jeff Sessions, Trump's son-in-law, Jared Kushner, and other members of the campaign meet with Russian ambassador to the United States Sergey Kislyak?** Of course they did. The special counsel concluded that their conversations had been "brief and non-substantive."[137] It's not a crime to be polite to or engage in casual banter, even with a Russian.

- **Did Michael Flynn, the incoming national security advisor, speak with the Russian ambassador during the transition about two pending foreign policy matters?** Absolutely. That is normal for any transition team, and nowhere in his report does Mueller state that it was criminal "collusion" or that it somehow violated the Logan Act, as so many Trump critics had alleged.[138]

The Mueller team's bias permeates nearly every page of the report. Facts were massaged in the way most detrimental to Trump without ever creating a basis to charge him. Other facts that would show that the president had been targeted with false evidence for political reasons were omitted.

The "dossier" was central to the FBI's investigation and surveillance. Yet the "collusion" volume made nary a mention of that document nor the bias of its author, Christopher Steele. Russia's involvement in feeding him disinformation in order to interfere in the election is unaddressed. The extent to which Steele appears to have invented information is disregarded. The role of his colleague Glenn Simpson and his company, Fusion GPS, are nowhere to be found. Though Mueller made a point of stating that paying money to a foreign national in exchange for information on an opposing candidate is a violation of the law, he ignored the fact that that was exactly what the Clinton campaign and the DNC had done.

These are not just deficiencies, they are fatal defects. It seems that the special counsel had no intention of producing a report that was fair, objective, or honest.

In reviewing the Mueller Report, Attorney General Barr penetrated the veneer of the special counsel's anti-Trump bias and spoke plainly about its ultimate conclusion: "Mueller has spent two and a half years. And the fact is, there is no evidence of a conspiracy. So it was bogus. This whole idea that Trump was in cahoots with the Russians is bogus." [139]

MUELLER KNEW TRUMP WAS INNOCENT OF COLLUSION LONG AGO—WHY DID HE DELAY?

All of this invites the question: When did Mueller *know* that there was no "collusion"? President Trump's lawyer John Dowd told me that by October/November 2017, the special counsel had everything he needed to conclude that there was no evidence of criminal "collusion." [140] More than a million pages of protected documents had been turned over voluntarily by the White House and the Trump campaign. Nothing had been withheld, and no privileges had been asserted, although they could have been. The president's transparency had been unprecedented. An endless parade of witnesses had all been made available and encouraged to speak.

According to Dowd, Mueller knew by the first week of December—slightly more than six months after his appointment—that "for sure there was no evidence of collusion." [141] That week, the special counsel had finished interviewing all "collusion" witnesses. Carter Page, who had never been charged with anything, confirmed to columnist Byron York that the special counsel had no longer been interested in him by the end of 2017. [142] The FISA wiretap warrant on him was abandoned; Mueller and the FBI did not seek a fourth renewal. Page was off the radar. Other witnesses, including George Papadopoulos, said that Mueller had closed the book on them the same year. That was not unexpected. Recall that senior FBI officials had admitted in private sessions with Congress that by the time Mueller was appointed, they had almost no evidence of "collusion" with Russia and had failed to corroborate the "dossier" allegations against Trump. [143] Still Mueller persisted.

Three months later, on March 5, 2018, a critical meeting took place between the president's lawyers and Mueller, in the company of several of the special counsel prosecutors, including James Quarles III, Michael Dreeben, and Andrew Goldstein. As before, Weissmann was not permitted to attend. John Dowd recited the following exchange:

> **DOWD**: Bob, does the president have any criminal exposure?
> **MUELLER**: No.
> **DOWD**: Okay, then what's his status?
> **MUELLER**: He's a witness-subject.[144]

The word "subject" is a Justice Department term of art. It means that a person's conduct is being examined, but there is no implication that he crossed the line into criminality. A "target," by comparison, means that there is reason to believe he may have committed a crime and there is substantial evidence in support of it. Dowd recalled vividly how "we laid out everything to him." There was no evidence whatsoever that the president had coordinated or conspired with any Russians. Nothing at all. Mueller had readily conceded this, said Dowd. Yet he had insisted that the president be interviewed. Dowd knew it was a trap. Mueller had done it to Flynn, Papadopoulos, and others. If a witness couldn't recall something precisely or if his memory deviated one iota from what documents showed, the special counsel team accused him of lying and applied enormous pressure to get him to implicate Trump or someone else in his campaign, even if it was a lie. Mueller was clearly angling for obstruction of justice and hoping to ensnare the president in the equivalent of a perjury trap if he consented to be interviewed.

> **DOWD**: I'm not going to have the president sit for hours and say, "I don't recall" over and over. Some of the subjects you want to ask him about are so insignificant that no one could remember them. Why do you want to humiliate the president?
> **MUELLER**: Well, I've got to square my corners about obstruction. I want to know if he [Trump] had a corrupt intent.[145]

Dowd explained that the firing of Comey had been an exercise of presidential authority under the Constitution and could not, by definition, constitute obstruction of justice. Even if it had been intended to end the investigation—and it was not—the president was legally and constitutionally empowered to do so as chief executive. It would not constitute obstruction "because that would amount to him obstructing himself," as Dowd had already pointed out to Mueller in a lengthy letter.[146] The special counsel would not relent.

MUELLER: I have to look at obstruction.

DOWD: You want to look at obstruction in the firing of Comey? I'll tell you who our first witnesses will be: the attorney general, the deputy attorney general, and the White House counsel. They all urged the president to fire Comey. He was off the rails in his handling of the Clinton email case.

MUELLER: We may pursue a grand jury subpoena to compel the president to testify.

DOWD: [pounds his fist on the table] You've got nothing! Go ahead! I can't wait for you to try. You've got no leg to stand on.[147]

Jay Sekulow, another of President Trump's lawyers, attended that meeting and confirmed to me this account of the exchange, which was also reflected in his handwritten notes. He said that the meeting with Mueller and his team "became heated," especially as Sekulow argued, "You don't have the right to interview the president. You are nowhere near establishing the legal threshold for questioning a sitting president."[148] Both Dowd and Sekulow maintained that Trump's actions had been constitutionally authorized and could not, as a matter of law, be considered obstruction of justice.

Mueller then pointed to other acts that might somehow be construed as obstruction, such as the president's public criticism of Attorney General Jeff Sessions and even the special counsel himself. This was now reaching a ludicrous level. It was silly to claim that a public expression of disapproval or anger by a president was a criminally obstructive act. Dowd felt that Mueller and his subordinates were living in an alternative universe where their version of the law bore no resemblance to statutes, Supreme Court decisions, and accepted constitutional law. "They acted like crybabies who were offended that someone dared to criticize them, so they instinctively labeled it obstruction—it was insane," said Dowd.[149] He paused their conversation, trying to restore some measure of calm, and attempted to reason with Mueller.

> **DOWD**: The president was raising hell with the attorney general, Bob, because he's doing a lousy job. So what? That's not obstruction of justice. He has every right to complain about the attorney general and every right to proclaim his innocence after being the victim of false accusations. Wouldn't you?[150]

Mueller seemed either stoic or confused. Dowd and Sekulow argued that there was no legal basis to justify interviewing the president. Citing the "Espy standard," based on a DC circuit court decision, they contended that there was no evidence that the information sought from the president could not be obtained elsewhere from the hundreds of witnesses questioned and more than a million pages of documents examined.[151] Mueller had submitted topics and questions in advance. The president's two lawyers explained that the special counsel had "already received the answers from the documents and testimony" on record. There was nothing more to ask the president.

Trump's lawyers held firm and refused to allow their client to be interrogated by Mueller. Eight months later, a compromise was reached in which the president answered questions in writing. They can be found in twenty-three unredacted pages of the Mueller Report and demonstrate that Dowd's argument was sound.[152] The information contained therein could be found elsewhere, and the president had had either no knowledge or little recollection of the events in question.

Dowd was unsparing in his criticism of the special counsel. "This entire report by Mueller is a fraud . . . it's an outrage," he said.[153] The president's lawyer was particularly incensed when incomplete and selectively edited quotes of a Dowd voice mail message left for Michael Flynn's attorney were misrepresented in the special counsel's report to make it appear as though Dowd had acted improperly. "Isn't it ironic that this man who kept indicting and prosecuting people for process crimes committed a false statement in his own report," he remarked.[154]

In my interview with Dowd, he said he had grown deeply disturbed by Mueller's mental state during the course of his investigation. "The environment surrounding Mueller was not healthy. There was this atmosphere of clicking heels and that Bob was God. He's used to people genuflecting, and we weren't. He was often sleeping there in the building where the investigation was headquartered." [155] During group meetings, only Mueller spoke. The other lawyers on the special counsel team were resolutely silent and behaved in complete deference to Mueller. Did he write any of the report himself? "Mueller didn't write it . . . it was written by committee," said Dowd. Giuliani agreed, "It was written by two groups—the James Quarles side, which was reasonable, and the Weissmann side, which was out of control, alleging collusion without any evidence." [156]

By December 2017, the evidentiary bucket was empty and bone dry. Instead of ending the investigation, Mueller kept it going for another fifteen months. That, of course, allowed Trump's detractors to continue to promote their vacuous claim that the president was a Russian marionette. It may also have been instrumental in helping Democrats win control of the House of Representatives in the November 2018 election. Was that the reason Mueller stubbornly pursued the case long after it was evident that there had been no criminal "collusion"? To pose the question is to answer it. With Democrats in charge, a path to impeachment of Trump was possible.

Dowd was disgusted by all of it. Even before Mueller's report was issued, he dismissed the entire investigation as a "terrible waste of time" and "one of the greatest frauds this country's ever seen." [157] He said he was shocked that Mueller hadn't had the backbone to tell his boss, Deputy AG Rod Rosenstein, that "this is nonsense . . . we are being used by a cabal in the FBI to get even." [158] At the very least, the special counsel could have issued an interim report a year earlier, stating that no "collusion" had been found. It would not have prevented Mueller from pursuing his quixotic quest for obstruction of justice, which, in the end, he decided not to decide. It was yet another colossal waste of time.

MUELLER'S FLAWED OBSTRUCTION THEORY

Once Mueller found that there had been no unlawful "collusion," he should have ended his investigation a year earlier. Why? Because obstruction of justice requires proof that a person has a corrupt or improper motive to interfere in a legal proceeding. But where that person has acted lawfully, there is no credible motive to obstruct an investigation into the lawful act. People are not motivated to cover up noncrimes. They have nothing to hide. Yes, it is legally *possible* to bring an obstruction case without an underlying crime, but it is exceedingly difficult to prove the required motive. For this reason, obstruction prosecutions in the absence of a predicate offense are quite rare.

There is, however, a more compelling reason the special counsel should never have considered obstruction against the president. Mueller's legal theory was fundamentally and egregiously flawed. In crafting Article II of the Constitution, the Framers granted to the president certain "plenary" powers.[159] They are absolute and unconditional. They are also discretionary. That is, the president may exercise them for any reason. Those reasons cannot be questioned, and they are not subject to review by anyone except voters every four years. Such powers include the authority to appoint, *remove*, and pardon individuals.

The president is also empowered to direct federal investigations and prosecutions under the "take care" clause of the Constitution to see that "the laws be faithfully executed." [160] This is an elemental duty. The president is the nation's top law enforcement official. To argue that he may not participate in or even direct decisions at the Department of Justice and the FBI is to disempower one of the president's core constitutional responsibilities. As former Harvard Law School professor Alan Dershowitz observed, "Presidents—from Adams to Jefferson to Lincoln to Roosevelt to Kennedy—played active roles in deciding who to investigate and prosecute." [161]

In addition, the "vesting clause" of the Constitution vests all enforcement power in the president.[162] This means he can direct, supervise, or *influence* legal proceedings. He is empowered to decide whether to initiate investigations and prosecutions or halt them, even if he has a personal interest in the outcome. This is appropriately referred to as "prosecutorial discretion," and the ultimate authority resides in the chief executive.[163] This power is accepted as absolute and has long been considered so by federal courts.[164] It may be unwise for a president to end an investigation or halt a prosecution. However, that is a political calculation, not a legal one. When a president exercises that power, he is not committing a crime. It is illimitable and not subject to a charge of improper motives or obstruction. It is presumed that he is acting lawfully. His decisions and subjective state of mind are not reviewable. In all of the instances in which Trump exercised these constitutional powers, Mueller had no right to question or investigate the president's motives. A special counsel, who is an inferior officer in the executive branch, cannot disempower what the Constitution has granted to the president.

To illustrate this, let's examine the firing of FBI director James Comey. This is what triggered immediate and enduring accusations of an attempt by Trump to obstruct the Bureau's investigation. Within days a special counsel was appointed. There is no question that Trump was authorized to fire the director under the "appointments clause" of Article II of the Constitution. It was a lawful act. He can appoint and terminate federal officials in the executive branch of government. They report to him and serve at his pleasure. (This principle was affirmed by the Justice Department years before Trump became president when it wrote, "The FBI Director is removable at the will of the President . . . no statute purports to restrict the President's power to remove the Director.") [165] Exercising this power cannot itself constitute obstruction of justice.

Comey also acknowledged this when, after his firing, he wrote to his colleagues that "a president can fire an FBI Director for *any* reason, or for no reason at all. [author's italics]" [166] Even if Trump's motive was improper—*and the evidence shows that it was not*—it would still not constitute obstruction of justice. Barr explained this during his congressional hearing when he admonished the special counsel for "trying to determine the subjective intent of a facially lawful act." [167] The attorney general was correct. The president's subjective intent was immaterial. A lawful act performed under constitutional powers cannot, by definition, be an obstructive act. Alan Dershowitz made the same argument when he wrote, "An improper motive cannot convert a lawful act into a crime." [168]

Whether we like it or not, presidents make constitutionally authorized decisions, such as the power to appoint, fire, or pardon, for reasons that are not always pure. Sometimes they are driven by self-interest. Dershowitz drew an important comparison with President George H. W. Bush's pardon of former secretary of defense Caspar Weinberger and other individuals during the Iran-Contra scandal in 1992. That had the effect of ending a troublesome investigation. Though the special counsel at the time claimed it was a "cover-up" of his probe, he did not accuse Bush of obstruction of justice because "the act of pardoning itself was authorized by the Constitution." [169] So, too, is the act of firing an executive branch officer. Motive is irrelevant when executing constitutional duties.

These unfettered powers granted to the president in the Constitution do not contain a limiting clause or a caveat that requires proof that the reason is not self-serving or otherwise improper. History abounds

with examples. President Gerald Ford pardoned his disgraced predecessor, Richard Nixon, for the express purpose of halting proceedings
that would have led to a criminal trial.[170] Ford was clearly interfering
in the judicial process, but it was not obstruction of justice because the
Constitution gave him that right. Did President Bill Clinton have an
improper intent when he pardoned his own brother, Roger Clinton?
What about his pardon of fugitive financier Marc Rich in 2001?[171]
Was it granted because Rich's ex-wife was a close friend of the president and had contributed generously to the Clintons? The answer to
both questions is "Probably," but under the law the pardons were not
obstruction of justice because they were constitutionally lawful acts.
A president is allowed to exercise his pardon and firing powers, even
if he has a personal stake in the result or if questions about his own
conduct are raised. If it were otherwise, an eternal array of special
counsels could investigate an endless number of presidential decisions
in search of hidden and subjective motives. The chilling effect on the
chief executive would be debilitating.

These constitutional principles should have prevented Mueller
from an obstruction investigation, but he pursued it anyway. That
created for him a serious dilemma. In composing his report, how
could he credibly convert Trump's lawful act under the Constitution
into something that would resemble unlawful obstruction? Mueller's
rationale, albeit a weak one, is buried deep in his report on page 333
in an extended discussion entitled "Constitutional Defenses." Here
the special counsel argued that there are no Supreme Court cases that
"*directly* resolved this issue. [author's italics]"[172] He then reasoned
that, in the absence thereof, he was perfectly free to reach his own
legal denouement. No, he was not.

Notice the use of the word "directly." That was dexterous sleight
of hand. It is true that the nation's highest court has never specifically and narrowly ruled that an Article II *firing* by the president can
or cannot constitute obstruction of justice. However, the Supreme
Court has twice ruled broadly that *general criminal statutes* do not
apply to the president when he is exercising his constitutional discre-

tionary powers, unless a statute expressly says so.[173] Obstruction is a general statute. Nowhere in the collection of obstruction laws does it expressly state that they apply to the president when executing an Article II function or prerogative. Hence, they do *not* apply to the president. The Supreme Court has referred to this as the "clear statement rule." [174] It means that the president's constitutionally conferred powers cannot be restricted or nullified by general statutes passed by Congress, such as those defining obstruction of justice, that do not contain a clear statement specifying that they apply to the president.

Understand what Mueller and his team of partisans seem to have done here. They spent numerous pages in their report writing *around* those adverse Supreme Court decisions. Lawyers are highly adept at creative composition. They revel in the challenge. Mueller and his prosecutors must have spent weeks and months trying to devise ways to bend and twist their report to circumvent those high court decisions, which threatened to ruin their suggestion that Trump might have obstructed justice. It's no wonder that it took them twenty-two months to conjure up such imaginative legal reasoning. Yet they paid no heed to what the Supreme Court had said nor to the opinion of their own Office of Legal Counsel (OLC) at the DOJ, which had written:

> The Supreme Court and this Office have adhered to a plain statement rule: statutes that do not expressly apply to the President must be construed as not applying to the President, where applying the statute to the President would pose a significant question regarding the President's constitutional prerogatives.[175]

The correct application of the law meant that Comey's firing could not be obstruction of justice since the president had been exercising his constitutional prerogative, regardless of his motive.

None of this is to imply that a president can *never* obstruct justice. He most certainly can if he acts "corruptly" *outside* his constitutional powers. On that, both Mueller and the president's lawyers agreed.[176]

It would clearly be obstruction of justice if, for example, a president were to suborn perjury, intimidate witnesses, fabricate or conceal evidence, destroy subpoenaed documents, lie to the FBI or Congress, or (as Nixon did) pay "hush money to witnesses in a criminal case."[177] All of these actions would be substantially beyond the bounds of a president's constitutional authority under Article II. However, that was not what Mueller and his team of prosecutors were investigating.

Some have argued that Mueller did not charge the president with obstruction because he was following another settled, but controversial, opinion of the Office of Legal Counsel at the DOJ that was first issued in 1973 and later updated. It provides that "The *indictment* or criminal *prosecution* of a sitting president would unconstitutionally undermine the capacity of the executive branch to perform its constitutionally assigned functions. [author's italics]"[178] Was that the reason for Mueller's indecision? Though it is true that his report made reference to the OLC opinion, he never said explicitly that it had prevented him from bringing an obstruction charge.[179] Then came his nine-minute news conference on May 29, 2019, which only compounded the confusion. He told the gathered media that his non-decision decision on obstruction had been "informed" by the OLC opinion.[180] "Informed" is a wonderfully elastic, vague, and ambiguous word. It can mean just about anything.

If Mueller was insinuating that he could not render a decision because of the OLC opinion, that was not what he told the attorney general and others during a meeting on March 5, 2019. Here is what Barr told senators during his May 1 testimony:

> We were frankly surprised that they were not going to reach a decision on obstruction and we asked them a lot about the reasoning behind this. Mueller stated three times to us in that meeting, in response to our questioning, that he emphatically was not saying that but for the OLC opinion, he would have found obstruction.[181]

Barr said there were others in the meeting who had heard Mueller say the same thing—that the OLC opinion had played no role in the special counsel's decision making or lack thereof. The attorney general repeated that in his news conference the day the special counsel report was released to the public:

> We specifically asked him about the OLC opinion and whether or not he was taking a position that he would have found a crime but for the existence of the OLC opinion. And he made it very clear several times that was not his position.[182]

Yet when Mueller stood before television cameras and reporters, he told a strangely different tale. He seemed to argue that he could not have accused the president of obstruction because he had been handcuffed by the OLC opinion. Why, then, had he also informed Barr that a special counsel can abandon the opinion if the facts merit it?[183] And why would the OLC opinion have deterred Mueller at all? It had not stopped him from making a decision on "collusion." It therefore follows that it should not have stopped him from rendering a decision on obstruction. His reasoning made no sense. Equally incomprehensible was his statement that "if we had had confidence that the president clearly did not commit a crime, we would have said so."[184] That was immediately contradicted by his statement "We concluded that we would not reach a determination *one way or the other* about whether the president committed a crime."[185] Which was it? Law professor Jonathan Turley called Mueller's reasoning "conflicted and, at points, unintelligible."[186]

Above all else, Mueller could have arrived at an obstruction decision without the need to either indict or prosecute. This was what independent counsel Kenneth Starr had done when he investigated President Bill Clinton. In his report, Starr had identified eleven criminal offenses, including obstruction of justice, that had allegedly been committed by Clinton and were supported by the evidence collected.[187] Starr knew—as Mueller surely knew—that the OLC opinion says nothing about drawing a *legal conclusion* that there is, or is not, sufficient evidence to support an obstruction offense. Starr stated that Clinton had lied and obstructed justice:

> The Office of Independent Counsel (OIC) hereby submits substantial and credible information that President Clinton obstructed justice . . . the President lied under oath to the grand jury and obstructed justice.[188]

There was nothing improper about Starr's legal findings. They did not in any way contravene the OLC opinion. Starr did what he was supposed to do—he found there was a basis for possible criminal charges and said so. Mueller surely knew of the Starr-Clinton precedent since he had been working for the Department of Justice at the time. The special counsel was stunningly mistaken when he asserted at his news conference that the OLC "opinion says that the Constitution requires a process other than the criminal justice system to formally accuse a sitting president of wrongdoing." [189] Mueller was obviously alluding to impeachment. However, as Professor Turley observed, "that is not actually what it [the OLC opinion] says." [190] The opinion addresses indictments and prosecutions, not legal conclusions. Mueller was informed of that fact by the attorney general and deputy attorney general.[191] Yet he ignored it. Even when he was instructed to reach a conclusion, he refused.

Two days after Mueller's news conference, Barr reinforced the point when he told CBS News, "He could have reached a conclusion. The [OLC] opinion says you cannot *indict* a president while he is in office, but he could have reached a decision whether it was criminal activity." [192] In fact, Barr made his own decision "without regard to, and not based on" the OLC opinion, as he specifically stated in a letter announcing that the president had not obstructed justice.[193] As for Mueller's cryptic reference to impeachment, Barr seemed to scold the special counsel: "The Department of Justice doesn't use our powers of investigating crimes as an adjunct to Congress . . . we are not an extension of Congress's investigative powers."[194] If impeachment was Mueller's goal, as it seems to have been, he abused his position and wasted taxpayer resources, not to mention tens of millions of dollars.

Take a look again at the language Mueller used in his carefully worded conclusion. It is actually a compound sentence conjoining two very separate conclusions. He wrote, "While this report does not conclude that the President committed a crime, it does not exonerate him." The second half of that sentence is an independent clause that is wholly irrelevant. You can toss it out because prosecutors don't exonerate. It was a gratuitous remark tagged onto the main conclusion that "this report does not conclude that the President committed a crime." If Mueller thought he had sufficient evidence that Trump had committed a crime, he could have said so without reservation or limitation. Adherence to the OLC opinion would not have precluded it. He might easily have stated, "This report concludes that there *is* sufficient evidence to support an obstruction offense against the President." Such a declaration is neither an indictment nor a prosecution. It simply draws a legal conclusion derived from the evidence. It would conform to the OLC opinion. Any decision to indict or prosecute would then be made by the attorney general. Barr would have been faced with three choices: he could overrule the OLC opinion and proceed against the president, he could defer legal action until Trump leaves office, or he could disagree with Mueller's conclusion and refuse to indict. But the special counsel didn't do that.

Mueller was well aware of the OLC opinion when he accepted the position of special counsel. If he thought he was barred from making a decision, why did he agree to take the job in the first place? The answer is obvious: he intended all along to decide the matter of obstruction. But when the evidence did not support it, he invented an excuse to avoid rendering the decision he had been hired to make. As he gathered facts and interviewed witnesses, he realized that evidence of obstruction was legally deficient. The president's constitutionally authorized decisions were not, by force of law, obstructive. Beyond that, there were plausible and legitimate explanations for Trump's actions.

The unabashed arrogance on display at Mueller's news conference was breathtaking. He refused to take any questions from the media, which is his right. But he then boldly announced that he would not answer questions from Congress "beyond our report." This is neither his right nor his prerogative. It was remarkably presumptuous for him to say he would not do so. Congress can command under subpoena that he appear to answer any and all questions about the evidence he gathered, the witnesses he interviewed, and the conclusions he reached in his report. There is no immunity for a special counsel. Does Mueller believe he is above criticism and reproach? For the better part of two years, he spent tens of millions of taxpayer dollars investigating "collusion" and obstruction, only to decide one but not the other. Congress had every right to demand that Mueller explain many of his inexplicable actions.

When Mueller eventually relented under a subpoena and appeared before Congress, he muddled his testimony on the extent to which the OLC opinion influenced, if at all, his decision not to decide obstruction of justice. During the morning session, he seemed confused and unsteady when he agreed with Representative Ted Lieu's (D-CA) statement that he "did not charge the president because of the OLC opinion." [195] During his opening remarks in the afternoon, he amended his earlier testimony by saying, "That is not the correct way to say it." He added, "As we say in the report and as I said at the opening, we did not reach a determination as to whether the president committed a crime." [196]

The whole of Mueller's disoriented testimony that day raised serious questions. He appeared detached and confused. It was obvious that he had not written the report that bore his name and may not have fully understood its contents. If so, Mueller may have been little more than a figurehead who was exploited by the partisans on his team.

EVEN ACCEPTING MUELLER'S FLAWED PREMISE, HE HAS NO OBSTRUCTION CASE

Having established here that Mueller's view of the law and regulations were distorted and wrong, let's assume for the sake of argument that he was correct—that a president can nevertheless obstruct justice while executing his constitutional authority. The *facts* presented in the special counsel report still do not support an obstruction offense under the law. The evidence of corrupt intent was woefully insufficient. Attorney General Barr explained, "For each of these episodes we thought long and hard about it, we looked at the facts, and we didn't feel the government could establish obstruction in these cases." [197]

Obstruction of justice is defined in a series of statutes in the criminal codes.[198] Proof requires that a person act "corruptly" to influence, obstruct, or impede a pending proceeding or a legal investigation.[199] An attempt or "endeavor" to obstruct is also criminal.[200] But the operative word is *corruptly*. What does it mean? It is defined as "acting with an improper purpose, personally or by influencing another." [201] Typically, it involves a lie, threat, or bribe, concealing evidence, or destroying documents. A prosecutor must prove not only that someone intended to obstruct but that it was done with a corrupt or improper purpose. Consciousness of wrongdoing must be shown. In the seminal case, *Arthur Andersen LLP v. United States*, the US Supreme Court further defined "corruptly" as acting with a "wrongful, immoral, depraved, or evil" intent.[202] This is an extremely high standard for prosecutors to sustain.

The firing of Comey was one of roughly a dozen actions and statements by Trump that the special counsel examined for potential obstruction. Although it is obvious that Mueller presented facts in a way that allowed others to misconstrue them for political reasons, his report was notable for what it did not show: provable corrupt intent or an improper purpose behind any of the president's actions. The report presents reasonable, logical, and legitimate explanations for every episode. The special counsel should, in good conscience, have stated that obstruction was unsupported. His refusal to do so was exactly what Attorney General Barr said was inappropriate when he testified during his confirmation hearing, "If you're not going to indict someone . . . you don't unload negative information about the person. That's not the way the department does business."[203] But that was the way Mueller did business. He delivered 183 pages disparaging Trump for actions that did not amount to obstruction of justice. It was a shameful smear. But a careful reading of the Mueller Report reveals both exculpatory evidence and analysis.

The Trump-Comey Conversation

The day after the February 13, 2017, resignation of National Security Advisor Michael Flynn, Comey visited the White House. According to the director's version of events, Trump spoke favorably about Flynn and emphasized that he "hadn't done anything wrong on his calls with the Russians, but had misled the Vice President."[204] That was true. Comey claimed that the president had then said, "I hope you can see your way clear to letting this go, to letting Flynn go. He is a good guy."[205] Trump insisted that he had "never asked Comey or anyone else to end any investigation, including the purported investigation of General Flynn."[206] The president characterized Comey's account as neither truthful nor accurate. But even if it were accurate, would it be obstruction of justice? The answer is found in Mueller's report.

The special counsel explained that under the Constitution, "The President has broad discretion to direct criminal investigations."[207] Comey acknowledged that when he told Congress that "the president is the head of the executive branch and could direct, in theory . . . anyone being investigated or not—he has the legal authority."[208]

Setting aside the accuracy of Comey's account in his meeting with Trump, did the president's alleged words about Flynn reflect a corrupt or improper purpose? Comey answered the question when he testified before the Senate Intelligence Committee. Asked if he'd ever heard of a case where a person has been charged for obstruction for "*hoping* something," Comey replied, "I don't as I sit here."[209] The director was correct. Hoping and wishing for an outcome is not an improper purpose; it is a deliberative statement. Comey knew that the president's words did not remotely establish an obstruction offense because he testified days before his firing that no one had attempted to interfere in any of his investigations.[210] So did his deputy, Andrew McCabe. That would have included the Flynn case. After he was terminated, Comey changed his story and testified that he had *thought* Trump was attempting to interfere in the Flynn matter. But Comey's interpretation of Trump's words is immaterial; the intent of the person speaking the words proves obstruction, not how the listener may construe them. No reasonable person would find the president's words obstructive, assuming he even said them.

If Comey truly believed the president was attempting to obstruct an investigation of Flynn, the FBI director would have said so in his self-serving memo. He did not. Moreover, he would have opened an obstruction investigation of Trump. He did not. In fact, in a March 30, 2017, phone call, Comey assured the president that he was not personally being investigated. If the FBI director felt that Trump's alleged remark about Flynn constituted obstruction, he would not have made that statement. If the president wanted Flynn cleared, why did he not raise the issue again with Comey? He did not. No one else at the White House broached the subject, which only undermines the legitimacy of the obstruction theory.

Documents produced by the special counsel also show that the FBI agents who interviewed Flynn "did not believe he intentionally lied to them."[211] He may have been confused or forgetful about his conversations with Russian ambassador Sergey Kislyak, but he was truthful. In fact, the FBI informed Flynn that it was "closing out" the case. That was confirmed by White House Counsel Don McGahn and memorialized in his notes.[212] Since Bureau agents concluded that Flynn had not lied, there was no investigation to obstruct. When the FBI appeared to have ended its investigation of Flynn, President Trump did not. He continued "gathering and reviewing the facts in order to ascertain whether Flynn's actions necessitated removal from office," which he thereafter ordered.[213] Far from obstructing justice, the president pursued the matter and fired Flynn.

Federal courts have consistently held that "The president may decline to prosecute certain violators of federal law just as the president may pardon certain violators of federal law."[214] Trump had the right to express his opinion, public or private, about the Flynn matter without being accused of obstruction. Recent historical precedent proves it. President Obama made numerous public comments about Comey's investigation into Hillary Clinton's mishandling of classified emails. Obama repeatedly opined that Clinton had not jeopardized national security and suggested that she had not broken the law. No one, including Comey, accused him of attempting to obstruct the FBI's investigation.

More than a year after the Trump-Comey conversation, when the special counsel began investigating Flynn for allegedly making false statements, there were several communications between the fired NSA and the Trump legal team. Some have suggested that the president was attempting to unduly influence Flynn's testimony. However, Mueller concluded that the president had never been "personally involved or knew" about any of this.[215]

Communications with Intel Leaders

In the first three months of Trump's presidency, media reports that the president had "colluded" with Russia escalated. That culminated in Comey's March 20, 2017, congressional testimony in which he refused to say whether Trump was personally being investigated. Trump was frustrated that the FBI director refused to make public what he had told the president privately. In a meeting on March 22 at the White House, Trump asked then director of national intelligence Daniel Coats and CIA director Michael Pompeo "whether they could say publicly that no link existed between him and Russia."[216] A similar request was made of NSA director Admiral Michael Rogers.[217] The president was asking them to tell the truth. The increasing number of media stories and concomitant condemnation by Democrats were making it difficult to govern, especially in matters of foreign relations with Russia. Coats and Pompeo declined, not because they considered Trump's request to be criminal obstruction of an investigation. They thought it was improper for the intelligence community to get involved. Mueller's report confirmed this when it stated, "The evidence does not establish that the President asked or directed intelligence agency leaders to stop or interfere with the FBI's Russia investigation."[218]

The Firing of Comey

Trump's firing of Comey did not constitute obstruction of justice for many of the same reasons that his alleged conversation about Flynn did not rise to the level of obstruction. His constitutional authority notwithstanding, there was no evidence of a corrupt or improper intent. The president had ample reason for discharging him. Comey's direct supervisor, Deputy Attorney General Rod Rosenstein, spelled out several incidents of serious malfeasance by Comey in his mishandling of the Clinton email case that more than justified the decision, and the deputy AG recommended to the president that the director be dismissed.[219] So, too, did the attorney general, Jeff Sessions, who endorsed the decision.[220] Trump cited those reasons and recommendations in his letter to Comey dated May 9, 2017.[221]

Mueller determined that that stated explanation for firing Comey had been a "pretextual reason."[222] The special counsel surmised that the real reason had been that Comey appeared to have been acting with duplicity and dishonesty. The director had repeatedly assured Trump in private that he was not personally under investigation by the FBI. Yet he had refused to say so publicly when the president had asked him to simply be candid with Americans. As the special counsel observed, "the erroneous perception he [Trump] was under investigation harmed his ability to manage domestic and foreign affairs, particularly in dealings with Russia."[223]

Following Comey's firing, the president sat down for an interview with NBC's Lester Holt. During the questioning, Trump referred to the Russia probe in the same sentence as the firing of his FBI director. Instantly, the president's opponents seized upon the televised interview as irrefutable proof of obstruction. It was not. A rigorous reading of what Trump said confirms that his intent had not been to interfere with or end the Russia investigation but to place someone who was neutral and competent in charge. "As far as I'm concerned, I want that thing to be absolutely done properly," he told Holt.[224] The president felt that Comey was "the wrong man for that position." Trump wanted him replaced with "a really competent, capable director," even if that meant that the Russia investigation might take *longer*.[225] If anyone on his campaign had coordinated with Russia in a way that had violated the law, Trump said, he wanted the FBI to find out.[226] The special counsel correctly cited those passages from the NBC interview, even though the media have consistently overlooked them and misrepresented the president's words.[227] Mueller even referred to a White House statement that made it clear that "the investigation would have always continued, and obviously, the termination of Comey would not have ended it." Removing the head of an executive agency does not halt or obstruct the work of the agency.

For the better part of two years following the firing of Comey, the mainstream media and Democrats who opposed Trump at every turn argued that that action, more than any other, by the president was definitive proof of obstruction of justice. They demanded his indictment, impeachment, and imprisonment. They assured Americans that Mueller would agree. Except that he did not. Mueller wrote, "The evidence does *not* establish that the termination of Comey was designed to cover up a conspiracy between the Trump campaign and Russia."[228] No conspiracy, no cover-up, and no obstruction.

Trump Tower New York Meeting

As noted earlier, the meeting between a Russian lawyer and the Trump campaign did not constitute "collusion." However, the president's critics, including many in the media, asserted that Trump's involvement in issuing official statements in late June and early July 2017 about what had occurred a year earlier at Trump Tower in New York had been false or misleading and therefore somehow constituted obstruction of justice. That was always a preposterous argument for two reasons. First, the statements were not false. The subject of the meeting had ended up being the Russian policy on adoption, which is what the White House communications team had conveyed. At worst, the statements were incomplete. Second, even if the statements were outright lies, it is not a crime to lie to the media or the public at large. If it were otherwise, most politicians in Washington would be behind bars. Mueller admitted this when he stated, "they would amount to obstructive acts only if the President, by taking these actions, sought to withhold information from or mislead congressional investigators or the Special Counsel."[229] And the report notes that Trump had never withheld emails or other information about the Trump Tower meeting from either Congress or the special counsel.

Trying to Get Sessions to Reverse His Recusal

Jeff Sessions was only three weeks on the job as attorney general when he held a news conference on March 2, 2017, to announce that he would recuse himself from any investigations involving Russia and the 2016 presidential campaign. He should never have done so. As explained in *The Russia Hoax*, neither the facts nor the law required his recusal, which applies to criminal investigations, not counterintelligence probes. Sessions set all of it into motion within hours after being sworn in and "basically recused" himself on his first day in office.[230] Trump was understandably livid. He considered it a serious betrayal. "If he was going to recuse himself, he should have told me prior to taking office," declared the disgusted and angry Trump at a Rose Garden news conference.[231] He was right.

The Mueller Report recapped several face-to-face meetings in 2017–2018 in which Trump tried to convince Sessions that he should reconsider his action, reverse his recusal, and take charge of the inquiry. Contemporaneous notes show that the president told his attorney general, "I just want to be treated fairly."[232] In a subsequent meeting, Trump repeated the statement. That was not an effort to halt or obstruct the investigation. The president had publicly accused Mueller of disqualifying conflicts of interest and frequently complained that the special counsel had assembled a team of partisans for his investigation. Trump wanted a fair, objective, and neutral investigation; he did not want a partisan "witch hunt," as he so often lamented in public comments and tweets. That was eminently reasonable from his point of view, and that expressed view is evidence of his intent under the law. Those statements, cited by Mueller in his report, are evidence that Trump's purpose was not improper or corrupt. To the contrary, it was sensible, proper, and lawful.

The Mueller Report spent several pages examining the president's intermittent efforts to force the resignation of Sessions and other steps that might "limit the scope of the Special Counsel's investigation."[233] However, since Sessions was recused, firing him would have had no impact on the special counsel investigation. Mueller's probe was never limited in any way.

Discussions to Remove the Special Counsel

Given Sessions's recalcitrance, Trump discussed with his staff the notion of removing Mueller as special counsel. The report identified Trump's intent as identical to one that had motivated his request that the attorney general reverse his recusal. That is, "The President told senior advisors that the Special Counsel had conflicts of interest."[234] At one point in June 2017, Trump allegedly telephoned White House counsel Don McGahn. According to the report, "The President called McGahn and directed him to have the Special Counsel removed because of asserted conflicts of interest."[235] McGahn never complied because he disagreed with the rationale, and "the President did not follow up with McGahn on his request to have the Special Counsel removed."[236] At roughly the same time, Trump received competing advice that terminating Mueller would be unwise politically. He then dropped the idea.

Those words and actions by Trump did not constitute obstruction of justice for several reasons. First, Mueller was not removed, so there was no obstructive act. The special counsel investigation proceeded uninterrupted. Second, it was not an *attempt* to obstruct because the law does not criminalize thoughts and discussions. Third, the president's intent was not improper, as the law requires. Conversations show that he sincerely believed Mueller was conflicted and could not render a fair conclusion to the investigation. Trump's expressed concerns seem legitimate and do not form a corrupt purpose. Fourth, the investigation uncovered evidence that Trump did not necessarily want to remove Mueller but "simply wanted McGahn to bring conflicts of interest to the Department of Justice's attention."[237] Finally, even if Mueller were removed, a replacement would be named. The report admits that the work of the special counsel would not have stopped.

When the story of the Trump-McGahn conversations was leaked to the media, the president denied their accurate portrayal and asked his counsel to deny them as well. Trump and McGahn disagreed over whether the reporting was factual. If the president genuinely believed that the story was incorrect, his intent was not improper. But this is largely irrelevant to any consideration of obstruction of justice, because media reports are not official investigations or legal proceedings. Obstructing the press is not a crime, although it tends to be a Washington preoccupation.

Trump Tower Moscow Project

As noted earlier, the concept of constructing a commercial and residential tower in Moscow was an undertaking that Trump considered and then abandoned in 2016. Mueller found no evidence of a criminal "collusion" conspiracy. But he did explore the lie that Trump's personal lawyer Michael Cohen had told to Congress about whether the president had directed him to give such false testimony. If true, it would have constituted a conspiracy to obstruct justice.

Cohen was the driving force behind the building proposal. He had personal financial incentives attached to the deal should it ever be consummated. But the obstacles proved to be insurmountable. The potential project fell apart in the spring and summer 2016. During the campaign, Trump insisted he had "no deals" and no investments in Russia. This was true. No deal had been struck. Discussions pursuant to a "letter of intent" had collapsed. On his own, Cohen began telling reporters that "the Trump Tower Moscow deal was not feasible and had ended in January 2016." That was untrue. It had ended months later, in May or June. But Cohen peddled that deception because, according to the Mueller Report, he thought "it limited the period when candidate Trump could be alleged to have a relationship with Russia to an earlier point in the campaign."[238] Cohen then repeated the same lie, and several others, to Congress when he submitted a statement in August 2017.

It is puzzling why Cohen would lie about something that is not a crime, until you consider that deceit is second nature to the disgraced lawyer who pleaded guilty to crimes including lying to Congress and fraud. Prosecutors portrayed him as a prodigious liar and cheat. However, the special counsel found that the president had not been a party to Cohen's mendacity. The report concluded that the evidence "does not establish that the President directed or aided Cohen's false testimony."[239] The specifics of this story will be scrutinized in a later chapter.

Evidence in Trump's Favor

In his report, Mueller recognized several factors that militated in favor of Trump and against any finding of an obstruction of justice offense.

First, many of "the President's acts were facially lawful" during the exercise of his constitutional powers.[240]

Second, "the evidence does not establish that the President was involved in an underlying crime related to Russian election interference."[241] Though not dispositive, it is compelling evidence of the lack of criminal intent. Why would a person obstruct or cover up a noncrime?

Third, many of the president's actions and statements had "occurred in public view."[242] People rarely obstruct in a conspicuous manner; discreet acts of interference are the norm.

Fourth, the FBI and special counsel investigations were not actually obstructed.[243] They continued to their conclusions. Trump had never asserted executive privilege. He had voluntarily handed over 1.4 million pages of documents and made all witnesses available to the special counsel. He had allowed his White House counsel to spend roughly thirty hours being interviewed, even though all of the McGahn-Trump conversations were privileged. The president's devotion to transparency was unprecedented.

A consistent underlying thread of evidence runs through the Mueller Report: the president was confident that he had not "colluded" with Russia and was being wrongfully targeted by investigators who were partisan and unfair. The probe itself was eroding his ability to govern. Quite often, Trump chose to avail himself of public forums to defend against what he perceived to be false accusations driven by political motivations. That was his right under the First Amendment. In fact, it was his only option. He was facing a one-sided investigation.

CONCLUSION

Mueller tried but failed to find evidence of a conspiracy with Russia. None existed, and the special counsel was forced to admit the obvious. Yet when it came to obstruction of justice, he turned the burden of proof upside down and refused to make the decision that he owed to the president and the American people. Inverting this time-honored standard and effectively forcing Trump to prove his own innocence is not how the rule of law works. Along the way, Mueller ignored rampant acts of gross misconduct at his former haunt, the FBI. Perhaps out of loyalty to the institution, he turned a blind eye to corrupt behavior by senior officials there, whom he knew all too well. Those misdeeds formed the basis of his probe. Yet his silence on those matters was deafening. Long after he realized that there had been no "collusion," he persisted in a feckless investigation of obstruction. Mueller came to symbolize the "witch hunt" that was originated by his longtime friend and colleague James Comey.

It was wrong of Mueller to hire a contingent of political partisans. It was a mistake to issue a nearly two-hundred-page report on obstruction of justice that alleged noncriminal wrongdoing. Selective inferences and ambiguities are not empirical evidence. It was equally inappropriate for him to try to influence congressional debate by composing what amounted to an impeachment referral. That was way beyond his authority and a shameful abuse of his power. Only when agitation to remove the president began to wane did Mueller elect to hold a news conference to fan the flames of impeachment. That was reprehensible.

Robert Charles, who once served as assistant secretary of state and is a former naval intelligence officer and litigator, posed an important question for Mueller:

> How is it that your report omits inquiry into origins of collusion allegations you were commissioned to investigate? And how do you explain your recent behavior—which appears aimed at clearing the FBI and intentionally encouraging impeachment?[244]

Mueller didn't want to talk about it, as he made plain in his brief statement to the media and the American public.

If the special counsel could not render a credible finding of obstruction, he should have said so and remained silent. Instead, Mueller repudiated established constitutional principles, adulterated the law of obstruction, and perverted due process. He produced a magnum opus that stands as an egocentric monument to the miscarriage of justice. Together with his partisan prosecutors, he rewrote more than two centuries of US jurisprudence by tainting Trump with the presumption of guilt. His tenure as special counsel was disgraceful.

To those who have long been suspicious of Mueller's tactics, dating back to his tenure as FBI director, his report came as no surprise. Representative Louie Gohmert, who for years has carefully tracked and questioned Mueller's "sordid history of illicitly targeting innocent people," produced and published a seventy-five-page compilation of what he described as chronic misconduct.[245] He rendered that verdict a full year before the conclusion of the special counsel investigation:

> Judging by Mueller's history, it doesn't matter who he has to threaten, harass, prosecute or bankrupt to get someone to be willing to allege something—anything—about our current President, it certainly appears Mueller will do what it takes to bring down his target, ethically or unethically, based on my findings.[246]

Former governor Chris Christie was prescient in his metaphorical description of Robert Mueller as "a trained assassin." I agree that Mueller's report reads like a "hit job."

CHAPTER 6

THE MEDIA WITCH HUNT

SCARECROW: I haven't got a brain . . . only straw.
DOROTHY: How can you talk if you haven't got a brain?
SCARECROW: I don't know. But some people without brains
do an awful lot of talking, don't they?
DOROTHY: Yes, I guess you're right.

—*THE WIZARD OF OZ*, 1939

*I think the media has been hurt by this political witch hunt hoax, maybe
more than anybody else. The Democrats and the media had a partnership.
They were deranged. They were totally deranged. They really did try to
take away an election. It's the greatest con job in the history of American
politics.*

—AUTHOR'S INTERVIEW WITH PRESIDENT DONALD J. TRUMP,
OVAL OFFICE, WHITE HOUSE, JUNE 25, 2019

Call him the "Typhoid Mary" of the Steele "dossier." David J. Kramer, a middle-aged man with a reddish beard, wire-rimmed glasses, and a serious demeanor, injected the virus of the Russia conspiracy into the bloodstream of the American mainstream media and triggered an infection so extreme that the patient nearly died.

A Harvard graduate and student of US-Russia geopolitics for thirty years, Kramer epitomized the Washington middle-level denizen dubbed "deep state." He'd worked in academia, in think tanks, at the US State Department, and at the McCain Institute. He sat on the board of directors of the prestigious Halifax International Security Forum, which hosts an annual conference attended by top civilian and military experts from around the world.

The mid-November 2016 conference held in Halifax, Nova Scotia, buzzed with excitement. Trump had just won the presidency and was viewed by many attendees, including Kramer, as a frightening wild card. During a break in proceedings on Saturday, November 19, Sir Andrew Wood, a former UK ambassador to Russia, pulled Kramer aside. The two men often shared articles and information about Russia by email with like-minded scholars.

"He said he was aware of information that he thought I should be aware of and that Senator McCain might be interested in," Kramer later said in a deposition.[1] Wood had not seen the material himself, but "he had been told that some information had been gathered that pointed to possible collusion" with the Russians by President-elect Trump.

Kramer had deep suspicions about Trump's ability to counter Putin's dangerous aggressions. Just a few days before the conference, he had coauthored a story for Politico titled, "How Trump's Victory Could Give Russia Another Win."[2] He contended that Putin posed an "existential threat" to Western democracies and that "controversy about ties to Russia shadowed Trump throughout the campaign."

Not only did he write about the threat, he took action. In July 2016, he had been one of about two dozen national security experts who had signed an open letter to Congress asking for an investigation into the hacking of the DNC by Russia.[3] And he had signed another open letter saying that Trump "lacked the character" to be president.[4] When approached by the esteemed Sir Andrew with secret dirt on Trump, Kramer was eager to help. "I believe that he felt that I was the best person he knew to act as a conduit to get to Senator McCain," Kramer said.

As senior director for human rights and democracy at the McCain Institute since 2014, Kramer had gotten to know the senator well. He was working on a book about Russia to be published by the McCain Institute in mid-2017. And it so happened that McCain was attending that year's annual meeting. Kramer sought out the senator and delivered Wood's tantalizing message. McCain was intrigued. Later that day, the senator and his top staffer, Christian Brose, met with Kramer and Wood in a private room. Wood explained that a London source with the "utmost credibility" had gathered evidence revealing Trump's ties to Putin and the existence of a sexually depraved video that could expose Trump to Kremlin blackmail. The senator, who had a long-standing feud with Trump, asked Kramer if he would go to London to meet the source. After Kramer agreed, Wood revealed that the informant was a former MI6 spy.

On Sunday, November 28, Kramer flew to London, so keen to oblige that he used his own air miles to book the flight and paid the airport fees himself. It felt a bit cloak and dagger. On Monday morning, he was met at Heathrow by Christopher Steele, who was wearing a blue coat and carrying a copy of the *Financial Times*, as had been prearranged. The Cambridge-educated former spy looked like his name: distinguished, with gray hair and a confident manner.

Steele took Kramer to his home in Surrey. He said that his firm had been engaged to look into Trump's business dealings in Russia to find any compromising material on the real estate tycoon. He did not reveal who had hired him but stressed several times that he did not tailor the information for the client or "sugarcoat it or shape it one way or the other." Just the facts. He handed over a lengthy document comprising sixteen or seventeen memos. Kramer took about an hour to read hair-raising, salacious details of Trump's corruption, sexual perversion, and cronyism outlined by Steele's numerous high-level Russian sources.

Steele assured him that although the memos needed to be corroborated and verified, "based on the sources and based on his own company's track record, he felt that at least he had the best sources possible to provide [the] information." In other words, this was solid material, not rumor or gossip. He assured Kramer that there was video proof of the "golden shower" allegation. However, despite his James Bond level of expertise, Steele had no idea how to obtain the video. Steele explained that he had approached an "FBI person" in Europe to share what he had found and hoped that the agency would take a serious look at it. If Senator McCain weighed in, that would give the FBI an "additional prod" to take this seriously. That's why he had reached out to Kramer through Wood.

Kramer felt honored to be singled out as someone with enough clout to provide the material to McCain, then the chairman of the Senate Committee on Armed Services. He had no idea that Steele had been trying for months to manipulate the FBI, the State Department, and media sources to get his "dossier" into the public domain. Nor did he know that although the FBI had used the "dossier" to obtain a FISA warrant on Carter Page, the agency had ended its relationship with Steele on November 1 because of his numerous conversations with the media and lying about them to the FBI.[5] And there were clear indications that his sources and methods were problematic.

During a meeting with Kathleen Kavalec, deputy assistant secretary of state, on October 11, 2016, Steele had expressed his intense desire to get the information into the public domain by his deadline: election day. In a memo she sent to the FBI, Kavalec wrote that Steele had cited a "technical/human operation run out of Moscow targeting the election" that recruited émigrés in the United States to do "hacking and recruiting." Steele had told her that they were paid through the Russian Consulate in Miami, which Kavalec noted did not exist.[6]

That was a red-flag warning to the FBI that Steele was unreliable as a source and that his "dossier" was likely a lie. If a State Department official, untrained in the techniques of uncovering deception, could accurately size up Steele as a fabricator during the course of just one brief meeting, the FBI must have known that its paid informant was selling false evidence. Yet ten days later the Bureau and the DOJ used Steele's suspect document as the basis for a warrant to spy on a Trump campaign associate.

During his meeting with Kramer, Steele refused to give him a copy of the "dossier" memos to carry back on the airplane. Too risky. But he promised that one would be given to him after he returned home by Glenn Simpson, who had hired him. They had never met, but Kramer knew of Simpson's reputation as a former writer for the *Wall Street Journal* who had written extensively about Russia. After fewer than twelve hours on the ground in London, Kramer flew back home to Washington—a whirlwind trip at considerable expense just so the supposed supersecret agent Steele could sell him on the reliability and grave national importance of the evidence against Trump.

On Tuesday, November 29, at 5:00 p.m., Kramer met with Simpson, who handed over two copies of the "dossier," one with more redactions than the other. Kramer still didn't realize that he was being set up. Simpson did admit to Kramer that he had spoken to the *New York Times*. But he did not disclose that he and Steele had also talked to the *Washington Post, Mother Jones*, CNN, ABC, and *The New Yorker*. His efforts thus far had yielded only a few stories.

Gradually, Simpson and Steele had discovered the key to spreading the hoax: journalists had trouble reporting out the allegations because they weren't true, but they could report truthfully that the government was looking into something. Of the tens of thousands of Russia stories about to push everything else off the front page for years, nearly all of them would be about the search for proof, rather than actual proof.

Former *Newsweek* reporter Michael Isikoff had published an article for Yahoo! News on September 23, 2016, about Carter Page's trip to Russia titled "U.S. Intel Officials Probe Ties Between Trump Adviser and Kremlin," relying on anonymous sources—clearly Steele and Simpson.[7] On the same day, Julia Ioffe wrote a story for Politico, "Who Is Carter Page? The Mystery of Trump's Man in Moscow."[8] On October 31, David Corn published a story in *Mother Jones* that was the first to surface Steele's existence, "A Veteran Spy Has Given the FBI Information Alleging a Russian Operation to Cultivate Donald Trump."[9] Steele was not named.

Corn passed a copy of the "dossier" to his longtime friend James Baker, then FBI general counsel, who later told Congress that the Bureau had been aware that Simpson was shopping the documents to many people in government and media in an effort to "elevate" the "dossier's" profile. "I know that David was anxious to get this into the hands of FBI," Baker said.[10] The three stories included quotes from a blistering letter to Comey, written by Senate minority leader Harry Reid, demanding an investigation of Trump-Russia ties, also instigated by the Steele "dossier." But the topic didn't catch fire in the wider media. Corn was known as a partisan activist, it was easy to dismiss the grandstanding Reid, and solid journalists who tried to corroborate the "dossier's" contents failed.

During his initial thirty-minute meeting with Kramer, Simpson said nothing about the identity of his client. Kramer didn't bother to ask. But he was hooked. Kramer became Simpson's single most important asset since the disinformation campaign crafted by Fusion GPS had begun.

The next day at about 5:00 p.m., Kramer met with Senator Mc-Cain and his staff aide, Brose, in order to physically hand over the memos. McCain read the "dossier," then asked Kramer his opinion. "I said that Mr. Steele himself seemed to be credible and believable, but that he himself had acknowledged he was not in a position on his own to verify everything in there." Perhaps that boilerplate lingo gave Kramer peace of mind. He would repeat it ad nauseam to those with whom he talked. "I suggested he provide a copy of it to the Director of the FBI and the Director of the CIA."

Over the next week or so, Kramer often spoke by phone to Simpson and Steele, who were desperate to know what, if anything, McCain was doing to advance the tale that Trump was a secret Russian asset. On December 9, Brose informed Kramer that McCain had met privately with Comey and had given him the document. Triumphantly, Kramer passed on the news. Mission accomplished.

That handoff—from Steele to Simpson to Kramer to McCain—gave the "dossier" a Republican patina. It wasn't merely Democrat Reid dishing a hated rival's dirty linen; it was a patriot and war hero on Trump's side of the aisle. "I think they felt a senior Republican was better to be the recipient of this rather than a Democrat because if it were a Democrat, I think that the view was that it would have been dismissed as a political attack," Kramer said. It was a clever con.

Reporters had been calling Kramer to ask if McCain had passed the "dossier" to Comey. They included Corn, Tom Hamburger and Rosalind Helderman with the *Washington Post*, Brian Ross at ABC, Peter Stone and Greg Gordon at McClatchy, Alan Cullison at the *Wall Street Journal*, Robert Little at NPR, and more. At Steele's request, Kramer met with Carl Bernstein of Watergate fame and Ken Bensinger, a reporter for BuzzFeed.

The Bensinger rendezvous proved to be a critical turning point. Bensinger visited Kramer at the McCain Institute on December 29 and revealed that he knew Steele from the FIFA investigation. Kramer declined to give him a copy of the "dossier" but said he could read it. Bensinger asked if he could take photos of the pages with his cell phone, but Kramer told him no. He was worried about the material being inadvertently spread. Bensinger agreed. After that pledge, Kramer conveniently excused himself from the room so that Bensinger could read in peace for thirty minutes, with his cell phone at the ready.

On January 10, 2017, Kramer sat near a TV in a lounge area at the McCain Institute, discussing the "dossier" with Julian Borger of *The Guardian*. At about 5:00 p.m., CNN reporters Evan Perez, Jim Sciutto, Jake Tapper, and Carl Bernstein broke the story that President Obama and President-elect Trump had been briefed on the "prostitute pee" part of the "dossier" in a two-page addendum attached to a report on Russian interference in the election.[11] Former director of national intelligence James Clapper had leaked news of the briefing to Tapper.

The CNN story was the hook Bensinger needed to run with the story. Within an hour, BuzzFeed published the full "dossier": "These Reports Allege Trump Has Deep Ties to Russia."[12] Despite his promise, Bensinger had photographed the documents. Using weasel words such as "potentially unverifiable," BuzzFeed posted the full thirty-five-page document online so that "Americans can make up their own minds about allegations about the president-elect that have circulated at the highest levels of the U.S. government."

"Holy shit!" Kramer said as he watched CNN report on the Buzz-Feed post. Shocked, he called Bensinger, demanding that the documents be taken off the internet. "You are going to get people killed."

"Why?" Bensinger asked. "How?"

"By posting this, you will put people's lives in danger," Kramer said. Steele's secret high-level Russian sources would be put at risk. Bensinger seemed unconcerned, as he had been blasé about first verifying the outrageous allegations.

BuzzFeed editor Ben Smith defended the decision, arguing that the "dossier" was worthy of being printed even without corroboration because its contents had been shared with Trump and others "at the highest levels of the U.S. government." Such transparency was "how we see the job of reporters in 2017."[13]

"FAKE NEWS—A TOTAL POLITICAL WITCH HUNT!" Trump tweeted later that day.

Kramer learned that the Wall Street Journal was going to publish Steele's name as the source of the "dossier." He tried to dissuade them but failed. Steele was frantic. "This wasn't supposed to happen this way," he said. Kramer told Steele he had no idea how Bensinger had gotten the documents—a lie, but he hoped to keep communications with Steele open.

After his name surfaced, Steele complained to Kramer that it "was causing considerable problems for him." He went into hiding. But Steele had no reason to be shocked. He and Simpson had been peddling the fabrications for months. Their scheme had worked splendidly. Kramer had talked to more than a dozen media people, assuring them of Steele's credibility, touting the involvement of Sir Andrew Wood and Senator McCain. They knew of Kramer's expertise and reputation.

After BuzzFeed splashed the whole thing onto the internet, the enraged Tapper emailed editor Smith to complain that BuzzFeed should have waited until the next day. (His emails became public after a Russian mentioned in the "dossier" filed a lawsuit against Steele and BuzzFeed.)[14] "No one has verified this stuff," Tapper told Smith. He had pointed out in his CNN report that a key allegation in the "dossier"—that Trump attorney Michael Cohen had traveled to Prague in August 2016 to meet with Kremlin operatives—was likely false, that government officials believed a different Michael Cohen had gone to Prague.[15] But Tapper was most angry about the timing. "Collegiality wise it was you stepping on my dick."[16] Tapper apparently had no problem with his own reporting about Comey briefing the president on the phony pee memo.

Within a few days, the Russia conspiracy erupted in the Washington and New York media. "What happened really quickly is that everybody committed to a narrative and they didn't examine or test the core hypothesis," Matt Taibbi of *Rolling Stone* later explained. That "impacted a lot of reporters whether they realized it or not. They believed this underlying story that Trump had been activated as an asset or just as a useful idiot for the Russians, but certainly at that point, none of this had been corroborated by actual reporting. But the reporting became increasingly maniacal and alarmist."[17]

The story was aided and abetted by timely leaks from FBI officials James Comey, Andrew McCabe, Peter Strzok, and Lisa Page, Representative Adam Schiff, the ranking Democrat on the House Intelligence Committee, and career intelligence officials James Clapper and CIA director John Brennan, now paid TV analysts with coveted security clearances.

The main reason the hoax worked so well and for so long is that many of the highest-ranking officials were willing to lie about and misrepresent the (lack of) proof they had had in secret for years. Cable news allowed them to go on air and promise, week after week, that we were just on the edge of a big reveal, of a scandal too horrible to even detail. However, none of it would have worked without top journalists allowing themselves to be played for fools over and over. Scoops are born by working sources, getting confirmations, and putting news into context. Top reporters allowed their inquiries to be shaped by sources that lied to them again and again. It is not surprising that newsrooms ran headline after headline over stories of partisan leaks from the Justice Department and Congress. That is the kind of thing that wins journalism prizes. What's utterly shocking is that no one in those newsrooms ever suggested delving deep into the other side. What if there was a partisan conspiracy holding sketchy meetings, lying under oath, meeting with Russians, and it was anti-Trump? What if the stories we were forced to retract weren't honest mistakes but our anonymous sources coordinating a massive scam? What if all our best stories are a kind of political pyramid scheme, where everyone is making accusations because they are convinced that someone else has uncovered a truth much more damning than the original deception?

The most reliable news sources in the world allowed themselves to be exploited by a gang of remorseless liars, but the truly scary part is that they would all do it again in a heartbeat.

Reporters for the *New York Times* and the *Washington Post* would go on to share the 2018 Pulitzer Prize for aggressive reporting on the "dossier," with this citation: "For deeply sourced, relentlessly reported coverage in the public interest that dramatically furthered the nation's understanding of Russian interference in the 2016 presidential election and its connections to the Trump campaign, the President-elect's transition team and his eventual administration."[18] Announcing its win, the *Post* bragged that it "helped set the stage for the special counsel's ongoing investigation of the administration."[19]

In reality they were duped into publishing fabrications paid for by the opposition candidate, Hillary Clinton. Brit Hume of Fox News would call the ensuing two-year hysteria the "worst journalistic debacle of my lifetime, and I've been in this business 50 years."[20] Reporters' reputations would be shredded, once-esteemed news organizations would be put to shame, the public's trust in the media would collapse.

The spectacular self-immolation—and it was indeed self-induced— occurred because journalists ignored the basics. They abandoned fairness and accuracy. They trusted sources with secret agendas and, in evaluating the Steele "dossier," failed to do what every student learns in Journalism 101: "Follow the money." Who was paying for the "opposition research" on which they relied so heavily?

One *New York Times* reporter, Ali Watkins, even carried on a sexual relationship with James Wolfe, the security director at the Senate Intelligence Committee, which she had covered for BuzzFeed and Politico.[21] Her "scoops," courtesy of his leaks, had led to her getting the job at the *Times*. The affair came to light when her much older paramour was indicted for leaking classified documents to her and other reporters. Her editors shrugged, declining to fire her.

The legacy print media, network and cable news, anchors, pundits, and analysts shrieked in unison, "Collusion, collusion, collusion." Fox News did not; across its many platforms, the network worked to maintain a balance. But even one network's not continually bashing Trump was too many. The venerable NBC newsman Tom Brokaw responded to a Trump tweet praising *Fox & Friends* for its success by slamming the president on MSNBC's *Morning Joe*. "[Trump] watches [F&F] because it reinforces what he believes," Brokaw said. "Fox News . . . is on a jihad right now." He cited Newt Gingrich, a Fox contributor, calling the FBI a corrupt organization. "So, we're at war here."[22]

The war was mounted against the duly elected president of the

United States. The media did it for ratings, they did it for the approval of their peers, and they did it because they hated Trump.

THE MEDIA ECHO CHAMBER DETERMINES
WHAT THE PUBLIC HEARS

Most Americans are not on Twitter. But the top stories of the day are dictated by it. Pundits and other journalists quickly identify which scoops are "important" and which can be ignored. Even if an outlet follows an unimpeachably nonpartisan agenda, journalists on Twitter are only going to point their hundreds of thousands of followers at the ones reinforcing the liberal narrative.

Over the past thirty years, the number of full-time journalists identifying themselves as Republicans has dropped significantly, from 25.7 percent in 1971 to 7.1 percent in 2013, according to a study by two Indiana University professors.[23] One analysis found that 88 percent of political contributions from television network executives, producers, and reporters went to Democratic candidates, while a comparative trickle of money went to Republicans.[24]

It's not just politics. Journalists at top media companies are less religious and far more likely to support issues such as government redistribution of wealth, alternative lifestyles, and abortion rights than are people in the country they cover.[25] In their abundant arrogance, reporters have never understood why most Americans do not embrace their liberal values. Most members of the media are too insular and dogmatic to conceive of any intelligent principles beyond their own. In their minds, they are the privileged elite possessed of total knowledge and ultimate wisdom.

Called out for their bias, they consistently protest it's not so. "*Hardball* is absolutely non-partisan," said MSNBC's Chris Matthews in 2010.[26] "I can see how the intensity of coverage on certain issues may, to some people, seem to reflect a liberal point of view," said Jill Abramson, the executive editor of the *New York Times* in 2013. "But

I actually don't think it does."[27] "It's true that journalists tend to be more 'liberal' than the average American," NPR's Brooke Gladstone said in a 2011 interview with CNN. "But hyper-awareness of that fact has caused some of our most respected mainstream media outlets to bend over backwards to compensate—offering far more conservative voices than liberal ones."[28]

Psychologists would call those laughable comments denial or groupthink. The problem is especially acute on the Washington–New York axis, where incestuous relationships prevail and conflicts of interest abound but are rarely disclosed. For example, CNN's Evan Perez, a former *Wall Street Journal* reporter, had coauthored stories with Glenn Simpson on national security. He repeatedly reported on the Trump-Russia story for CNN, usually relying on anonymous sources, without disclosing his close relationships with Simpson and other Fusion GPS executives such as Tom Catan, Peter Fritsch, and Neil King, who had also worked at the *Wall Street Journal*.[29] In addition, King is married to Shailagh Murray, another former *Journal* reporter, who was a top communications adviser to President Obama.[30]

The relationships prompted the *Wall Street Journal* to publish an editorial in October 2017 accusing the media of complicity for attacking House Intelligence Committee chairman Devin Nunes after he issued subpoenas for Fusion GPS principals to testify before Congress: "Americans don't need a Justice Department cover-up abetted by Glenn Simpson's media buddies."[31] Politico's Jason Schwartz in turn attacked the *Journal* for not following the media herd on the Trump-Russia "collusion." Schwartz quoted "former *Journal* editor Neil King" as saying "I don't know a single WSJ alum who's not agog at where that edit page is heading."[32] Schwartz failed to mention that King worked for Fusion GPS.

Media people move into and out of jobs with administrations. They marry government employees and political strategists. They live in the same neighborhoods, vacation in the Hamptons, all the while congratulating themselves on being the elite. They value the advice of established experts and rarely watch what they considered lowbrow prime-time network fare such as *The Apprentice*. They knew nothing about Trump's massive fame among middle Americans, or how well the fact that he listened to their political concerns went over. Hubris is the reason journalists never imagined that Trump would be elected president. They thought it was inconceivable that a man they judged to be so unrelievedly vulgar and unsuited for the presidency could possibly be elected to the highest office in the land.

By 2016, Americans' trust in journalists "to report the news fully, accurately and fairly" had plummeted to its lowest level in the history of Gallup polling.[33] During the first one hundred days of Trump's administration, a study by Harvard University's Shorenstein Center on Media, Politics, and Public Policy showed that coverage by CNN and NBC of the president had been 93 percent negative. In contrast, Fox News was more balanced, at 52 percent negative and 48 percent positive.[34] A study by the Media Research Center, a conservative media watchdog, examined statements made by reporters and nonpartisan sources on evening newscasts in 2017 and determined that 90 percent had been negative of the president.[35] Stories of alleged "collusion" with the Russians "comprised almost exactly one-fifth of all Trump News."[36]

But the mainstream media in the twenty-first century have insisted that their reporting is "objective" despite their own biases. This is a myth. Their political favoritism and personal dislikes influence the stories they choose to cover and the way they tell those stories. They shape the content to conform to their own beliefs and, often, misconceptions. Bias doesn't creep into a reporter's story; it hits the viewer and reader over the head with it.

During the 2016 campaign, Jim Rutenberg of the *New York Times* encapsulated many of his colleagues' views about covering Trump: "If you're a working journalist and you believe that Donald J. Trump is a demagogue playing to the nation's worst racist and nationalistic tendencies, that he cozies up to anti-American dictators and that he would be dangerous with control of the United States nuclear codes, how the heck do you cover him?"[37] Well, "you have to throw out the textbook American journalism has been using for the better part of the past half-century," Rutenberg wrote. "You would move closer than you've ever been to being oppositional."

Why the old standards of accuracy, fairness, and solid sourcing didn't apply were not explained except to blame Trump for being Trump. His candidacy was "extraordinary and precedent-shattering" and to "pretend otherwise is to be disingenuous with readers," said Carolyn Ryan, the *Times* senior editor for politics.[38] Thus, Rutenberg explained, "it would also be an abdication of political journalism's most solemn duty: to ferret out what the candidates will be like in the most powerful office in the world. It may not always seem fair to Mr. Trump or his supporters. But journalism shouldn't measure itself against any one campaign's definition of fairness. It is journalism's job to be true to the readers and viewers, and true to the facts, in a way that will stand up to history's judgment."[39]

In running an overwhelming number of stories that focused negatively on Trump, the media exhibited its "selectivity bias." That is, they made a conscious decision to report on matters that conformed to their ideological sympathies and against those of the president. Then, in the body of those stories, there was a distinct "presentation bias" in which viewers were treated to a slanting of the news intended to emphasize an unfavorable viewpoint. Headlines emphasized the negative. Often the storytelling contained no countervailing information that would otherwise balance the narrative—lies by omission. Stories became agenda driven, not knowledge driven.

When the incomprehensible happened on election day, the media reacted to Trump's victory by using their lofty perch to try to undo the results—to drive the president from office in a slow-motion coup. The drip-drip of leaks became a flood, with favored reporters receiving information from employees of the FBI, the DOJ, and other insiders who considered themselves members of the Trump "resistance."

On a daily basis TV and Twitter mocked, ridiculed, and demeaned the president. They proclaimed that he had conspired with Russians by committing acts of lawlessness and perfidy akin to treason. They repeatedly gave a platform to Trump's political enemies and made little effort to hold them accountable for the condemnation they so freely disgorged on air.

After a Trump speech in August 2017, former DNI Clapper spouted off about Trump's "fitness to be in this office"—a lot of nerve for someone who should have been indicted for criminal perjury.[40] In 2013, while testifying before Congress, Clapper was asked, "Does the NSA collect any type of data at all on millions or hundreds of millions of Americans?" The DNI responded, "No, sir . . . not wittingly."

When the story broke soon thereafter that the NSA had been doing exactly that, Clapper first claimed he hadn't realized what the question was about. Then he told a reporter who called him on the lie, "I responded in what I thought was the most truthful, or least untruthful manner by saying 'no.'" Absurd. Later he apologized for his "clearly erroneous" answer but explained that he had simply forgotten about the massive government operation to secretly collect metadata on hundreds of millions of American citizens. That was like saying Christmas had slipped his mind.

Lying to Congress is a felony, but Obama's attorney general, Eric Holder, made sure the case was tossed into a broom closet. Clapper later lied to Congress again, flatly denying that he had discussed the Steele "dossier" or any other intelligence involving Russian hacking of the election with journalists. Asked specifically about his conversations with CNN's Jake Tapper, Clapper acknowledged that he had done so and "might have spoken with other journalists about the same topic."[41] (A few months after the leak to Tapper, Clapper was rewarded with a contract as a CNN contributor.)

Possibly the most egregious offender was John Brennan, President Obama's director of the CIA. On MSNBC's *Morning Joe*, when asked if Trump was afraid of Putin, Brennan opined that the "Russians may have something on [Trump] personally . . . [they] have had long experience of Mr. Trump, and may have things they could expose." No one on the panel asked Brennan if he had proof or was just speculating. Surely the former director of the CIA wouldn't make something up? But when asked by the *New York Times* to respond to written questions about his wild claim, Brennan admitted, "I do not know if the Russians have something on Donald Trump that they could use as blackmail."[42]

Instead of exposing Brennan as a liar, the *Times* published a story with this headline: "Ex-Chief of the CIA Suggests Putin May Have Compromising Information on Trump," burying Brennan's vacillation in the eleventh paragraph. This is not journalism; it's a political smear aided and abetted by those catering to the clicks of readers desperate to hear the worst about Trump.

Many viewers sensed what was going on. A Monmouth University poll released in April 2018 showed that public trust had dropped precipitously from just a year earlier, when the media's Russian obsession had erupted. Of 803 Americans who responded to a survey that March, 77 percent agreed that major traditional television and newspaper outlets report "fake news," compared to 63 percent the year before. That response wasn't limited to those on the right; 61 percent of Democrats believed that media outlets peddled misinformation.[43]

That belief was reinforced by a strange phenomenon that gave new meaning to the term "echo chamber." On certain networks, catchphrases were repeated by pundits, reporters, and guests, as if Hillary had handed out talking points at dawn. For example, the phrase "the walls are closing in" around President Trump was repeated fifty times during a ten-day period on CNN and MSNBC in December 2018.[44] The repetition of the mantras reinforced the idea of fake news being dished out by actors hired to repeat lines in a play.

RUSSIA HYSTERIA: THE MEDIA LOSES ITS MIND

Political essayists such as "Typhoid Mary" Kramer began painting Trump as beholden to Russia's president well before the election, not long after Simpson had put his plan into operation. He had company.

In July 2016, Franklin Foer penned a pulpy piece of partisan fancy entitled "Putin's Puppet: If the Russian President Could Design a Candidate to Undermine American Interests—and Advance His Own—He'd Look a Lot like Donald Trump," which opened thus: "Vladimir Putin has a plan for destroying the West—and that plan looks a lot like Donald Trump."[45] Weeks later, Jeffrey Goldberg, the editor in chief of *The Atlantic*, wrote, "Donald J. Trump has chosen this week to unmask himself as a de facto agent of Russian President Vladimir Putin."[46] Not to be outdone, David Remnick, the editor of *The New Yorker*, published a column in August 2016 called "Trump and Putin: A Love Story." Employing journalistic superpowers of precognition, Remnick declared that "Putin sees in Trump a grand opportunity. He sees in Trump weakness and ignorance, a confused mind. He has every hope of exploiting him."[47]

The scribblings of those three journalists likely left readers convinced that Trump was a "Manchurian Candidate." (The original movie, not the remake.) All Putin had to do was activate Trump by flashing a Queen of Hearts card. Ludicrous? Obviously. But that did not deter Abigail Tracy from publishing a hit piece in *Vanity Fair* just days before the election with the titillating title, "Is Donald Trump a Manchurian Candidate?"[48] As evidence, she relied on Senator Reid, who, she wrote, claimed to have seen " 'explosive information' linking Donald Trump and his top aides to the Russian government"—in other words, the Steele "dossier."[49] However, clueless citizens could be forgiven for assuming that those well-placed, intelligent thought leaders and politicians had seen some of the explosive proof.

Other news outlets, such as Huffington Post, ran similar articles with nearly identical titles.[50] The *New York Times* printed a column by Ross Douthat with the headline "The 'Manchurian' President?"[51] Above the headline was a photograph from the 1962 film in which a Communist "sleeper agent" is programmed to take over the US government. Douthat, a purported conservative voice, wrote that it is not "impossible to believe . . . that Trump's inner circle was actually colluding with Russian intelligence . . . or that Trump himself, for reasons financial or personal, was really a Russian asset of some sort."[52]

The irony, of course, is that the media paid little attention to Hillary Clinton's prodigious and lucrative connections with Russia. Incisive reporting that Clinton might have used her office for profit rarely appeared in the pages of the press or on television news.

No one had seriously accused President Barack Obama of "colluding" with Putin when an active microphone picked up his quiet words about missile defense as he leaned over to confide in then–Russian president Dmitri Medvedev, "This is my last election. After my election, I have more flexibility." Medvedev responded, "I will transmit this information to Vladimir."[53] The press concluded that Obama was not a Russian asset but a cunning negotiator who was scamming Kremlin apparatchiks.

Journalists tend to occupy a bubble of their own making. Their thoughts and words reverberate against the walls and bounce back to themselves. They hear only their own voices in a self-perpetuating circle, thrilled to be retweeted by one another, Hollywood celebrities, and "legal analysts" such as Benjamin Wittes, a senior fellow at the Brookings Institution, editor of the Lawfare blog, and close friend of Comey. According to the book *Shattered: Inside Hillary Clinton's Doomed Campaign*, Hillary's closest allies decided in the days immediately after the election to push two narratives to explain her devastating loss: Comey's handling of her email investigation and Russia hacking to assist Trump. A Clintonite confided to the authors that Hillary "wants to make sure all these narratives get spun the right way."[54]

Clinton campaign official Robby Mook, who had been getting briefings from lawyers at Perkins Coie, had spread the "dossier's" rumors to friendly reporters, even saying publicly right before the Democratic National Convention that Putin wanted to help Trump win.[55] A lawyer with Perkins Coie was also behind a lengthy story published by Franklin Foer just before the election on October 31, 2016, linking the Russian-based Alfa Bank to Trump: "Was a Trump Server Communicating with Russia?"[56] Steele had written a memo about Alfa Bank's connection to Trump in September, misspelling it as "Alpha."

Clinton tweeted the same day: "Computer scientists have apparently uncovered a covert server linking the Trump Organization to a Russian-based bank." Attached to her tweet was a statement from Jake Sullivan, senior policy adviser to Clinton: "This could be the most direct link yet between Donald Trump and Moscow. . . . This secret hotline may be the key to unlocking the mystery of Trump's ties to Russia."[57] The story was debunked; the alleged communications turned out to be spam.[58] Much later it emerged that the lawyer pushing the story for Clinton had likely been Michael Sussmann, an attorney for the Clinton campaign and the DNC.[59]

Even as the janitors were sweeping up the popped balloons at Hillary's watch party at the Jacob Javits Convention Center, her allies in the media took up the refrain. *That's how we got the run-up to the election so wrong! The Russians fooled everybody.* "We may be living through the most successful Russian intelligence operation since the Rosenbergs stole the A-bomb," wrote David Frum, the senior editor of *The Atlantic.*[60] His colleague senior editor Adam Serwer wrote, "Congratulations to Vladimir Putin, the Ku Klux Klan, and the Federal Bureau of Investigation."[61] Veteran media analyst Howard Kurtz of Fox News observed that by refusing any attempt to "normalize" the new president, reporters were saying "Trump is not a legitimate president and doesn't deserve to be treated as such."[62]

The condemnation of Trump for "colluding" with the Russians was just getting started. He was guilty, to be sure. The media just needed to prove it. And if they couldn't prove it, no matter; public denunciations and smears would be sufficient. Because Trump was a public figure, they were safe from libel lawsuits filed in US courts. Often during television interviews, anchors and reporters would cite some encounter or conversation between a Russian and a person in the Trump campaign and then ask a guest the leading question "Isn't this evidence of collusion?" as if some heinous crime had been committed by the mere act of a communication, association, or handshake. But no one would actually state what law or statute had been violated.

An example of this occurred on October 31, 2017, when Trump's lawyer Jay Sekulow was interviewed on ABC by George Stephanopoulos about a conversation a peripheral Trump adviser had had with a professor who had spoken to someone in Moscow. Even though the adviser had not talked with the Russian directly and the content of the discussion was twice removed, the anchor seemed to imply that it was somehow illegal.

"There is no crime of collusion," Sekulow said. "What is a violation of the law here?"

"Collusion is cooperation," Stephanopoulos said.[63]

Viewers were left with the impression that the simple act of talking with someone constituted "collusion," which equaled the commission of a crime. Yet there had been no secret agreement between the adviser and the professor to achieve a fraudulent or illegal or deceitful purpose. Was there a conspiracy? If so, a conspiracy to carry out what crime? It was left unstated, while viewers were fed the *appearance* of criminality.

Others took to the airwaves to either imply or outright pronounce that criminal "collusion" by Trump and/or his campaign was a foregone conclusion:

CARL BERNSTEIN (CNN): I think this is a potentially more dangerous situation than Watergate. We're at a dangerous moment. And that's because we are looking at the possibility that the president of the United States and those around him during an election campaign colluded with a hostile foreign power to undermine the basis of our democracy—free elections.[64]

DAN RATHER (MSNBC): Donald Trump is afraid. He's trying to exude power and strength. He's afraid of something that Mueller and the prosecutors are going to find out. A political hurricane is out there at sea for him. We'll call it hurricane Vladimir, if you will, the whole Russian thing. It is approaching Category Four.[65]

PAUL KRUGMAN (*NEW YORK TIMES*): There's really no question about Trump-Putin collusion, and Trump in fact continues to act like Putin's puppet. The only question is how high the indictments will reach, and how much damage they'll do. But it won't be good.[66]

JAKE TAPPER (CNN): This is evidence of willingness to commit collusion. That's what this is on its face.[67]

NICK AKERMAN, FORMER WATERGATE PROSECUTOR (MSNBC): There's outright treason. I mean, there is no question that what he's doing is giving aid and comfort to the enemy.[68]

LAWRENCE O'DONNELL (MSNBC): Donald Trump now sits at the threshold of impeachment.[69]

At the time of those legal and political proclamations by the media, Trump was only a few months into his presidency. The media obsessed over "collusion" without defining it. They either declared or implied that it was a crime that surely must exist in the vast body of law books. Just dust them off and you'll find it there, buried somewhere. And Trump had surely committed that crime, they reasoned. Be patient; the incriminating evidence would eventually emerge, proving the media's prescience.

At one point, MSNBC's Joy Reid was so giddy over the prospect of Trump's arrest, she even dreamed aloud on air about the day Trump would barricade himself inside the White House as federal marshals banged on the door to take him into custody.[70] In Reid's distorted universe, wishing was believing.

There was no credible evidence of any crime called "collusion." Yet those individuals were waxing recklessly about such matters as a Russian conspiracy, treason, and impeachment in an attempt to convict Trump without proof, charges, or the benefit of that quaint constitutional protection called "due process of the law."[71] Their rants were the equivalent of "Off with his head, trial to be had later."

The hysteria was fanned a few weeks before Trump's inauguration by the *Washington Post* when it claimed that the Russians had hacked the US power grid through a utility company in Vermont. It was untrue, but it planted in readers' minds the thought that those dastardly Russians were at the gate, with Trump ready to let them in unchallenged.[72]

In March 2017, Jennifer Palmieri, the director of communications for Hillary's 2016 presidential campaign, penned a column for the *Washington Post* calling for Democrats to "fight back" against the new administration. The Clinton campaign had tried to raise the alarm about Russia hacking stolen emails from the DNC to help Trump and hurt Hillary, she said. Alas, nobody had listened.

"The lessons we campaign officials learned in trying to turn the Russia story against Trump can help other Democrats (and all Americans) figure out how to treat this interference no longer as a matter of electoral politics but as the threat to the republic that it really is," she said. "If we make plain that what Russia has done is nothing less than an attack on our republic, the public will be with us. And the more we talk about it, the more they'll be with us."[73] The nation was at a tipping point, more serious than Watergate, a constitutional crisis, she declared. "We all have a role to play in stopping it." Meaning you Democrats in the media, bang that drum!

The smug Rachel Maddow took up the challenge on her prime-time show on MSNBC. Citing a *New York Times* story about Donald Trump, Jr.'s, emails, she proclaimed, "They're not even six months into this administration and they're confessing to colluding with the Russians during the campaign."[74]

On the same network, Mika Brzezinski predicted the imminent incarceration of various members of the Trump family who had dared to meet or talk with a Russian: "I think they're shocked that the noose is tightening and that people might go to jail for the rest of their lives."[75]

Chris Hayes managed to surpass his MSNBC colleagues when he was asked by Stephen Colbert on CBS, "Are you all in on collusion?" In response, Hayes declared unqualified guilt: "The simplest explanation is that everyone's running around acting guilty because they're guilty. They're acting super guilty because they're guilty."[76] It was a guaranteed applause line, but guilty of what specifically? That vexing information was conveniently left out, although the fuliginous topic of "collusion" had been broached.

It should be noted that Maddow, Brzezinski, and Hayes do not have law degrees. Their collective experience in courtrooms appears to be meager, if not nonexistent. That did not dissuade them from offering confident judicial pronouncements. Yes, they are hosts of opinion-driven programming. But shouldn't their opinions on matters of criminality be informed by the content of criminal statutes or case law?

Therein lies a serious problem. Journalists or hosts who are not schooled in the law felt no reticence in drawing conclusions on air about criminal conduct without having the requisite education, proficiency, or experience in the law. When they rendered legal opinions with scant foundational knowledge, they disserved the viewing public while debasing themselves. Did any of them research the criminal codes? You don't have to be a lawyer to comprehend them. They are sometimes dense, but they are mostly understandable even for laypeople. Thus a reference by journalists to a statute might have lent some credibility to their arguments. If their searches proved fruitless, they might have reformed their points of view. But personal biases blinded their minds to their fundamental ignorance about legal matters.

Nothing illustrated bias more than the way the press covered Comey. When he went against Clinton in the email investigation, he was the devil who cost her the election. When he testified before Congress in June 2017, after Trump had fired him, Evil Comey became Saint Comey trying to stand up to Mean Mr. Trump.

"Comey: Trump Lied," blared the *Dallas Morning News* in a gigantic font on June 9, 2017. Buried in the story was Comey's complaint that Trump had lied in a tweet, saying that the FBI was in chaos and people had lost confidence in his leadership. Trump's assertion was true, but the misleading headline would be what people took away from the story.

Ultimately, not only were the media hysterical, but they were willing to commit malpractice to support their anti-Trump narrative.

MEDIA MALPRACTICE IN THE AGE OF TRUMP

In their zeal to bring down Trump, reporters revealed not only their bias but their sloppiness and sheer incompetence. Phone records and intercepted calls showed that Trump had had "repeated contacts with Russian intelligence," the *New York Times* reported in February 2017, citing "four current and former American officials."[77] Comey later testified that the story wasn't true: "The challenge, and I'm not picking on reporters, about writing on classified information is: The people talking about it often don't really know what's going on, and those of us who actually know what's going on aren't talking about it. . . . And we don't call the press to say, 'Hey, you got that thing wrong about this sensitive topic.' We just have to leave it there."[78] The paper responded with no correction, no explanation, no asking of the question: Why did our sources tell us something that was false?

Bloomberg News published a piece on July 20, 2017, entitled "Mueller Expands Probe to Trump Business Transactions." The president's attorney John Dowd had checked with his contacts within Mueller's independent counsel team, as he often did. They had confirmed that the story was false. "But I couldn't get the press to accept it," Dowd said. "They were off on their own toot."[79]

In December 2017, the chief investigative correspondent for ABC News, Brian Ross, reported that General Michael Flynn was prepared to testify that Trump had ordered him to make contact with the Russians *during the campaign*, contradicting all the president had said to date.[80] You could almost hear the sound of champagne corks popping across America's newsrooms. That had to be the long-awaited "smoking gun" evidence of "collusion" with Moscow. The alarmed stock market took a precipitous plunge on the news. Impeachment seemed certain. Forget the fact that talking to Russians during a campaign is not, by itself, against the law.

Alas, Ross's report was wrong. Trump had directed Flynn to speak with the Russians *during the transition*, as was common practice for most incoming administrations. The red-faced ABC News first "clarified" the story. When that was greeted with guffaws, the network officially issued a "correction." Ross was suspended for four weeks without pay and banned from covering the president. But the damage was done. The network's tweet about the story had been shared more than 25,000 times before it was deleted.[81] It's a safe bet that few of the people who had originally circulated the story tweeted the correction.

Less than a week later, CNN's Manu Raju breathlessly hyped— for twelve straight minutes—his "exclusive" report that Trump, his son Don Jr., and others in the campaign had received an email giving them special access to a decryption key for WikiLeaks documents concerning stolen DNC emails.[82] The email, from "Michael J. Erickson," whom CNN did not identify, was dated September 4, more than a week *before* WikiLeaks had uploaded the trove of stolen documents on September 13, 2016.[83] Chasing its competitor's scoop, CBS News claimed it had confirmed CNN's story through an independent source.[84] MSNBC joined the chorus. Trump critics, including many in the media, crowed that this was credible proof of "collusion" with the Russians, who were suspected to have been behind the theft.

Hyped by CNN for hours, Raju's stunning report was sucked up into the maelstrom of social media, traveled far and wide via Twitter, then crumpled with a whimper after being debunked by the *Washington Post*. From an ordinary Trump fan, the email was actually dated September 14, a day *after* WikiLeaks had published the stolen documents. Erickson, who identified himself as president of an aviation management company, had simply told the Trump family to check out the public posting. No secret early access. No conspiracy.

CNN insisted that "multiple sources" had provided the network with the false date, but Raju admitted that he'd never seen the email. CNN offered this laughably weak correction: "The new information indicates that the communication is less significant than CNN initially reported." CNN and CBS corrected their stories only after the *Post* published its slap down.[85] Most of the people who had tweeted the scoop moved on to something else without correcting the falsehood, leaving a trail of slime.

That would have been the perfect moment for newsrooms everywhere to ask if two anonymous leakers is enough to run a story, now
that government officials across the branches had shown themselves
willing to lie maliciously.

As Glenn Greenwald of The Intercept pointed out, "No Russian
Facebook ad or Twitter bot could possibly have anywhere near the
impact as this CNN story had when it comes to deceiving people with
blatantly inaccurate information."[86] Calling for transparency, he speculated that the "multiple sources" peddling the story to three different
networks at the same time were Democrat members of the House
Intelligence Committee, which had obtained Donald Trump, Jr.'s,
emails. CNN did not fire or discipline Raju, who remained mum on
how he had been fooled.

The push for speed and the thrill of being the one to nail Trump's
hide to the barn door compelled many media personnel to violate
long-standing journalistic norms. Raju's embarrassment came on the
heels of earlier reporting by CNN of Russian "collusion" involving
an anonymous source who claimed that Trump Communications director Anthony Scaramucci was involved with a $10 billion Russian
investment firm. The story was later retracted.[87] The mistake about
"The Mooch" resulted in three of the network's journalists being
shown the door for failure to follow editorial procedures. CNN issued
a rare apology.

Numerous errors involved reporting on the president's attorney
Michael Cohen. Here are just two.

CNN reported in July 2018 that, according to two sources, Cohen
would confirm that the president had had advance knowledge of the
Trump Tower meeting between his son and Russian intelligence operatives. Collusion! Cohen's lawyer Lanny Davis—the real originator
of the lie—finally confessed, suggesting that the network had gotten
"mixed up," but CNN continued to pump it, rationalizing that Davis
was trying to protect his client.[88]

BuzzFeed claimed in early 2019 that Special Counsel Mueller had
evidence that Trump had ordered Cohen, who had recently been sen-

tenced to prison, to lie to Congress about when negotiations ended for a potential Trump Tower construction project in Moscow, an allegation given credibility when both Cohen and Davis refused to comment. The thinly sourced story by Jason Leopold and Anthony Cormier, which had been "months in the making," relied on "two federal law enforcement officials." [89] Cormier admitted that he hadn't seen the evidence but said that "two officials we have spoken to are fully, 100 percent read into that aspect of the Special Counsel's investigation." He claimed that the sources had been working on the "Trump Moscow tower portion of the investigation" before Mueller's appointment.

"So they had access to a number of different documents, 302 reports which are interview reports," Cormier said. "That stuff was compiled as they began to look at who the players were speaking with, how those negotiations went, who all from the Trump organization and outside the organization were involved in getting that tower set up." [90] Wow. FBI 302 reports, extensive collection of documents. Despite the sketchy background of his fellow reporter, Jason Leopold, who had botched several other blockbuster stories, Cormier defended their sourcing as "rock solid."

"This is stunning," proclaimed CNN's Don Lemon. For twenty-four hours, frenzy erupted. Trump had suborned perjury, a federal crime. ABC, CBS, and NBC spent more than twenty-seven minutes on three evening shows on the BuzzFeed report but provided meager critical analysis. [91] Legislators called on Trump to resign. Former attorney general Eric Holder tweeted that "Congress must begin impeachment proceedings." Harvard law professor Laurence Tribe predicted prison for Trump. All of this was predicated with the meaningless phrase "If it's true." Denials by the president's lawyer were largely ignored or contemptuously dismissed. [92]

The BuzzFeed-triggered hysteria prompted Mueller's team, usually silent, to do something extraordinary. Spokesman Peter Carr issued a public statement: "BuzzFeed's description of specific statements to the Special Counsel's Office, and characterization of documents and testimony obtained by this office, regarding Michael Cohen's Congressional testimony are not accurate." Why Mueller waited twenty-four hours before deflating the swollen egos of TV anchors, hosts, and reporters remains a mystery. All the while, Trump was being battered like a human piñata.

"The larger message that a lot of people are going to take from this story is that the news media are a bunch of leftist liars who are dying to get the president and they are willing to lie to do it," said CNN legal analyst Jeffrey Toobin.[93] His comment was salient and correct; then he blew it by adding "And I don't think that's true." Sadly, he failed to clarify that government operators might be a bunch of leftist liars and the news media continued to be willing to print their lies if it led to traffic and praise.

CNN anchor Chris Cuomo whined about the criticism of Buzz-Feed and by extension his network. "Reporting is hard," he said on *The View*. "The idea that anonymous sourcing is somehow weaker sourcing is BS, OK?"[94] Except that the story had just been destroyed by the office of the special counsel. BuzzFeed had allowed lying sources to hide behind anonymity. It was media malpractice on steroids. BuzzFeed refused to concede. "We are confident that our reporting will stand up," said editor Ben Smith. We're still waiting.

The sourcing problem was pervasive. In May 2017, a *Washington Post* headline screamed, "Trump Revealed Highly Classified Information to Russian Foreign Minister and Ambassador."[95] Specifically, he had divulged the name of a city in ISIS-held territory where an aviation terror threat had been detected. Media madness ensued. Except, as National Security Advisor H. R. McMaster repeatedly said, "it was a totally appropriate conversation" and a part of routine information sharing in cases of common interest, such as fighting terrorism.

The story was built on anonymous sources. "It's all kind of shocking," said a "former senior official who is close to administration officials." But who are the "officials" the media are fond of quoting? How "senior" are they? In which branch of government are they? How politically biased might they be? Do they really know what they're talking about? The *WaPo* story was editorializing disguised as straight news reporting.

Behind the scenes, Glenn Simpson and Christopher Steele were still working their agenda, goosing the Trump "collusion" story when it lagged. Court documents obtained by the *Washington Examiner* in November 2017 showed that Fusion GPS had made payments to three journalists from June 2016 through February 2017.[96] Though their names were withheld, all were known to have reported on the Russia allegations. The attorney for Fusion GPS argued that the reporters had been paid for research, not to publish stories, but that explanation was unconvincing given the tactics employed by Simpson and his dirty tricks team.

Every week brought a new bombshell, only to be debunked:

- Relying on an anonymous source, CNN reported that Comey, during congressional testimony in June 2017, would refute Trump's claim that the former director of the FBI had told the president multiple times that he was not under investigation. CNN was forced to retract its story after Comey's testimony, which confirmed exactly that.[97]

- According to Bloomberg and the *Wall Street Journal*, in late 2017 Mueller's team had served Deutsche Bank with a subpoena for President Trump's financial records.[98] Bloomberg claimed that the independent counsel had "zeroed in" on Trump. Wrong.

- CNN reported that Attorney General Jeff Sessions had failed to disclose meetings with Russian officials when he had applied for his security clearance, portraying him as dishonest. It walked back the story after it emerged that an FBI employee had told Sessions that he didn't need to list dozens of meetings with foreign ambassadors that had occurred while he was a senator.[99]

- In April 2018, McClatchy relied on two anonymous sources to resurrect Steele's claim that attorney Michael Cohen had made a secret trip to Prague to collude with the Russians during the run-up to the 2016 presidential election and that the independent counsel had proof of that.[100] The story was false and remained false when McClatchy hauled it out again in December 2018, now with four anonymous sources claiming that Cohen's phone had "pinged" cell towers in Prague.[101]

- CNN sent reporters to Thailand to interview the incarcerated Anastasia Vashukevich, a model and "sex coach" who claimed she had "top secret" information about Trump and Russian billionaire Oleg Deripaska that she would divulge if someone would rescue her from the Thai pokey.[102] Spicy, yes, but completely bogus.

- Paul Manafort, according to *The Guardian*, sneaked into the Ecuadorian Embassy in London three times to meet with Julian Assange of WikiLeaks. Spectacular, if true, but Manafort denied it and Mueller did not charge him with a crime related to consorting with WikiLeaks or colluding with Russia.[103]

Using anonymous sources, relying on material not seen but "shared," reporters chased around the world looking for the elusive proof of Trump-Russia "collusion." "Gossip treated as gospel," in the words of Joe Concha, a contributor to Fox News. "Sources providing information to reporters all too willing to accept it, like seagulls at the beach."[104]

Along the way, the urge to post snarky comments on their personal Twitter accounts sucked reporters into making errors that traveled the globe in an instant. In December 2017, to mock Trump's tweet that a Florida rally "was packed to the rafters," *WaPo* reporter Dave Weigel posted a picture of a half-empty arena—a picture taken well before the event had started. Weigel deleted it and apologized after Trump called him out.[105] MSNBC's Joy Reid tweeted that an Amtrak derailment that had killed three people was the result of the GOP putting tax cuts for the wealthy ahead of funding for infrastructure. But the accident had actually happened on a new track built for high-speed rail.[106] Journalists spread disinformation in the blink of an eye. When challenged, they delete their tweets, often without correcting or apologizing for their errors.

If journalists couldn't convict Trump of "collusion," they sought to have him sentenced for the crime of obstruction of justice. CNN's chief legal analyst, Jeffrey Toobin, called the firing of Comey "a grotesque abuse of power by the president of the United States. This is the kind of thing that goes on in non-democracies."[107] Of course, Comey himself admitted that the president had been constitutionally empowered to fire him.

But Toobin didn't stop there. Days later, it was reported that Trump had allegedly told Comey that fired national security advisor Michael Flynn "is a good guy" and "I hope you can see your way clear to letting this go." Toobin brayed, "Three words: obstruction of justice."[108]

Granted, Toobin is entitled to his legal opinion. But don't you dare disagree with him, as former Harvard Law School professor Alan Dershowitz did on air regarding Comey's firing, which he called "not illegal today under the law."[109] In one joint appearance, Toobin, who was once a student of Dershowitz, scolded his former professor: "Alan, I don't know what's going on with you. How has this come about that, in every situation over the past year, you have been carrying water for Donald Trump?"[110]

Toobin seemed clearly exasperated:

TOOBIN: This is not who you used to be, and you are doing this over and over again in situations that are just obviously rife with conflict of interest. And it's just, like, what's happened to you?

DERSHOWITZ: I'm not carrying water. I'm saying the exact same thing I've said for 50 years. And, Jeffrey, you ought to know that, you were my student. The fact that it applies to

Trump now rather than applying to Bill Clinton is why people like you have turned against me.[111]

Dershowitz, a strong Hillary supporter, has insisted that his fidelity is to the rule of law and the Constitution, not politics. In his denunciation, Toobin oozed contempt. Dershowitz had committed ideological treason. How could a liberal professor not abide by the liberal doctrine?

For his integrity, CNN banned Dershowitz.[112] "It would confuse CNN's viewers at a time when one had to be either for or against Trump," Dershowitz said. "Today, everyone has to pick a team—Trump or anti-Trump—and picking the side of the Constitution and civil liberties just doesn't do it." Instead of a constitutional scholar, CNN viewers were subjected to more than a hundred appearances by the ethics-challenged lawyer Michael Avenatti, who crowed, "I guarantee Trump will not serve out his term." [113]

Dershowitz learned that the directive to keep him off CNN had come straight from the top. "The brass didn't want their viewers' minds to be confused by the law or the facts," he said. "Trump was guilty; that's all they needed to know." [114] Why? CNN president Jeff Zucker had looked at his failing network's bottom line and realized that the Trump-Russia fantasy was ratings gold.

The highly respected liberal journalist Glenn Greenwald, who had won a Pulitzer Prize in 2014 for his coverage of Edward Snowden and the NSA, was also banned by CNN. Instead, Zucker put on relics from the Watergate years such as Bernstein and former Nixon official and convicted felon John Dean to diagnose the president as mentally ill. Those has-beens had to keep the lie going if they wanted to be on the air.

But behind the scenes, the network knew the truth. CNN star Van Jones was caught on an undercover camera in 2017 by Project Veritas in its *American Pravda* series admitting that "the Russia thing is just a big nothingburger." Jones later claimed that quote was taken out of context.[115] But an undercover reporter also filmed CNN producer John Bonifield saying that the network's Russia coverage was "mostly bullshit" but good for ratings and was being pushed on Zucker's order.

What else could explain the network's fascination with Avenatti, the lawyer for porn star Stormy Daniels and Julie Swetnick, the fantasist who accused Supreme Court nominee Brett Kavanaugh of participating in gang rape? Over a one-year period, Avenatti appeared on CNN 121 times, making the man Tucker Carlson had dubbed the "Creepy Porn Lawyer" a household name.[116]

Avenatti, celebrated by CNN as a possible 2020 candidate for president, photographed partying with CNN's Don Lemon and April Ryan, saw his star crash when in March 2019 he was indicted by federal authorities on multiple counts, including attempting to extort $20 million from Nike, stealing from clients, committing bank and wire fraud, cheating on his taxes, and other crimes.[117] The bad news multiplied when Avenatti was indicted two months later for stealing from his cohort in media hijinks, Stormy Daniels, who had propelled him to international fame. Accused of forging Daniels's name to two payments for her book *Full Disclosure*, he pleaded not guilty to the charges.[118] There was no mea culpa and accompanying apology from CNN to its viewers.

All along, the so-called legacy media were feeding the beast, from news reporters to opinion writers to editorial boards. The *New York Times* deliberately abandoned fairness and objectivity in its zeal to push the Russia story. Anonymous sources ran rampant, opinion bled into news stories, and always in the anti-Trump direction. That compelled others to follow. As Michael Goodwin of the *New York Post* put it, the *Times* "is the bell cow, and it led the media over the cliff by getting the big story wrong."

BASHING TRUMP IS GOOD FOR BUSINESS

On March 7, 2019, the legendary TV newsman Ted Koppel stunned his colleagues during a televised discussion with colleague Marvin Kalb decrying the erosion of journalistic standards at the *Times* and *WaPo*. "[We're] not talking about the *New York Times* of 50 years ago," the former ABC star said. "We're not talking about the *Washington Post* of 50 years ago. We're talking about organizations that I believe have, in fact, decided as organizations that Donald J. Trump is bad for the United States." [119]

A few months earlier, Koppel had pointed out in a public forum that CNN's ratings would have been "in the toilet" without Trump. Now he argued that Trump was "not mistaken" that the media was out to get him. "We are not the reservoir of objectivity that I think we were."

Even Jill Abramson agreed. In *Merchants of Truth: The Business of News and the Fight for Facts*, the former executive editor of the *New York Times* criticized her successor, Dean Baquet. "Though Baquet said publicly he didn't want the *Times* to be the opposition party, his news pages were unmistakably anti-Trump," she wrote. "Some headlines contained raw opinion, as did some of the stories that were labeled as news analysis." [120]

By abandoning the foundation of news reporting hewn to by the legendary publisher Adolph Ochs, "the more anti-Trump the *Times* was perceived to be, the more it was mistrusted for being biased. Ochs's vow to cover the news without fear or favor sounded like an impossible promise in such a polarized environment, where the very definition of 'fact' and 'truth' was under constant assault." [121]

A generational rupture at the *Times*, she said, manifested by the split in print and digital reporters, had created an environment in which younger staffers favored attacking the president. "The more 'woke' staff thought that urgent times called for urgent measures," she wrote. "The dangers of Trump's presidency obviated the old standards." [122]

And bashing the president was great for the bottom line; subscriptions to the *Times* jumped by 600,000, to more than 2 million, in the first six months of Trump's presidency.[123] "Given its mostly liberal audience, there was an implicit financial reward for the *Times* in running lots of Trump stories, almost all of them negative," Abramson wrote. "They drove big traffic numbers and, despite the blip of cancellations after the election, inflated subscription orders to levels no one anticipated."[124]

Even as stories blew up like clowns' exploding cigars, the media clung to the belief that Special Counsel Robert Mueller would vindicate them by squeezing Trump's colleagues, such as Paul Manafort and Michael Cohen, to get to the truth about Trump-Russia "collusion," setting up Trump's removal from office. "An impeachment process against President Trump now seems inescapable," wrote Elizabeth Drew, a journalist who had covered Watergate, in an op-ed for the *Times*.[125] *The Atlantic* featured a cover story in March 2019: "Impeach: It's Time for Congress to Judge the President's Fitness to Serve."[126]

THE ANTI-TRUMP NARRATIVE IS FUNDED BY MILLIONS OF DOLLARS FROM "NONPARTISAN" AND LEFTIST NONPROFITS

Many of the stories were being pushed by the Democracy Integrity Project (TDIP), a Washington-based nonprofit founded after the 2016 election by Democratic operative Daniel J. Jones, which hired Fusion GPS and Steele to continue the stealth campaign, according to an investigation by journalist Paul Sperry.[127] TDIP is linked to several offshoots tasked to manipulate social media. The supposedly "nonpartisan" TDIP had raised $50 million from George Soros, the Hollywood activist Rob Reiner, a nonprofit linked to billionaire Tom Steyer, and other sources primarily in New York and California to go after Trump, according to a declassified congressional report.[128]

Jones was a former FBI analyst and top staffer for Senator Dianne Feinstein, who had unilaterally released a transcript of a closed interview of Glenn Simpson without disclosing her aide's connection to Fusion GPS.[129] For TDIP, Jones set up an email distribution list for a newsletter of anti-Trump gossip, rumor, and innuendo blasted five days a week to reporters and Democratic aides on Capitol Hill. "What's significant about them is they're totally one-sided," a veteran reporter with a major newspaper told Sperry. Working the national security beat, he insisted on anonymity. "It's really just another way of adding fuel to the fire of the whole Russian collusion thing."

According to Sperry, leaked texts from March 2017 revealed that Jones had boasted to a lawyer working with Senator Mark Warner (D-VA) that he had planted several anti-Trump articles with Reuters and McClatchy. "Jones has been chumming out his own share of garbage stories," a Republican legislative assistant told Sperry.[130]

Similar pieces were promoted by Media Matters, the liberal activist group founded by David Brock that former *60 Minutes* correspondent Lara Logan called "the most powerful propaganda organization" in the United States during an appearance on *Fox & Friends* on April 2, 2019. "And they describe themselves as a propaganda organization, that's what they do," she said. "They're pushing their agenda. And one of the main pillars of their strategy is to make sure that Donald Trump is seen as the most unpopular president in history."[131]

Media Matters targeted conservative media as well. In 2011, Brock had announced a campaign of "guerilla warfare and sabotage" aimed at the Fox News Channel.[132] With a generously funded budget, Brock's staff of ninety people "smear, manipulate and invent false narratives driven by their well-funded political agenda," Logan wrote in an op-ed for the *New York Post*. "With armies of bots and a stable of journalists that parrot their talking points, they silence and intimidate. They use our criticism of unfairness and bias to falsely accuse us of being conservative."[133]

The phony narratives being promoted about Trump-Russia "collusion" prompted Logan to give a podcast interview that she called "professional suicide for me." [134] She slammed reporters who rely on a single, anonymous government source. "That's not journalism, that's horseshit," she said.

ELITE COMMENTATORS MAKE LUDICROUS CLAIMS

The articles by supposedly serious journalists reached ludicrous levels. Jonathan Chait of *New York* magazine wrote a cover story in July 2018 about the upcoming summit between Trump and Putin with a teasing subtitle "What If Trump Has Been a Russian Asset Since 1987?" [135] Chait suggested that "it would be dangerous not to consider the possibility that the summit is less a negotiation between two heads of state than a meeting between a Russian-intelligence asset and his handler." No evidence, just wild conjecture.

During his appearances on MSNBC, former CIA director turned pundit John Brennan used his former position to suggest that he'd seen collusive material that ordinary journalists had not. After Trump ordered that his security clearance be revoked, he penned an op-ed for the *New York Times* saying that "Trump's claims of no collusion are, in a word, hogwash." [136] After the summit, Brennan pronounced his verdict in a tweet: "Donald Trump's press conference performance in Helsinki rises to & exceeds the threshold of 'high crimes & misde-meanors.' It was nothing short of 'treasonous.' Not only were Trump's comments 'imbecilic,' he is wholly in the pocket of Putin." [137]

Presidential historian Douglas Brinkley followed Brennan's lead on CNN. "The spirit of what Trump did is clearly treasonous. It's a betrayal of the United States. He threw our U.S. intelligence services, flushed them away and it came off as being a puppet of Putin." [138] A historian might be more circumspect about the times the intelligence service has been wrong or simply lied to the people and the president. But not this historian.

Treason, of course, was legally preposterous, under both Arti-cle III of the Constitution and statutory law, 18 U.S.C. § 2381. [139] But the word became a favorite cudgel for Trump critics to bludgeon the president for policy statements and conduct that do not fall under the law of treason.

The media loved booking the most flagrant huckster of hysteria in Congress, Representative Adam Schiff, who, as the ranking Democrat and later chairman of the House Intelligence Committee, received classified intelligence briefings. "I can't go into the particulars, but there is more than circumstantial evidence now," he said on *Meet the Press* in March 2017.[140] He repeatedly trumpeted that he had seen incontrovertible "evidence" of collusion but repeatedly failed to produce any such evidence. In May 2018, Schiff told ABC that the nefarious connection between Trump and Russia was of "a size and scope probably beyond Watergate."[141] If pressed for further details by any reporter, he hid behind the supposedly "classified" nature of the evidence he claimed to have seen.

Members of the Senate Intelligence Committee were just as bad if not as ubiquitous. "Enormous amounts of evidence" exist of collusion, said Senator Mark Warner. "There's no one that could factually say there's not plenty of evidence of collaboration or communications between Trump Organization and Russians."[142] Leaks from the House and Senate intelligence committees fueled numerous phony stories. Saying that Donald Trump, Jr., "went into the Senate Intelligence Committee, took an oath to tell the truth, and lied his butt off," former federal prosecutor Paul Butler predicted that Mueller would indict the son on perjury charges to create leverage against his father.[143] Except that there was no evidence that Don Jr. had lied, and he was not indicted.

The Senate hearings to confirm William Barr as attorney general brought another opportunity for media madness. Jonathan Turley, a commentator for Fox News, testified to the Senate that his longtime friend Barr was extraordinarily qualified to become attorney general again; he'd served in the same post under President George H. W. Bush. But the media attacked Barr over a well-reasoned twenty-page memo he'd written for the DOJ regarding executive power and Mueller's investigation of Trump for obstruction of justice over his firing of Comey.[144] Because it was favorable to Trump, they called his opinion "bizarre" and "strange," alleging that it was "based entirely on made-up facts."

"These criticisms were wrong," Turley wrote. "Many of these same critics have explored the same possible use of the law against the obstruction of justice before and after Barr's memo came out." But now that Barr was likely to be confirmed as attorney general and might not be a rabid "Never Trumper," he had to be attacked.[145]

A good example of the Twitter echo chamber came in August 2018 after the man who had investigated both Clinton and Trump was discharged. Trump tweeted: "Fired FBI Agent Peter Strzok is a fraud, as is the rigged investigation he started. There was no Collusion or Obstruction with Russia, and everybody, including the Democrats, know it. The only Collusion and Obstruction was by Crooked Hillary, the Democrats and the DNC." [146]

Chuck Todd, NBC's *Meet the Press* moderator, condemned Trump with borderline hysteria, tweeting "I wonder if other civil servants who believe they have seen wrongdoing are watching how POTUS and his echo chamber can character assassinate so viciously and get second thoughts about doing their job given the risk POTUS is showcasing to anyone who crosses him?" [147] Perhaps Todd had been asleep for a year and was unfamiliar with the loathing and scheming by Strzok and his lover, Lisa Page, explicitly revealed in their texts, to take Trump down. It seems as though only those conspiring to lie to the media and their colleagues have something to worry about.

Strzok had been justifiably fired by FBI deputy director David Bowdich, the most senior career official in the agency. Legal analyst and former FBI official Chuck Rosenberg told MSNBC's Andrea Mitchell that "as much as I like Peter Strzok, he exhibited remarkably bad judgment. . . . It's not a crazy decision. In fact, I understand why Bowdich did it." [148]

But Todd saw only what he wanted to see. He should have taken the time to read Strzok's texts, which had been included in the report by Inspector General Michael Horowitz, released a few months earlier. On August 8, 2016, Lisa Page texted Strzok, "[Trump's] not ever going to be president, right, right?" Strzok, the upstanding civil servant, texted back, "No, no he won't. We'll stop it."[149]

The media's endless supply of anti-Trump stories—often accompanied by the qualifier "if true!"—reached a fever pitch by early 2019. From January 20, 2017, through March 21, 2019, the evening news shows of ABC, NBC, and CBS produced a combined 2,284 minutes of "collusion" coverage, roughly three minutes a night for 791 days.[150] And the coverage was consistently over 90 percent negative.

A few people tried to slam on the brakes. In February 2019, NBC News reporter Ken Dilanian reported that the Senate Intelligence Committee had, after two years and interviews with more than two hundred witnesses, found "no direct evidence" of collusion between Trump and Russia.[151] Though guests on an MSNBC panel expressed skepticism, Trump could claim vindication, Dilanian warned, "and he'll be partially right."

Terry Moran, ABC TV's senior national correspondent, predicted a "reckoning" for those who had staked their credibility on Mueller validating their reporting. "That's a reckoning for progressives and Democrats who hoped that Mueller would essentially erase the 2016 election, it's a reckoning for the media, it's a reckoning around the country if, in fact, after all this time, there was no collusion."[152]

As it became clear that Mueller was closing up shop, anonymous Democratic lawmakers told Politico their prediction that the special counsel's report "will be a dud."[153] But the mania among the media who had invested so much in the narrative for more than two years didn't diminish.

Bombshell after bombshell, night after night: the "noose is tightening around Trump's neck . . . the walls are closing in . . . the beginning of the end . . . the tipping point . . . Trump will resign . . . like the last days of Nixon . . . his presidency is crippled . . . Trump's going down . . . I do not think the president will serve out his term . . . it's over . . . another bombshell!"

THE MUELLER REPORT SHATTERS
THE MEDIA NARRATIVE

No one bit on the Russia "collusion" bait with more gusto than Rachel Maddow, MSNBC's star anchor, paid $7 million a year. She pounded the story relentlessly, using wild graphics to connect the dots of bizarre "collusion" theories, like Mel Gibson's character in the movie *Conspiracy Theory*.

Her ratings surged to 4 million viewers who tuned in to hear their Trump hatred reinforced. With her help, MSNBC reaped a 62 percent bump in viewership during 2017.[154] By February 2019, Maddow's show had been number one in her time slot for three straight months in the coveted demographic of adults ages twenty-five to fifty-four. During the weeks before the anticipated release of the Mueller Report, Maddow gleefully predicted that Trump would finally be exposed by the intrepid special counsel and summarily frog-marched out of the White House.

On Friday, March 22, when the Mueller Report dropped, Maddow was trout fishing in Tennessee and raced to a local newsroom for the big reveal.[155] As she reported the news that her hero Mueller had found no collusion and would file no more indictments, her eyes seemed to tear up and her mouth twisted into a grimace.

"This is the start of something, not the end of something," she insisted, gasping like a fish hooked and plopped onto a pier. However, there was no escaping the fact that the narrative she'd hyped for over two years had been an elaborate hoax. She had relished every twist and turn, staking her credibility and career on the outcome. And now it was in tatters. Her show shed a half-million viewers the following week. Though she plays one on TV, Maddow is no journalist; she's a pundit, an entertainer weaving elaborate tales that defy facts and logic.

But somehow the crack Pulitzer Prize–winning uber-objective reporters at the *New York Times* and *Washington Post* had missed the story of the biggest political hoax in the history of the United States.

They had swallowed Hillary's Big Lie, with the help of Fusion GPS, the FBI, and former intelligence officials. Michael Goodwin of the *New York Post* wrote a column titled "The New York Times Owes Americans a Big, Fat Apology." [156]

The announcement by Attorney General Barr that Mueller had found no "collusion" came as a "thunderclap to the mainstream news outlets and the cadre of mostly liberal-leaning commentators who have spent months emphasizing the possible collusion narrative," wrote *WaPo* media critic Paul Farhi.[157] The *New York Times*' "conservative" columnist David Brooks, who had hyped the hoax with the rest, had the grace to suggest Trump was owed an apology in a column titled "We've All Just Made Fools of Ourselves—Again." [158]

"It's clear that many Democrats made grievous accusations against the president that are not supported by the evidence. . . . If you call someone a traitor and it turns out you lacked the evidence for that charge, then the only decent thing to do is apologize." But Brooks meant people like Brennan and Schiff, not himself or his newspaper.

Charles Hurt of the *Washington Times* singled out Michael Isikoff and David Corn for well-deserved scorn. Early promoters of the hoax, they had coauthored a book entitled *Russian Roulette: The Inside Story of Putin's War on America and the Election of Donald Trump*, which had regurgitated the Steele "dossier." [159] "No two people have done more than these two buffoons to peddle fantasy claims, unfounded conspiracy theories and outright lies regarding President Trump's supposed collusion with the Kremlin," Hurt wrote.[160] Their book title had been "entirely fitting considering who wound up getting their heads blown off in the whole misadventure."

The once-respected Isikoff didn't apologize, but he found a spine at last. On an MSNBC panel with Corn and Chris Hayes, he acknowledged that Mueller's findings had obliterated the reliability of the "dossier," which he had so avidly promoted for more than two years. "You know, it was endorsed on multiple, multiple times on this network, people saying it's more and more proving to be true, and it wasn't," he said.[161] But Corn was still reluctant to give up the Holy

Grail of Russian "collusion." He probably never will. A few days later, he published a story with this headline: "Trump Aided and Abetted Russia's Attack. That was Treachery. Full Stop." [162]

LEFTIST JOURNALISTS REFUSE TO ADMIT THEY WERE WRONG

What can a cult leader do when his predictions about the world ending fail to come true? Pick a new date and prophecy. Television producers found people willing to come on and engage in the most brazen attempts at goalpost moving ever seen on the little screen.

John Dean of Watergate infamy claimed that the Mueller Report hadn't resolved whether or not the president was a Russian agent. [163] Jonathan Chait was at it again, contending that Trump could still be guilty. "People who want to demonstrate their innocence make displays of cooperation with investigators," he wrote. "His flamboyant refusal to cooperate deprives Trump of any claim to have been cleared." [164] Except that Trump and his lawyers had provided unprecedented access to mountains of documents; staffers and lawyers had submitted to hours of interviews with Mueller's team. Dowd, President Trump's attorney, had given a remarkable interview to Byron York outlining his interactions with the special counsel.

Despite Trump blasting Mueller on Twitter and decrying the investigation as a "witch hunt," behind the scenes, Trump's people were cooperating. "Bob [Mueller] was a big boy about the political side," Dowd told York. "He understood the president had to address the politics of it. He couldn't just say nothing. People were pounding him about this thing every day, both privately and publicly." [165] The special counsel's office even used "code" to let the president's lawyer know when a press report about something allegedly emanating from Mueller's team was false. [166]

With the help of useful tools like Corn, Dean, and Chait, the press had played an indispensable role in promulgating the largest disinformation operation in US history. The media "rolled right into an absolutely deranged conspiracy territory that held the country hostage for two years," said Mollie Hemingway, a senior editor at The Federalist. "It is shocking how many people believed this crazy theory about Russia collusion, but many people lacked the courage to speak against it in the face of hysteria." [167]

Among them, the *New York Times*, the *Washington Post*, CNN, and MSNBC websites had published 8,507 articles mentioning the Mueller investigation, according to the Republican National Committee.[168] The RNC is a pro-Trump source, but no one could seriously dispute the sheer magnitude of the resources poured into promulgating the hoax.

In *Tablet*, an American Jewish online magazine, writer Lee Smith called the debacle an "extinction level event" for the media. "The farce that passed for public discourse the last two years was fueled by a concerted effort of the media and the pundit class to obscure gaping holes in logic as well as law," Smith wrote. "And yet they all appeared to be *credible* because the institutions sustaining them are *credible*." [169]

Instead of apologies and soul-searching *mea culpas*, the self-serving excuses began. "I'm comfortable with our coverage," *New York Times* executive editor Dean Baquet said. "It is never our job to determine illegality, but to expose the actions of people in power. And that's what we and others have done and will continue to do." [170]

Sure, line by line, most of what the newspaper had reported was true. But how can it feel comfortable with missing the bigger story: a former CIA director attempting to take down a sitting president? On *Meet the Press*, John Brennan had boldly thrown down the gauntlet: "I called [Trump's] behavior treasonous, which is to betray one's trust and aid and abet the enemy, and I stand very much by that claim." [171] Now revealed as a fabulist who had had a large hand in spreading the conspiracy theories, Brennan blamed his despicable denunciations of Trump on "bad information," as if he had gotten a few details wrong.[172]

"I am relieved that it's been determined there was not a criminal conspiracy with the Russian government over our election," he said on MSNBC. "I think that is good news for the country. I still point to things that were done publicly, or efforts to try to have conversations with the Russians that were inappropriate, but I'm not at all surprised that the high bar of criminal conspiracy was not met." In spook speak, that's called CYA. That that sordid man was once the director of the CIA is a shocking indictment of Barack Obama's administration.

Lester Holt of NBC tackled the other prevaricator who had feasted on the rotting corpse of the collusion narrative. On tour to promote his self-aggrandizing book, James Comey was repeatedly asked if the Russians had something on Trump. He offered the same incendiary answer, knowing it would lead news reports: "I think it's possible." [173] Comey had led the media down the "collusion" path with classified memo bread crumbs that had triggered the special counsel appointment. Now, asked by Holt if he saw the results of Mueller's investigation as a rebuke of his leadership of the FBI and its role in the catastrophe, Comey shamelessly tried to spin: "No, I actually see it the other way. It establishes, I hope, to all people, no matter where they are on the spectrum, that the FBI is not corrupt, not a nest of vipers and spies, but an honest group of people trying to find out what is true. That's what you see here."

No, what we see is an unscrupulous and unprincipled leaker who, in the words of "Jason Beale," a pseudonym for a retired US Army interrogator, created a "$25 million, two-year hostage crisis, wherein the president of the United States' political and foreign policy agenda was handcuffed by an investigation he repeatedly told us was unnecessary and unjust." [174]

During an interview with White House chief of staff Mick Mulvaney, CNN's Jake Tapper defended his network: "I'm not sure what you're saying the media got wrong. The media reported that an investigation was going on. Other than the people in the media on the left, not on this network, I don't know anybody who got anything wrong. We didn't say there was conspiracy. We said that Mueller was investigating conspiracy." [175]

Not much evidence of soul-searching there.

Tapper had conveniently forgotten CNN's long record of screw-ups and fake scoops. Viewers remembered. The network lost one-third of its prime-time audience and more than half of its viewers in the coveted younger demographic.[176]

Anchor Chris Cuomo also refused to accept that CNN had had any responsibility for its distorted coverage and the collapse of the public's trust in media. "For the first time in our history you have a president who tells everybody that the media is their enemy and that the institutions of our democracy can't be trusted," he insisted in a heated interview with Representative Sean Duffy (R-WI).[177] "But if the media was doing its job they would be far more skeptical of some of the details that they got," Duffy shot back. That was exactly what Trump had been saying since the election—and it had turned out that he was correct.

CNN president Jeff Zucker didn't apologize. He puffed out his chest and proclaimed, "We are not investigators. We are journalists and our role is to report the facts as we know them, which is exactly what we did. A sitting president's own Justice Department investigated his campaign for collusion with a hostile nation. That's not enormous because the media says so. That's enormous because it's unprecedented."[178] In other words, we accepted bogus information from manipulative dirty cops and partisan hacks attempting to destroy a duly elected president and never noticed because we didn't investigate it. CNN wasted no time ginning up new angles: "Trump Moves to Weaponize Mueller Findings."[179]

Over at MSNBC, anchor Katy Tur dismissed the fuss over the result of the Mueller investigation by stressing that several dozen people, including Paul Manafort, had been indicted for crimes. "Let's just put on the screen of everybody who's been found guilty or already indicted in the Mueller investigations," Tur said. "Lots of faces, lots of pleas, and lots of indictments." Of course, none of them was guilty of the crime Mueller was supposedly appointed to investigate. That bothered Tur not at all.[180]

The *New York Times* had the audacity to print an op-ed titled "Trump's Shamelessness Was Outside Mueller's Jurisdiction," by Bob Bauer, who was identified as a "professor of practice and distinguished scholar in residence at New York University School of Law."[181] The *Times* failed, however, to point out that until May 2018, Bauer had been a partner at Perkins Coie, in the firm's "political law practice," and was still representing certain of the law firm's clients.[182] Hiding the fact that the op-ed had been written by an attorney with the firm involved in originating the hoax was dishonest.

But honest reporters who had remained skeptics refused to let them off the hook. "Check every MSNBC personality, CNN 'law expert,' liberal-centrist outlets and #Resistance scam artist and see if you see even an iota of self-reflection, humility or admission of massive error," tweeted Greenwald.[183] "While standard liberal outlets obediently said whatever they were told by the CIA and FBI, many reporters at right-wing media outlets which are routinely mocked by super-smart liberals as primitive and propagandistic did relentlessly great digging and reporting."

He singled out Maddow for criticism, saying she "went on the air for 2 straight years & fed millions of people conspiratorial garbage & benefited greatly."[184] Greenwald praised Tucker Carlson of Fox News, in particular, for letting his audience hear both sides of the story. "MSNBC did the exact opposite," Greenwald said on Carlson's show. The network "should have their top hosts on prime time go before the cameras and hang their head in shame and apologize for lying to people for three straight years, exploiting their fears to great profit. These are people who were on the verge of losing their jobs. That whole network was about to collapse. This whole scam saved them. And not only did they constantly feed people, for three straight years, total disinformation, they did it on purpose."[185]

Tim Graham, the director of media analysis for the conservative-leaning Media Research Center, also had harsh words for the mainstream media: "So now it's apparent the news channels merely channeled their wishful thinking. They had a grand denouement in

mind and it didn't happen. They mocked Trump for saying 'no collusion,' and that ended up being the truth. . . . The voters should feel
punked, swindled." [186]

"Nobody wants to hear this, but news that Special Prosecutor
Robert Mueller is headed home without issuing new charges is a
death-blow for the reputation of the American news media," wrote
Matt Taibbi. "Nothing Trump is accused of from now on by the press
will be believed by huge chunks of the population." [187]

Though former CBS investigative reporter Sharyl Attkisson had
not joined the anti-Trump bandwagon, on behalf of her colleagues
she wrote "Apologies to President Trump," published by The Hill,
which is worth quoting at length: [188]

> Whatever his supposed flaws, the rampant accusations and
> speculation that shrouded Trump's presidency, even before it
> began, ultimately have proven unfounded. Just as Trump said
> all along. Yet each time Trump said so, some of us in the media
> lampooned him.
>
> We treated any words he spoke in his own defense as if they
> were automatically to be disbelieved because he had uttered
> them. Some even declared his words to be "lies," although
> they had no evidence to back up their claims. We in the media
> allowed unproven charges and false accusations to dominate
> the news landscape for more than two years, in a way that was
> wildly unbalanced and disproportionate to the evidence. We
> did a poor job of tracking down leaks of false information. We
> failed to reasonably weigh the motives of anonymous sources
> and those claiming to have secret, special evidence of Trump's
> "treason."
>
> As such, we reported a tremendous amount of false infor-
> mation, always to Trump's detriment. And when we corrected
> our mistakes, we often doubled down more than we apolo-
> gized. We may have been technically wrong on that tiny point,

we would acknowledge. But, in the same breath, we would insist that Trump was so obviously guilty of being Russian President Vladimir Putin's puppet that the technical details hardly mattered. So, a round of apologies seem in order.

The Donald J. Trump for President Campaign released a memo to "Television Producers" after the news that Mueller had found no collusion regarding the "credibility of certain guests." [189] It named Schiff, Senator Richard Blumenthal, Representative Jerrold Nadler, Representative Eric Swalwell, DNC Chairman Tom Perez, and John Brennan as among the most aggressive spreaders of false information on Trump-Russia "collusion." The memo suggested that producers ask a basic question: "Does this guest warrant further appearances in our programming, given the outrageous and unsupported claims made in the past?" If they do reappear, "you should play the prior statements and challenge them to provide the evidence which prompted them to make the wild claims in the first place." Of course, TV producers were highly offended—and continued to book the same guests, especially Schiff, who didn't miss a beat, peddling the same discredited nonsense about "collusion."

"But the media question of the day is why members of the press corps aren't deciding on their own to reject the source who seems to have been misleading them for years," wrote James Freeman of the *Wall Street Journal*.[190] "One might expect any reasonable person, journalist or not, to stop providing a platform to someone who had gone two full years without backing up a sensational claim." That would include people such as Malcolm Nance, the author of *The Plot to Destroy Democracy: How Putin and His Spies Are Undermining America and Disarming the West*[191] and a former intelligence officer, who went on MSNBC the day after Barr's statement with a hysterical rant: "Everyone repeat after me—single most serious scandal in the history of the United States—was the president of the United States an agent of an enemy of the United States? Look. This—it could technically eclipse Benedict Arnold, who at least did it for money." [192] Nance will always be welcome on MSNBC as long as he keeps up the anti-Trump rhetoric, no matter how nutty he sounds.

Maddow's reputation crumbled. She won "The Worst!" crown in the *New York Post*'s "Mueller Madness" bracket game, getting more than twice as many votes as other serial offenders such as Brennan and Avenatti.[193] "I'm a liberal who loved her show a few years back," wrote one reader, "but the lack of skepticism and the unfortunate shift from fact-based reporting to hysteria was shocking." Within days, Maddow was back to whipping up new conspiracy theories. Fans who hate Trump will forgive her.

"Maddow viewers absolutely had the expectation that Mueller would find that Trump had colluded with Russia," wrote Caleb Howe for Mediaite. "And now, they don't feel that she led them astray. They believe that, because reality did not meet expectation, reality must have been tampered with. Because that's how conspiracy theories work. It's dot-connecting and innuendo and ultimately more theory to explain the theory." [194]

Like many journalists, Maddow believes in her own righteousness. "They buy their own PR about being the last line of defense, the noblest and most self-sacrificing profession, speaking truth to power," Howe wrote. "Being a check on power is the distilled value of the profession. . . . That belief in one's own nobility and righteousness, and unquenchable thirst to display it and be seen as it, is the biggest problem of all." [195]

THE MEDIA'S FUTURE

Bizarrely, the evening newscasts on ABC, CBS, and NBC ramped up talk of "impeachment" after Mueller's report announced that it had found no evidence of collusion, according to a Media Research Center report. "Despite the Mueller report's lack of an anti-Trump smoking gun, the broadcast networks actually became more invested in the Russia story," wrote MRC senior editor Rich Noyes. "TV news viewers heard three times as many references to 'impeachment' compared to before Mueller concluded there was no campaign conspiracy with Russia, while making no determination about potential obstruction of justice."[196]

Although *New York Times* executive editor Dean Baquet didn't acknowledge that his paper had erred, he tried to stanch the erosion of its reputation by blocking his reporters from appearing on such partisan programs as *The Rachel Maddow Show* and *CNN Tonight with Don Lemon*, according to *Vanity Fair* magazine.[197]

So far, this is the only hint that the biggest figures in journalism see a need for remedies, a small step in the right direction.

It's worth noting that they got off the gravy train only when the gravy ran out. By late May 2019, Maddow's weekly ratings had plummeted to a Trump-era low, an average 2.6 million compared to the 3.1 million she had averaged in the first quarter of 2019, during the peak hysteria.[198]

"For two years, this grifter strung her audience along with the promise that tonight's show will be the Trump killshot," wrote media critic John Nolte. "But it was all bullshit, all lies, all desperate paranoid delusions of collusion invented by a con woman/conspiracy theorist who got high sniffing her own press clippings. . . . She looks foolish now, exposed, and that is something almost impossible to overcome or to live down. Her intellectual bravado comes off a little hollow now, a little shrill. . . . She's not the old tried, true, and confident Rachel Maddow anymore and she never will be; no one is after they grift away their credibility and integrity."[199]

Accuracy in Media national editor Carrie Sheffield said that Maddow had long "ignored basic journalistic, fact-checking practices and the presumption of innocence in our legal system by relentlessly pushing unproven conspiracy theories about supposed Russian collusion," and her "lapse in journalistic balance" had contributed to Americans' declining trust in the national media.[200]

"This was a long time coming, and we hope MSNBC will allow

for greater balance moving forward," Sheffield said. "We hope that Maddow's programming will include substantive fact-checking, balanced debate and dialogue, rather than an echo-chamber monologue that further divides Left and Right. Americans deserve better."

One can hope. However, Cornell Law School professor William A. Jacobson, who writes a conservative blog, has his doubts. MSNBC will stand by Maddow, he contended, "because she still has a large viewership emotionally invested in bringing down Trump. That some reporters refuse to go on her show is important, but is unlikely to change her behavior. Maddow long ago carved out her faux-intellectual paranoid niche, and she's stuck in it."[201]

Can Maddow and the rest of the Trump-hating media survive the Mueller fiasco? Audiences will continue to erode unless they reevaluate their coverage and concede their mistakes.

"The one incontestable fact was that a paid advocate who was trading on his previous profession as a British intelligence agent, in the middle of a presidential election, was being shepherded around Washington by a notorious PR schemer and promoting allegations whose truth he was unwilling to vouch for and whose source he was unwilling to reveal," wrote Holman W. Jenkins, Jr., in the *Wall Street Journal*. "Unless you are an exceptionally dim journalist, whenever somebody peddles a salacious story to you, a question naturally and unbidden leaps to mind: Is the real story the one I'm being peddled? Or is the real story the fact that I'm being peddled it?"[202]

He pointed out the "ludicrous overkill" of the raids on Manafort and Stone. Those tactics had signaled Mueller's desperation as he prepared to release a disappointing report that would prove that the media had been chasing their tails for two years. Those guys—and their former boss, Trump—must be really bad if dozens of FBI agents bust into their homes at dawn. Few questioned Mueller's tactics because they fit their narrative.

"Can you imagine the horror right now in the network planning rooms?" Matt Taibbi asked.[203] "This is the greatest story that has ever existed for the news media business. And you can't separate that out from the coverage, because the financial incentive to keep hammering this home was tremendous.

"I mean this was the greatest reality show in history. It had everything. It had sex, it had cloak-and-dagger intrigue, it had the shadowy British spy [Christopher Steele], it had obscure meetings on remote islands. And it had the added advantage of being able to tell audiences, 'You can't turn us off, because this thing could blow up any minute. That bombshell could be coming at any time, so keep tuning in.'"

As Taibbi inquired: "What will they sell now?"[204]

What's more, who will buy? At every turn, the media demonized President Trump by declaring him guilty of a multiplicity of crimes. In the process, journalists squandered their only currency: their credibility.

And they proved that the Scarecrow was right.

CHAPTER 7

CROOKED COHEN COPS A PLEA

I have lied, but I am not a liar. I have done bad things, but I am not a bad man.

—Michael Cohen, testimony to Congress, February 27, 2019

The problem with liars is that, for them, truth has no meaning. They spin so many lies they lose track. They lie to themselves about their own lies. Over time, they cannot distinguish fantasy from reality. People stop listening. Credibility is lost. That's when the *serious* lying begins.

And so it was with Michael Cohen when he testified before the House Oversight and Reform Committee in February 2019. President Trump's onetime personal attorney admitted that he had lied to Congress before.[1] But now, he insisted, he was telling the truth. He asked the committee to believe that his lies did not make him a liar. It's what rhetoric experts call "dissociation."[2] He then promptly blamed Trump for his lies. Liars, after all, are always blameless.

Having squandered his believability, Cohen had every reason to continue inventing stories. Facing a substantial prison term, maybe he thought he could curry favor with prosecutors and the sentencing judge if he trashed the president of the United States before the cell door clanged shut. Or perhaps revenge was his motive. Trump had refused to give him a job after the election, causing him to feel discarded and humiliated. Payback would be in order.

Whatever his motivations, Cohen's newly revised storytelling was greeted with abiding skepticism—and for good reason. Even the chief prosecutor, who had charged him with a variety of felony crimes, portrayed him as a prodigious con artist who had peddled "lies and dishonesty over an extended period of time."[3] Though Cohen could never be called as a reliable witness against anyone in a trial, that didn't stop Democrats in Congress from mining his skills as an expert prevaricator in their relentless effort to destroy Trump. In retrospect, it had all been carefully choreographed by Special Counsel Robert Mueller more than a year earlier.

Mueller's spectacular dumpster dive into building a case against Trump's personal attorney came to the world's attention on April 9, 2018. News broke of dramatic early-morning FBI raids on Cohen's home, office, safe-deposit box, and hotel suite, where he and his family were staying while his apartment was undergoing renovation.

Federal agents, at the direction of the US Attorney's Office for the Southern District of New York, scooped up 4 million electronic and paper files, a dozen mobile devices and iPads, and twenty external hard drives, flash drives, and laptops. A later release of government documents revealed that the extraordinary warrant had been triggered by Mueller's team, which had referred its case involving Cohen to Manhattan prosecutors.[4]

Included in the haul were attorney-client privileged documents, including some regarding Cohen's negotiations of hush money "payoffs" to two women who claimed they'd had a sexual relationship with Trump, adult film star Stephanie Clifford, better known as Stormy Daniels, and former *Playboy* model Karen McDougal.

The privilege is sacrosanct, one of the bedrocks of legal jurisprudence. The raids prompted Trump to tweet: "Attorney-client privilege is dead!" Cohen's own attorney called the action "completely inappropriate and unnecessary."[5] The move was in sharp contrast to the treatment of Cheryl Mills, Hillary Clinton's lawyer: the FBI had given her partial immunity in return for nothing and even allowed her to be present during its interview of her client.[6]

DOJ rules require prosecutors to first consider less intrusive alternatives than no-knock raids before seeking records from lawyers. Cohen had already turned over thousands of documents and sat for depositions under oath. The SDNY prosecutors justified the unusual breach of ethics by claiming that the intrusions had been necessary because they hadn't been able to obtain access to Cohen's Trump Organization email account or phones with encrypted apps.[7]

The explanation of the breach was that Mueller believed that Cohen, as a central figure in the Trump empire, could be pressured to flip on his former boss. He had avoided DOJ rules by enlisting the SDNY to carry out his dirty work. If Manhattan prosecutors found information related to his investigation of Russia collusion, they could pass it on.[8]

Geoffrey Berman, the Trump-appointed US attorney for the SDNY, recused himself from the case. His deputy, Robert Khuzami, a Republican, approved the raid and supervised the SDNY prosecution team, which included three highly partisan lawyers from the special counsel's office: Jeannie Rhee, personal attorney of Obama adviser Ben Rhodes, as well as the Clinton Foundation; Andrew Goldstein, a former chief of public corruption for the SDNY who had donated $3,300 to Obama in 2008 and 2012; and Rush Atkinson, a former DOJ prosecutor under Andrew Weissmann.[9]

The raids on Cohen's properties galvanized the New York media, which had watched his posturing as Trump's devious troubleshooter and keeper of secrets for more than a decade. Cohen was involved in everything Trump. His voluminous files would surely provide abundant documentation of Trump's corruption involving taxes, real estate deals, the Donald J. Trump Foundation, and bankruptcies.

Trump would be forced to resign as president and would soon be in the dock facing charges. It was delicious to imagine what might be in the truckloads of documents hauled away by the FBI.

The government's tawdry dumpster dive ended in a dumpster fire. Cohen, after pleading guilty to violating campaign finance laws, tax evasion, and bank fraud, repeatedly testified to Congress. He exposed himself as a duplicitous liar who couldn't tell the truth for a twenty-minute stretch. He had committed crimes, then embroidered stories about Trump to give the prosecutors hounding him what they wanted in return for a reduced sentence. The irony is that even when he flipped, Cohen torpedoed Mueller's grand scheme to prove that Trump had conspired with the Russian government to steal the presidential election.

COHEN AND THE STEELE "DOSSIER"

The name of Michael Cohen was sprinkled throughout the Steele "dossier," which claimed that Cohen had journeyed to Prague in August 2016 in order to conspire with Kremlin officials to make non-traceable payments to hackers for the Trump campaign. Cohen's involvement in the collusion scheme was the tent pole holding up all the other wild claims of election collusion. Steele's memo, dated October 19, 2016, said:

> TRUMP's representative COHEN accompanied to Prague in August/September 2016 by 3 colleagues for secret discussions with Kremlin representatives and associated operators/hackers. Agenda included how to process deniable cash payments to operatives; contingency plans for covering up operations; and action in event of a CLINTON election victory. Some further details of Russian representatives/operations involved; Romanian hackers employed, and use of Bulgaria as bolt hole to 'lie

low.' Anti-CLINTON hackers and other operatives paid by both TRUMP team and Kremlin, but with ultimate loyalty to head of PA, IVANOV and his successor/s.[10]

You can almost see Boris twirling his mustache while Natasha pouts. The memo had been written by Steele the same day the *New York Times* had reported the arrest of a Russian hacker in a Czech hotel. A bit of artistic inspiration?

The Steele allegations were denied by Cohen, who said that during the time period mentioned—August 23 to 29—he had been in Los Angeles with his daughter, not in Europe. In fact, he'd never been to Prague. Czech intelligence said there was no evidence putting him in Prague; it had not monitored any such meeting between Cohen and Russian intelligence officers.[11] But the story wouldn't die. The journalist Paul Sperry identified Glenn Simpson as the source behind the discredited Cohen-Prague story, whispering in the ears of gullible journalists to keep the story alive.[12] Because if Michael Cohen had not been in Prague, Steele's credibility was torched.

Another line of investigation pursued steadily by Mueller related to a Russian-born real estate developer in New York named Felix Sater, who had pitched Cohen on building a Trump-branded tower in Moscow in late 2015. "Our boy can become president of the USA and we can engineer it," Sater wrote. He then elaborated:

> I will get all of Putin's team to buy in on this, I will manage this process. . . . Michael, Putin gets on stage with Donald for a ribbon cutting for Trump Moscow, and Donald owns the Republican nomination. And possibly beats Hillary and our boy is in. . . . We will manage this process better than anyone. You and I will get Donald and Vladimir on stage together very shortly. That's the game changer.[13]

Cohen told investigators that he had regarded Sater's discussions as mere puffery. Though Sater had pushed vigorously to set up a deal and to connect Trump with Putin, it had never happened. The tower project had not come close to fruition.[14] Trump had entertained the notion and granted permission to explore it, but it had not been his idea. Sater had been the driving force and had been unremitting in his attempts to lure Cohen into the plan.

Sater was not prosecuted. The Mueller Report, not unintentionally, created the distinct impression that Sater might somehow be working for the benefit of Russia. There is no evidence of this, and Mueller managed to omit the critical fact that Sater had "done extensive work for *American* intelligence agencies [author's italics]"[15] and was a "cooperator, confidential source, and intelligence asset" for the U.S.[16] Far from being a Russian asset, Sater was an *American* asset. In fact, he had been working for the FBI and US intelligence since 1998, according to a federal judge. Sater had given the government Osama bin Laden's telephone number before 9/11.[17] (You will not find this anywhere in the 448 pages of the report.)

It was none other than Andrew Weissmann, who later became Mueller's top prosecutor on the special counsel team, who had signed off on that cooperation agreement,[18] negotiated after "Sater pleaded guilty to a federal charge of racketeering for his role in a Mafia-linked $40 million stock fraud scheme," according to the *Los Angeles Times*.[19] Had he or someone at the FBI directed Sater's interactions with Cohen after Trump had announced his candidacy? Was Sater yet another informant/spy, like Stefan Halper, pushing to enmesh Trump associates in the phony Russia "collusion" story? There is strong evidence that the genesis of the Trump-Russia hoax began as early as 2015 and that Sater's Moscow venture set it in motion.

Weissmann surely knew at the outset that the Trump Tower project in Russia had been instigated by an FBI informant, but he allowed allegations that it was somehow evidence of Trump-Russia collusion to be used to bludgeon Trump. With their dumpster full of documents, Mueller's team homed in on Cohen's lucrative foreign lobbying work and found material they could exploit for the purpose.

COHEN AND "THE BOSS"

Cohen had grown up on Long Island, part of a moneyed crowd. He had driven to law school in a Jaguar but cultivated a streetwise image. He had married Laura Shusterman, whose father, Fima, had built a fortune after emigrating from Ukraine. After getting his start as a taxi driver, Fima Shusterman had gone on to invest in taxi medallions.[20] At the wedding of a friend, Cohen bragged that he belonged to the Russian mob. His friend found the boast unconvincing. It was as if Cohen wanted to be seen as a tough, don't-mess-with-me guy despite his pampered background.[21]

Court documents described how Cohen had become an attorney for Trump after an unspectacular early career:

> Cohen is a licensed attorney and has been since 1992. Until 2007, Cohen practiced as an attorney for multiple law firms, working on, among other things, negligence and malpractice cases. For that work, Cohen earned approximately $75,000 per year. In 2007, Cohen seized on an opportunity. The board of directors of a condominium building in which Cohen lived was attempting to remove from the building the name of the owner [Trump] of a Manhattan-based real estate company. Cohen intervened, secured the backing of the residents of the building, and was able to remove the entire board of directors, thereby fixing the problem for [Trump]. Not long after, Cohen was hired by the Company to the position of "Executive Vice President" and "Special Counsel" to [Trump]. He earned approximately $500,000 per year in that position.[22]

Cohen's sharp elbows in the condo dispute had impressed Trump. He had said, "Who is this guy? My lawyers that I give thousands of dollars to couldn't do it. I'd like to meet him," remembered Dr. Morton Levine, Cohen's uncle.[23]

Cohen's income had more than quadrupled almost overnight, as had his prestige. In his own estimation, Cohen became Trump's "Ray Donovan," the TV character who acts as a fixer for his tycoon boss.[24] He'd threaten lawsuits and berate reporters who maligned Trump. But in many ways, he was "always just at the edges" of Trump's world.[25]

Cohen had no official role in Trump's presidential campaign, though he had a campaign email address. He never thought Trump would win, but after the election, he told CNN's Chris Cuomo that he "certainly" hoped he'd be asked to join the administration and would "100 percent" move to Washington, DC, if asked.[26] His loyalty was so deep, so abiding, he pledged that he would "take a bullet" for Trump. "My sole purpose is to protect him and the family from anyone and anything."[27]

When Trump did not offer him a post, Cohen took it badly, as if spurned by a lover. "Boss, I miss you so much," he told the president in a phone call after Cohen had breakfast with an outspoken Trump critic, the billionaire Mark Cuban; the meeting had been photographed by paparazzi. "I wish I was down there with you." The amused Cuban said, "I think he does it to piss off Trump when Trump is ignoring him."[28] But Trump had concluded that bringing Cohen to Washington would be risky, describing him privately as a "bull in a china shop."[29] Cohen grew frustrated, believing he deserved the White House chief of staff job.

Two days before the inauguration, Cohen announced that he was leaving the Trump Organization to serve as Trump's personal lawyer. Despite his efforts to build relationships with others who joined the administration, he had effectively been sidelined.[30]

After the inauguration, Cohen set up a solo practice, taking an office on the twenty-third floor of Rockefeller Center with the law and lobbying firm Squire Patton Boggs. They agreed to a "strategic alliance," with the firm paying Cohen $500,000 a year plus a percentage of fees from five clients he had brought in. But he "maintained complete independence" from the firm, according to a search warrant, locking his office with a key no one else had, doing business on a separate computer server.[31]

His financial situation was precarious due to severe losses in the taxi business. Cohen's posh Manhattan lifestyle was at risk. He approached major US corporations with a blunt pitch: "I have the best relationship with the president on the outside, and you need to hire me," a person "familiar with Mr. Cohen's approach" told the *Wall Street Journal*.[32]

Touting his status as "the president's lawyer," uniquely able to help clients understand how Trump operated, he snagged Ford, AT&T, and the pharmaceutical company Novartis AG. Uber turned him down because of his ownership of taxi medallions, an obvious conflict of interest. He buttonholed Cuban, saying "I want you to know I'm out looking for deals." Cohen urged Cuban to hire a health industry firm he'd taken on as a client. "Michael is a hustler," Cuban said. "That's who he is, that's what he does."[33]

For a while his aggressive approach worked. AT&T paid him $50,000 a month and Novartis AG paid him $100,000 a month, both for a year. A lobbying deal with the investment management firm Columbus Nova netted him seven monthly payments totaling $583,332 for January through August 2017. He also earned $600,000 from a South Korean aircraft manufacturer. The payments were for "political consulting work, including consulting for international clients on issues pending before the Trump administration."[34] The money was routed through a shell company called Essential Consultants LLC, which Cohen had created to route money to Stormy Daniels.[35]

Cohen's contact with Trump dwindled after the inauguration. He complained that the boss "was not calling him and not helping him." His efforts to build a robust base of clients stagnated; some canceled their deals, unimpressed with his services. Squire Patton Boggs ended its relationship with Cohen in March 2018. The raids on his home, office, and hotel room took place on April 9.[36]

The next day, legal scholar Jonathan Turley sounded a warning for Cohen and Trump, saying the involvement of the SDNY prosecutors was a move "as cunning as it is hostile. The timing and manner of the raid have all the characteristics of a wolf pit and Trump—not

Cohen—could be the prize." A wolf pit, he pointed out, is a hole dug by farmers with sharp spikes at the base, branches covering the hole, and a juicy piece of meat on the top.[37] Cohen was the bait. The wolf lunges for the steak, falls in, and the more he thrashes about, the more he injures himself.

"Like any good wolf trap, this set-up, first and foremost, protects the hunters," Turley said. "By referring the matter, Mueller and Rosenstein protected themselves from criticism of expanding the investigation."[38]

Mueller had no compelling case against Trump for collusion; perhaps he could entice the enraged president to obstruct justice by rescuing his longtime lawyer. The wolf trap nearly worked. Enraged, Trump blasted the raids as "a disgraceful situation" and "an attack on our country in a true sense." He thundered about the possibility of firing the special counsel: "This is really now on a whole new level of unfairness."[39]

Cohen had relished his various roles with Trump—friend, fixer, business associate, attorney—but his "bizarre concept of representation is so convoluted and conflicted that, after months of litigation, we still are unclear as to whether Cohen was acting for himself or his client or his shell company," Turley said.[40] Mueller, using the SDNY by proxy, had stripped away the attorney-client privilege that should have attached to their relationship, especially regarding possible payments to Daniels and McDougal. Cohen's only hope was to make a deal with Mueller or obtain a pardon from Trump.

COHEN'S SECRET TAPES

Mueller's investigation into Cohen had actually begun nine months before the raids, using the Steele "dossier" as its basis. On July 7, 2017, federal agents had obtained warrants allowing prosecutors to access Cohen's Gmail account for messages from the start of 2016 through July 2017. Mueller got the SDNY involved, and they obtained a warrant for the same email account through the end of February 2018. In addition, they used a "pen register" to log phone calls he made and received and a "triggerfish" cell phone surveillance device to locate him while he was staying at the Loews Regency hotel in New York.[41]

The investigation of collusion now morphed into two prongs: campaign finance violations and Cohen's complex business interests.

He owed $22 million on taxi medallion loans; in an effort to avoid the debt, he had claimed his net worth had cratered, going from $75.9 million in 2014 to negative $12.2 million by 2017, due to the depreciation of the taxi medallions and real estate. But he was allegedly hiding his assets. Court documents quoted FBI forensic accountants saying that Cohen had had approximately $5 million in cash as of September 30, 2017.[42]

The alleged campaign finance violations got the most attention from the press, since they involved sex. Porn star Daniels claimed she had had a single encounter with Trump in 2006. *Playboy* model Karen McDougal claimed she had had a ten-month affair with Trump that had ended in April 2007. Trump denied having had a sexual relationship with either woman, and Cohen tried to buy their silence.[43]

The puerile melodrama created by Daniels and her seedy attorney, Michael Avenatti, took cable TV by storm in March 2018, after Daniels filed a lawsuit against Trump, alleging that Cohen had "intimidated and coerced" her into signing a nondisclosure agreement just a few weeks before the election in return for a payoff of $130,000. Soon after came the raid, which yielded documents involving Daniels and McDougal that Mueller twisted into campaign finance violations.[44]

Cohen had the remarkably poor judgment to trust the leadership of his defense team to attorney Lanny Davis, a longtime Clinton sycophant, still vexed at the trouncing his beloved Hillary had received at the hands of Trump in the 2016 election. Cohen told the press that Davis was representing him free of charge. That should have been Cohen's first clue that he was expendable.

Cohen and Davis made the perfect couple: two conflicted and

compromised lawyers, loyal to a fault until it was not in their best interests. Both are like the pilot fish that hang around sharks, happy to nibble on the debris that attaches itself to the skin and teeth of the apex predators, providing valuable services without being eaten themselves. For decades, Davis had sucked up to the Clintons and had fully expected to be an adviser to the president when Hillary moved back into the White House. Cohen had attached himself to Trump, who had various lawyers but gave him the chores that needed the finesse of a New York fixer, not a legal scholar. The ticklish problems presented by Daniels and McDougal were perfect examples.

The media embraced the sex and campaign finance story with zeal, making the dirty duo of Daniels and Avenatti round-the-clock circus stars. In an example of collective delusion, CNN anchors promoted Avenatti as Trump's rival for the presidency in 2020.[45]

But Mueller's tactics of seizing privileged documents should have sent shudders down the spine of every lawyer in America. Among the items confiscated were hundreds of audiotapes with clients and reporters who had not known they were being recorded; they included a dozen tapes of conversations with Trump, released by a judge to federal investigators after Trump withdrew his objections. Davis claimed that Cohen had secretly recorded people in lieu of taking notes and had had no intention of "publicizing" them, "nor any intention to ever deceive anyone." Trump certainly felt deceived. He tweeted: "What kind of lawyer would tape a client? So sad!" Most clients expect that their conversations with their lawyers are privileged. Though not illegal—New York is a "one-party" state, which allows recording as long as one party to the conversation knows—it was an egregious breach of trust. Cohen had violated a fundamental legal ethic.[46]

The tapes revealed that in September 2016, Cohen had suggested to Trump that they buy the rights to McDougal's story, which had been purchased for $150,000 by *National Enquirer*, owned by American Media, Inc. (AMI). The chairman of AMI, David Pecker, had been Trump's friend since the 1990s. However, there was no indication by Trump in the recorded conversation that he knew she'd sold her story to AMI. To carry out a possible purchase, Cohen created a shell company called Resolution Consultants LLC in Delaware. However, Pecker declined to sell, and the story was never published.[47] Cohen dissolved that company on October 17, 2016, and created a second shell company, Essential Consultants LLC, which he used to pay Daniels, drawing on his home equity line of credit.[48]

Cohen had also surreptitiously recorded conversations with reporters, including a nearly two-hour conversation with CNN's Chris Cuomo talking about negotiating the payment to Daniels in October 2016. "I did it on my own," he told Cuomo. "It wasn't for the campaign. It was for him [Trump]."[49] That comment would be exculpatory in any potential trial regarding violation of campaign finance laws. Cohen would later complain to friends that Trump had never repaid him. That wasn't true; he had been reimbursed through a monthly retainer. The unethical taping and willingness to go into debt to pay off Trump's accuser revealed two sides of Cohen's character: loyalty undermined by deceit.

By May 2018, Cohen no longer represented Trump. When attorney Lanny Davis entered the picture, he was the perfect symbol of Cohen's alienation from Trump's circle of protection. Davis had become friends with Hillary Clinton at Yale Law School and worked as a spokesman and special counsel for Bill Clinton in the 1990s. Known as a shill for whoever anted up his fees, injecting himself into controversies, masquerading as a political analyst, Davis has been a Washington fixture for decades.[50] A story by the *New York Times* once described him as a "front man for the dark side," representing dictators in Equatorial Guinea and Ivory Coast. In 2017, Davis was lobbying on behalf of Ukrainian oligarch Dmitry Firtash, who had been fighting extradition to Chicago to face charges of racketeering and money laundering in relation to Russian organized crime.[51]

"Lanny Davis is just a face that reflects the grime and sleaze that lies at the core of our political culture," said journalist Glenn Greenwald. "He's presented by numerous media outlets as an independent analyst who opines on the news of the day—yet does so almost exclusively in order to promote the interests of those who are paying him, relationships which are often undisclosed."[52]

Davis's most famous clients are the Clintons. A 2010 email released by the State Department from Davis to "my dear friend" Hillary describes their close relationship. Asking Clinton to speak with a reporter on his behalf, Davis concluded, "Aside from Carolyn, my four children, and my immediate family, I consider you to be the best friend and the best person I have met in my long life."[53] Another email uncovered by Judicial Watch was just as cringe inducing: "Thank you H for who you are and what you do," Davis wrote, followed by another exchange: "PS. I swear you look younger and better every time I see you. Good night, dear Hillary. Lanny."[54]

Perhaps in representing Cohen, Davis saw an opportunity to humiliate Trump and avenge Hillary's loss. He announced that his new client had "turned a corner in his life and he's now dedicated to telling the truth." A few weeks later, Davis gave CNN the tape of Cohen speaking to Trump, a sign he was serving the media's interests, not those of his client.[55] The muddled audio lasted less than two minutes, with Trump heard to say that, if he was going to make any kind of payment, to use a check so there would be a paper trail—not the kind of thing someone would do if he was conspiring to commit a crime.

Cohen was indicted August 21, 2018, on eight charges: five counts of tax evasion from 2012 through 2016; one count of making false statements to a bank regarding a loan; and two counts of violating federal campaign finance laws for paying a combined $280,000 in hush money to McDougal and Daniels. By paying the women himself, the government said, Cohen had exceeded the maximum allowable amount of contributions of $2,700 per candidate.[56] The next day, Cohen pleaded guilty to all eight counts.

Significantly, there was no "collusion"-related charge. The information filed by the government revealed new details about the payoffs.

McDougal had sought to sell her story to the *National Enquirer* in June 2016. An editor and Pecker had notified Cohen. At Cohen's urging and after his promise of reimbursement, the publication had bought McDougal's rights to the story for $150,000 in early August. The rights were to be assigned to Cohen for $125,000.[57]

The difference in the amount was related to additional rights AMI acquired for McDougal's appearance on two magazine covers as a model and a health and fitness column, which she hoped would resurrect her modeling career. The deal by AMI editor Dylan Howard was made despite the fact that McDougal had offered no documentation that proved her alleged relationship with Trump.[58]

But Pecker contacted Cohen, told him the deal was off, and asked him to tear up the agreement. Thus, Cohen never paid either McDougal or Pecker. The documentation was later found in the search of Cohen's office.[59]

On October 8, 2016, Cohen was approached by an agent and an attorney (not Avenatti) representing Daniels, who alleged that she had had sex with Trump at a celebrity golf tournament in Lake Tahoe in 2006. Daniels's agent had also approached a *National Enquirer* editor, demanding $200,000. The editor had passed, pointing out that Daniels had called the report "bulls—" when approached by a TV show in 2011.[60] A second statement provided to the media by her manager said, "I am denying this affair because it never happened."[61]

But after the *Access Hollywood* tape surfaced, putting Trump's campaign into jeopardy, Daniels resurfaced. Cohen negotiated a $130,000 agreement to buy her silence in return for a nondisclosure agreement (NDA). In the tumult of the campaign, Cohen hadn't paid her by October 26. Her representatives threatened to take Daniels's story elsewhere. Cohen arranged a line of credit against his home mortgage and wired $130,000 to Daniels's attorney. In so doing, he agreed in court documents, he had intended to "influence the 2016 presidential election." He billed Trump for $130,000, plus a $35 wire fee and $50,000 for "tech services" performed during the campaign. He would end up being promised $420,000, paid in monthly installments of $35,000, for legal work performed over the course of twelve months for the Trump Organization.[62]

At the same time as the Cohen raids, Pecker and the Trump Organization were subpoenaed for records related to the payoffs. Pecker was granted immunity in return for his testimony; he was not charged with a campaign finance violation.[63]

Cohen faced a potential of sixty-five years in prison. By pleading guilty, he narrowed that to a range of three to five years. Davis described Cohen as feeling a sense of "relief" and "liberation" upon entering his plea. The "straw that broke the camel's back" for Cohen was Trump's supposedly disastrous summit with Putin in July, Davis said. After the press conference, Davis claimed, Cohen had become "very emotional," a ludicrous embellishment dreamed up by a Clintonite to fit the news cycle.[64]

Davis pointed out that Cohen had stood up in court and read a statement written by the prosecutors. "He stated under oath that Donald Trump directed him to commit a crime, making Donald Trump as much guilty of that felony as my client, Mr. Cohen," claimed Davis.[65] Perhaps the government had Cohen dead to rights on tax evasion and bank fraud. But the payments to Daniels and McDougal were not illegal. Cohen had pleaded guilty to two noncrimes, an indication that Davis was serving Mueller's interests, not his client's.

Why were they noncrimes? Although Cohen initially had used his own funds, Trump had reimbursed him. Under the law, a candidate is allowed to spend an unlimited amount of his or her own money on his or her campaign. Moreover, paying money to someone in exchange for an NDA is not considered a campaign contribution at all, as long as there is a secondary or collateral reason for the payment. Trump's reasons would include protecting his business and avoiding personal embarrassment to his wife and other family members.

As former Federal Election Commission chairman Bradley Smith explained, "Not everything that might benefit a candidate is a campaign expense."[66] Such a payment can be a crime only if there's a showing that the person making it "knowingly and willfully violated" the law. Campaign laws are exceedingly complex; few candidates understand them sufficiently to break them knowingly. Indeed, not many lawyers fully understand these complicated finance laws.

It should also be remembered that the vast majority of campaign violations are civil cases with penalties assessed. For example, Obama's 2008 campaign paid a $375,000 fine for taking $2 million in illegal contributions.[67] Prosecutors would have to show that Trump had *known* the campaign laws and specifically intended to break them—something that would be exceedingly difficult, if not impossible, to prove.

Smith, though not a Trump supporter, was concerned about the prosecutors' actions: "I find the whole thing kind of disturbing— what seems to be almost a determination to use whatever laws we can find to get Trump because a lot of people know Trump's such a bad guy, he must be violating the law, and even if he's not, we have to get him anyways."[68]

If Trump committed a crime in his payments, so, too, have many members of Congress. On Capitol Hill, the department that handles sexual harassment and discrimination claims, the Office of Compliance, has paid 268 victims more than $17 million in claims since the office was established in the 1990s.[69] Taxpayers have footed the bill. In most cases, the information was kept quiet. It appears that none of the 268 payments was ever reported as a campaign contribution. Didn't those payments either directly or indirectly benefit the reelection campaigns of members of Congress? Of course they did. But they also served a dual purpose in protecting members from disclosures that would personally embarrass themselves and their families. Hence, they were not considered campaign contributions under the law. Compare this to the case of Hillary Clinton, who somehow received legal dispensation from paying a foreigner for the fabricated "dossier" from Russian sources during her campaign. As noted in an earlier chapter, her payment established it as a "thing of value," which is arguably illegal under campaign finance laws.

"That money was not campaign money," Rudy Giuliani said regarding Daniels's compensation. "No campaign finance violation. They funneled through a law firm, and the president repaid it."[70]

Daniels offered evidence that the payment had had little to do with the campaign. "Daniels herself has said that years before Trump declared for president, she was threatened about not disclosing any affair, suggesting, if she's telling the truth, that her silence was desired long before Trump became a candidate," said former FEC chairman Smith.[71] Mueller knew the case was weak. Former senator John Edwards (D-NC), who ran for president in 2008, was charged with a similar violation for paying his mistress to keep silent during his campaign, but federal prosecutors failed to obtain a conviction at trial.[72]

"Lanny Davis had his client plead guilty to a crime that isn't a crime," said David Warrington, an attorney who had worked with the Trump campaign. "Who's Lanny Davis really working for, Michael Cohen or the Clintons?"[73]

But Cohen wanted to curry favor with the prosecutors, which would help him during sentencing, and to strike back at Trump, whom he blamed for all of his woes. The pilot fish had lost his shark and deeply resented Trump's leaving him behind.

Davis shamelessly put out a crowdfunding request for a Michael Cohen "truth fund," hitting every TV and radio network in a flurry of appearances. His pitch on Cohen's behalf so galled the audience at *Megyn Kelly Today* that they guffawed and even booed the former Clinton attorney.[74]

As it turned out, the pro bono tale was nothing more than a facade. Davis would later acknowledge to two Republican members of Congress that he was getting paid, but not by Cohen.[75]

THE LANNY DAVIS SPIN MACHINE

The ink was barely dry on Cohen's plea agreement before Lanny Davis was revealed as the anonymous source behind two "bombshell" allegations made against Trump by a CNN story that aired on July 27, 2018.[76]

Carl Bernstein, Jim Sciutto, and Marshall Cohen reported, based on unnamed sources, that Michael Cohen could tell Mueller that he had witnessed Donald Trump, Jr., inform his father beforehand about the infamous Trump Tower meeting with a Russian lawyer to get "dirt" on Hillary Clinton. Trump had publicly denied that. In addition, Cohen could confirm to Mueller that Trump had been aware of and encouraged Russian hacking before it became publicly known. The story specifically said that Davis had declined comment, implying he wasn't a source. *Lie number one.*[77]

CNN hosts discussed the allegations for days, crowing that the report was the strongest evidence yet of Trump-Russia "collusion" and a cover-up. Nearly a month later, days after Cohen pleaded guilty, Davis batted down the bombshell about the meeting. "I think the reporting of the story got mixed up in the course of a criminal investigation," he told CNN's Anderson Cooper. "We were not the source of the story."[78] *Lie number two.*

However, that night, according to the editorial board of the *New York Post*, Davis again confirmed the story—anonymously—to a *Post* reporter and to the *Washington Post*.[79] *Lie number three.*

Then he changed his mind and said he could not independently confirm it.[80] Finally forced to admit that he had been CNN's source, Davis told BuzzFeed that he'd "made a mistake" and "unintentionally misspoke" about the Trump Tower meeting. In fact, he'd done so three times.[81]

"I should have been much clearer that I could not confirm the story," Davis said. "I regret my error."[82] He also hedged on his allegation that Cohen could confirm to Mueller that Trump had been aware of and encouraged Russian hacking before it had become publicly known. "I am not sure," he said. "There's a possibility that is the case. But I am not sure." He told the *Washington Post*, "I was giving an instinct that he might have something to say of interest to the special counsel" about hacking. In other words, Davis had made it up.[83]

Davis's flurry of deceptions—designed to turn up the heat on Trump—undermined his client's previous closed-door testimony to the Senate Select Committee on Intelligence (SSCI), exposing Cohen to perjury charges. The committee's leadership contacted Davis after the CNN story broke and asked if Cohen wanted to amend his testimony. The new version of events wasn't lining up with his prior statements under oath. Cohen declined.

CNN, without shame or apology, declined to retract or modify the report, according to the editorial board of the *New York Post*, "because it suspects Davis changed his story only because Cohen could face perjury charges. . . . The network offers no justification for its own deception of reporting that Davis hadn't commented when he actually had, though anonymously." [84]

Whatever his reasons for the seesaw lies, Davis had eroded his client's credibility and value to the government in cooperating with ongoing investigations. "To the extent that Michael Cohen knew that Lanny Davis would make (or had made) that representation and took no steps to repudiate it, that would create a credibility issue that could be used on cross-examination," said Mitchell Epner, a former federal prosecutor. "Special Counsel Mueller would consider that information as part of the baggage that comes with Michael Cohen." [85]

Roland Riopelle, another former federal prosecutor, concurred: "Mr. Davis has done a disservice to his client, in my opinion, by presenting his case in the press and then contradicting himself and making grandiose claims about what Mr. Cohen knows or can do. These sorts of statements may very well benefit the Clintons . . . but they do not benefit Mr. Cohen." [86]

Doing damage control, Davis told Bloomberg News that the thirteen references to Cohen in the Steele "dossier" were false, "and he has never been to Prague in his life." [87] The Trump-Russia "collusion" story was dead, kept on life support by gullible members of the media. Davis was looking down the road, planning to use Cohen to attack the president in the political arena. Impeachment had always been the suspected goal of the "insurance policy." Now someone intimately familiar with Trump, with a sword of Damocles hanging over his head, could be deployed against him.

After his first guilty plea, Cohen entered a second guilty plea in November 2018, admitting he had made false statements to the

SSCI in 2017, when he had claimed that the negotiations involving the Trump-branded tower in Moscow had ended in January 2016. In truth, the talks had ended in June 2016. He had also stated that his discussions with Trump regarding the project had been limited when in fact they were more extensive than he portrayed them.[88]

The plea marked the first time that Mueller had charged Cohen—or anyone else associated with Trump—as part of his investigation into Russia meddling and alleged "collusion."

As Fox News political reporter Brooke Singman wrote, "Cohen made the false statements to minimize links between the Moscow Project and Individual 1 [Trump] and give the false impression that the Moscow Project ended before the 'the Iowa caucus . . . the very first primary,' in hopes of limiting the ongoing Russia investigations."[89] Whether Cohen deliberately misled Congress or got his dates wrong was a moot point, given his plea. At the very least, he should have been more forthcoming about the extent of the discussions. But his plea deal made it clear that he was cooperating, and as a result he "will not be further prosecuted criminally by this Office."[90]

Since Cohen was an admitted liar and tax cheat, we have to be skeptical about how honest he was being in the second plea. "Michael Cohen is lying and he's trying to get a reduced sentence for things that have nothing to do with me," President Trump told reporters. "This was a project [in Moscow] that we didn't do, I didn't do. . . . There would be nothing wrong if I did do it."[91]

As the old saying goes, guilty pleas are like paper currency: some are more valuable than others. But insofar as proving some amorphous crime of Trump-Russia "collusion" to win the 2016 presidential election, with Cohen's second plea Mueller got something about as valuable as a crumpled dollar bill. It gave Democrats and members of the Trump-hating media—who know little about the law—something to howl about. Beyond that, it had zero value in proving that the president and his campaign had somehow conspired, coordinated, or "colluded" with Russia. And that was what Mueller was supposed to be investigating.[92]

Amid the hysteria over Cohen's guilty pleas, many journalists overlooked one important and immutable fact: nothing Cohen said about the never-consummated Moscow real estate project had established an election conspiracy. It is not a crime to develop real estate in Russia, which made it all the more mystifying why Cohen lied about it. How would Russia gain favor or leverage over Trump through a deal that had never materialized? Recall that among the specious claims in the Steele "dossier" was that Trump had rejected offers for "various lucrative real estate business deals in Russia."[93]

A week later, prosecutors with the SDNY filed a sentencing memo outlining Cohen's serial lies in both his business and private life, revealing how he'd monetized his relationship with Trump to rake in more than $4 million from corporations that had hired him as a consultant even though he had provided little or no real service under the contracts. Driven by greed, Cohen had avoided paying taxes for five years, had made private loans at double-digit interest rates, and had not reported the income the loans generated. He had lied to his accountant, using unreported income to finance a lavish lifestyle. The SDNY prosecutors asked the judge to sentence Cohen to about four years in prison.[94]

Mueller also filed a memo regarding the lies Cohen had told Congress regarding the Moscow tower project. But Cohen was not charged with collusion regarding Russia. He had never gone to Russia and had had no contact with the Kremlin or Putin. Instead he was accused of—get this—having a "telephone call about the project with an assistant to the press secretary for the President of Russia." That torpedoed the idea that the Trump campaign had a secret "back channel" to the Russian government.[95]

But the media heralded Cohen's plea deal as a sign that Mueller was closing in on Trump. "Well into the 2016 campaign, one of the president's closest associates was in touch with the Kremlin on this project, as we now know, and Michael Cohen says he was lying about it to protect the president," said CNN's Wolf Blitzer. "Cohen was communicating directly with the Kremlin." Legal analyst Jeffrey

Toobin announced that the plea deal was so "enormous" that Trump "might not finish his term."[96] Such declarations were fanciful at best; gibberish at worst. The plea deal read more like an exoneration. If he'd had any information about "collusion" to offer, Cohen would have laid it on the table. It had never happened.

Cohen's negotiations with the Russian immigrant who had come up with the idea for the Trump Moscow tower project had come to nothing because Felix Sater had not had the clout to pull it off. All the FBI informant had offered Cohen was access to "one of his acquaintances, who was an acquaintance of someone else who was partners in a real estate development with a friend of Putin's," as the journalist Paul Sperry aptly described it.[97]

Despite claiming that Cohen's payments to Daniels and McDougal had been illegal campaign contributions, the prosecutors knew that with his guilty plea the issue would never be argued in court. Trump had not violated campaign finance laws. The FEC law specifically states that campaign-related expenses do not include any expenditures "used to fulfill any commitment, obligation, or expense of a person that would exist irrespective of the candidate's election campaign."[98] Like many billionaires, Trump had made payments out of his own funds—which was perfectly legal—to protect the reputation of his brand and save his family embarrassment.

Pleading guilty to the campaign finance counts added little to Cohen's exposure in sentencing. "Cohen's defense team perceived that the SDNY is trying to make a case on President Trump, so pleading guilty to two extra felonies paradoxically improved his chances for sentencing leniency," said Andrew McCarthy, former federal prosecutor. "I think Cohen's rolling over on the campaign-finance allegations made the SDNY more amenable to leniency."[99]

That proved to be the case. The sentencing took place on December 12, 2018, in a lower Manhattan courtroom mobbed with reporters and spectators, including the odious Avenatti. Three members of Mueller's team sat with the SDNY prosecutors. Backed by a large group of family members, Cohen expressed his remorse and declared war on Trump.

"Today is the day that I am getting my freedom back," he read in a prepared statement. "I have been living in a personal and mental incarceration ever since the day that I accepted the offer to work for a real estate mogul whose business acumen that I deeply admired." Exactly how Trump had compelled Cohen to commit tax evasion and bank fraud was not explained. The judge sentenced Cohen to three years in prison, forfeiture of $500,000 in assets, and payment of $1.393 million to the IRS, plus an extra $50,000 fine for lying to Congress.[100]

One of Cohen's lawyers told the judge that Cohen and his family had been receiving "threats" because of his decision to "come forward to offer evidence against the most powerful person in our country."[101] Davis promised that his client planned to reveal "all he knows about Mr. Trump" after the Mueller investigation was complete, with testimony to congressional committees "interested in the search for truth and the difference between facts and lies." Clearly revenge was on the agenda.[102]

One clue was his deferred prison sentence; Cohen was told he didn't have to report to prison until March 6, 2019. Representative Elijah Cummings (D-MD), the chairman of the House Oversight and Reform Committee, appeared on CNN and said he wanted Cohen to testify. "This is a watershed moment," he averred, comparing Cohen to John Dean, who had testified about Watergate and "changed the course of America." Of course, Dean's testimony had led to President Richard Nixon's resignation.[103]

McClatchy added to the melodrama, reporting in December 2018 that, according to unnamed sources, a mobile phone traced to Cohen had "sent signals ricocheting off cell towers in the Prague area" in late summer 2016, and "electronic eavesdropping by an Eastern European intelligence agency picked up a conversation among Russians, one of whom remarked that Cohen was in Prague."[104]

That story, based on triple hearsay, proved to be false as well. Cohen and Davis again denied it. As the Mueller Report would subsequently put it, "Cohen had never traveled to Prague and was not concerned about those allegations, which he believed were provably false." [105]

However, as far as the media were concerned, Cohen's upcoming appearances before Congress put collusion, with the added dollop of campaign finance violations, on the table. And Davis was there to assist, shaking up Cohen's legal team for the second time, bringing in attorneys to focus on "Washington and Congress." [106] If "collusion" had collapsed, impeachment over some tortured allegation of obstruction of justice would do just fine.

THE CAPITOL CIRCUS

When Cohen agreed to testify before three different congressional committees before reporting to the penitentiary, Trump foes rejoiced. But Cohen and Davis couldn't play it straight; his testimony needed building up. All eyes must be focused on Cohen's epic takedown of Trump.

So someone leaked to BuzzFeed, which had gotten the ball rolling by publishing the phony Steele "dossier" two years earlier. On January 27, 2019, BuzzFeed reported that Trump had instructed Cohen to lie to Congress in his statements about the "Moscow Project," directly accusing Trump of committing a felony. "The special counsel's office learned about Trump's directive for Cohen to lie to Congress through interviews with multiple witnesses from the Trump Organization and internal company emails, text messages, and a cache of other documents," wrote BuzzFeed reporters Jason Leopold and Anthony Cormier. "Cohen then acknowledged those instructions during his interviews with that office." [107]

Cohen and Davis made no official comment. The poorly sourced story, citing "two law enforcement officers involved in an investigation of the matter," made little sense. But for twenty-four hours, rabid media piranhas and feverish Democrats speculated that the president would resign. Many predicted Trump's imminent impeachment for suborning perjury and obstruction of justice.

Then the special counsel's office issued a rare public statement. "BuzzFeed's description of specific statements to the special counsel's office, and characterization of documents and testimony obtained by this office, regarding Michael Cohen's Congressional testimony are not accurate," said Mueller's spokesman Peter Carr.[108] The frenzy fizzled out.

On February 5, 2019, BuzzFeed tried again. The same reporters released a trove of internal Trump Organization documents about efforts by Cohen and Sater to advance the Moscow project. But the bottom line remained the same: no tower had been built. The deal hadn't made financial sense and—spoiler alert—Trump had been elected president.[109]

Cohen announced that he would not appear to testify before the Senate Intelligence Committee on February 7, 2019, as planned. Perhaps with the flood of new documents, he needed to get his story (or lies) straight. Davis announced that he was postponing his testimony for several reasons, including his continued cooperation with the special counsel regarding ongoing investigations and "threats against his family from President Trump and Mr. Giuliani." [110]

Hysteria about obstruction of justice ensued. "By referencing threats, Cohen can let people run wild with speculation of witness tampering without ever having to actually accuse anyone of tampering," said Ronn Blitzer, legal analyst at Law & Crime. Twitter obliged.

"The worst thing that could happen for [Cohen] or Mueller would be to get trapped and make the mistake of lying," said Blitzer. "Not only would it expose Cohen to additional charges, it would compromise Mueller's investigation by casting significant doubt on the credibility of a key witness."

Cohen then postponed his appearance before the House Intelligence Committee. Chairman Adam Schiff gave him a pass, attrib-

uting the postponement to "the interests of the investigation."[111]
Then Cohen delayed an appearance before the SSCI, claiming "post-surgery medical needs." But photographs of Cohen dining out on the town with five buddies appeared on social media, infuriating Senator Richard Burr (R-NC), who said, "I can assure you that any good will that might have existed in the committee with Michael Cohen is now gone."[112]

The big showdown arrived in the last week of February. On February 26, Cohen met behind closed doors with members of the SSCI and apologized to them for lying to them in 2017. Senators grilled him over his previous prevarications. During the hearings eighteen months earlier, Cohen had defended Trump. Staffers reported that he now took the opportunity to eviscerate his former client. After more than eight hours of questioning, he told the press, "I look forward to tomorrow to be able to, in my voice, tell the American people my story. I am going to let the American people decide exactly who is telling the truth."[113]

Cohen was playing for time. On February 20, 2019, a federal judge had approved delaying the start of his three-year prison sentence until May 6, due to physical therapy and his need to prepare to testify regarding Trump's business practices, campaign, charitable foundation, and payments made to "influence" the election. Questions about Mueller's investigation were off limits.[114]

As Davis and Cohen intended, Cohen's three days of testimony coincided with Trump's second historic meeting in Vietnam with North Korea's supreme leader, Kim Jong-un, in an attempt to prevent war on the Korean Peninsula. The strategy suited the *New York Times*, which ran fifteen stories on Cohen's testimony, including three op-eds. There was nothing on the front page about Trump's summit with the North Korean leader; three stories on that subject were buried on pages 8 and 9.

On February 27, 2019, Cohen took an oath to tell the truth before the House Government Reform and Oversight Committee. He began with a written statement that included this gem: "Mr. Trump did not directly tell me to lie to Congress. That's not how he operates. In conversations we had during the campaign, at the same time I was actively negotiating in Russia for him, he would look me in the eye and tell me there's no business in Russia and then go out and lie to the American people by saying the same thing. In his way, he was telling me to lie."[115] Huh? Cohen was channeling Johnny Carson's psychic character, "Carnac the Magnificent," who could divine unknown answers to unseen questions.

The disgraced lawyer, now disbarred, accused Trump of having committed a number of criminal infractions during the campaign. But Cohen did not deliver a scintillating takedown of the president as promised, nor did he provide evidence of collusion. "I have no direct evidence that Mr. Trump or his campaign colluded with Russia. I have my suspicions," he said. He did annihilate the key element of the Steele "dossier" that wouldn't die. "I've never been to Prague," he said in answer to a question. "I've never been to the Czech Republic."[116]

For more than five hours, attorney Cohen blamed his former client for his woes, calling Trump a "racist," a "conman," and a "cheat."[117] Representative Jim Jordan (R-OH) accused him of turning on Trump out of bitterness. "That's the point, isn't it? You wanted to work in the White House, but you didn't get brought to the dance." Cohen rejected that portrayal: "Sir, I was extremely proud to be personal attorney to the president of the United States. I did not want to go to the White House." That was demonstrably untrue; he had desperately wanted the chief of staff position.[118]

Cohen testified about his belief that the president had known about the Trump Tower meeting with the Russian lawyer to get "dirt on Hillary" ahead of time: "I remember being in the room with Mr. Trump, probably in early June 2016, when something peculiar happened. Don Jr. came into the room and walked behind his father's desk—which in itself was unusual. People didn't just walk behind Mr. Trump's desk to talk to him. I recalled Don Jr. leaning over to his father and speaking in a low voice, which I could clearly hear, and saying: 'The meeting is all set.' I remember Mr. Trump saying, 'Ok good . . . let me know.'"[119] What meeting? This allegation had no evidentiary value whatsoever.

Cohen offered up sexy tidbits, saying he'd pursued buying a supposed tape of Trump striking his wife, Melania, in a Moscow elevator in order to destroy it. But he had determined that the tape didn't exist—nor did he believe that the event had occurred. He'd also pursued the allegation that Trump had had a "love child," only to become convinced that that claim was phony as well. The "pee tape" was another fabrication. "I have no reason to believe that that tape exists," Cohen said.[120]

The most provocative claim Cohen offered was that he'd been listening on speakerphone in July 2016 when Trump had taken a call from Roger Stone, who said "he had just gotten off the phone with Julian Assange and that Mr. Assange told Mr. Stone that, within a couple of days, there would be a massive dump of emails that would damage Hillary Clinton's campaign." The drop occurred on July 22. Even as Cohen testified, WikiLeaks responded via its Twitter account: "WikiLeaks publisher Julian Assange has never had a telephone call with Roger Stone. WikiLeaks publicly teased its pending publications on Hillary Clinton and published >30k of her emails on 16 March 2016."[121]

Cohen insisted that his false written statement about the negotiations of the Trump Moscow project ending in January 2016 had been edited and reviewed by Trump's attorneys, including Jay Sekulow and Abbe Lowell. (This was refuted by the attorneys as "completely false." Documents showed that the date mentioned in the original draft written by Cohen was January 2016.)[122]

Among his false statements during the hearing: "I have never asked for, nor would I accept, a pardon from Mr. Trump."[123] But in the spring of 2018, after the raids, Cohen had instructed one of his attorneys at the time, Stephen Ryan, to ask the president's attorneys about a possible pardon. (He had been rebuffed, according to Giuliani.)[124]

The world saw a liar, lashing out at "the boss" he'd once idolized. Cohen knew that no pardon would be forthcoming, so he pandered to the Trump haters. *Washington Post* columnist Marc Thiessen called Cohen's testimony the "bombshell that didn't explode." Cohen's statement that Trump hadn't thought he was going to win the election "kind of undermines the argument" that Trump wanted to pay off Stormy Daniels to help his chances.[125]

The Cohen dumpster fire raged for seven hours, convincing viewers that the attorney had schemed and lied and peddled his connection to Trump to rake in millions of dollars. But if Mueller had hoped that Cohen would prove Russia collusion, it was a humiliating fail. Cohen had every motive to lie, to say he had been to Prague and had negotiated with Russians. He didn't. Instead he offered exculpatory evidence that the entire Mueller investigation was a fraud.

In his closing remarks, Cohen joined hands with the #Resistance: "Given my experience working for Mr. Trump, I fear that if he loses the election in 2020 that there will never be a peaceful transition of power." His concern for the future was "why I agreed to appear before you today."[126] That was the dumbest, most partisan statement he could have made. Perhaps Lanny Davis, who had had experience with Hillary's unwillingness to face her 2016 election loss, wrote it for him.

Cohen's performance led Representatives Jim Jordan and Mark Meadows to write a letter to the newly sworn-in attorney general, William Barr, alleging that Cohen had committed perjury. They cited Cohen's statements that he had "never defrauded any bank" and that he hadn't sought employment in the White House. His claim that he had not sought a pardon had also been false. Cohen's testimony "was a spectacular and brazen attempt to knowing and willfully testify falsely and fictitiously to numerous material facts," the letter said. "Mr. Cohen's prior conviction for lying to Congress merits a heightened suspicion that he has yet again testified falsely before Congress."[127]

Cohen wasn't finished. The following day, he met with the House Intelligence Committee in a closed-door session, where, Representative Adam Schiff said, the committee was able to "drill down" on certain issues in more detail. It emerged that Cohen had met several times with Schiff's staff before the public hearing without telling the Republicans, raising the possibility that he had been "coached."[128] Or had Schiff's team written some of his anti-Trump comments?

Schiff and his subordinates wanted ammunition to use in a Trump impeachment, but Cohen had lied to lawyers, the IRS, his accountants, bank officials, and Congress, making him useless as a witness against the president.

Cohen testified a final time behind closed doors to the House Intelligence Committee in early March 2019. Representative Devin Nunes, the ranking member, called that pointless exercise just part of the Democrats' "building out a narrative." Cohen had no classified information to impart. "They want to keep it behind closed doors so that they can conveniently say, well, we can't talk about what happened behind closed doors. . . . [The] guy has no classified information. It's ridiculous." He said that Cohen's public testimony had been "great for Republicans, it was great for Donald Trump, because we now know that the dossier was total bunk."[129]

On May 6, Cohen reported to federal prison in Otisville, New York, to begin serving his sentence.[130] His Shakespearean fall from Trump's inner circle was complete. George Sorial, a Trump Organization executive vice president and its chief compliance officer, told the *Wall Street Journal* that Cohen had lasted a decade at the company despite his shortcomings as a lawyer because "he was loyal. . . . Some very bad people with an anti-Trump agenda led him astray."

That included Davis, the Mueller team, and the SDNY prosecutors, who had cleverly turned a noncrime into a crime and convinced Cohen to go along with it. He had been particularly ill served by an attorney/adviser long afflicted with Clinton-itis. Either Davis had manipulated Cohen, or the two had worked in concert to target Trump. Why Cohen trusted Davis remains a mystery.

Maybe his bitterness at being left behind by "the boss" whom he'd once pledged "to take a bullet for" explains it. But in the end, Cohen's lies finally caught up with him.

CHAPTER 8

COLLATERAL DAMAGE

Throughout the interview, Flynn had a very "sure" demeanor and did not give any indicators of deception. Strzok and [redacted FBI agent] both had the impression at the time that Flynn was not lying or did not think he was lying.

—FBI REPORT ON INTERVIEW WITH
LIEUTENANT GENERAL MICHAEL FLYNN, JULY 19, 2017

Four days after Trump's inauguration in January 2017, FBI director James Comey did something he knew was out of line: he told his second in command, Andrew McCabe, to send two FBI agents to interview Trump's new national security advisor, retired army Lieutenant General Michael Flynn, about his phone conversations with the Russian ambassador, Sergey Kislyak, during the transition.

"I sent them," Comey admitted to MSNBC's Nicolle Wallace. By so doing, he had gone around standard practice, which he knew well from years of experience, as he pointed out: "In the George W. Bush administration or the Obama administration, if the FBI wanted to send agents into the White House itself to interview a senior official, you would work through the White House counsel, and there would be discussions and approvals and who would be there."[1]

THE PLOT TO IMPLICATE MICHAEL FLYNN

Comey wanted to avoid going through official channels. His move was "something I probably wouldn't have done or maybe gotten away with in a more . . . organized administration." The key phrase is "maybe gotten away with."[2]

He and McCabe had a scheme in mind: laying a trap for the retired general, who had been forced from his role as head of the Defense Intelligence Agency by President Barack Obama in 2014 for expressing opinions at odds with the administration's stance on Iran and other national security issues. They wanted to disrupt the new administration, get revenge on a man who had been a thorn in Obama's side, and supercharge the Trump-Russia collusion investigation.

Before Trump took office, on January 5, 2017, Comey had met with Deputy Attorney General Sally Yates, National Security Advisor Susan Rice, Vice President Joe Biden, and Obama at the White House. That was the meeting Rice memorialized in an email to herself on Inauguration Day: "The President stressed that he is not asking about, initiating or instructing anything from a law enforcement perspective. He reiterated that our law enforcement team needs to proceed as it normally would by the book. . . . From a national security perspective, however, President Obama said he wants to be sure that, as we engage with the incoming [Trump] team, we are mindful to ascertain if there is any reason that we cannot share information fully as it relates to Russia."[3]

Joseph diGenova, a former US Attorney for the District of Columbia, has charged that this unusual meeting was held to discuss targeting Flynn.[4] They despised and feared him, as they despised and feared Trump. They would use Flynn's perfectly legal and appropriate communications with the Russian ambassador during the transition period and twist them into something nefarious.

Highly decorated, with thirty-three years in the army, Flynn had served multiple deployments to Iraq and Afghanistan, building a reputation as a maverick "known for his candor and his unorthodox sensibilities about intelligence and military operations."[5]

He had led the Defense Intelligence Agency from 2012 to 2014, trying to overhaul the way the military treats intelligence.[6] And he had repeatedly challenged Obama's foreign policy related to military action.[7]

In his last interview as director of the DIA, Flynn had said he was being forced into retirement for questioning Obama's public statements that al-Qaeda was near defeat.[8] Under Flynn, the DIA had sent numerous classified warnings about the "dire consequences of toppling Syrian President Bashar Assad" that had been ignored.

Former DIA official W. Patrick Lang summarized it thus: "Flynn incurred the wrath of the White House by insisting on telling the truth about Syria . . . they shoved him out. He wouldn't shut up."[9]

Flynn opened Flynn Intel Group, a private consulting firm, with his son, Michael G. Flynn.[10] Building on contacts in Russia, he became a regular guest on RT, formerly Russia Today, a media outlet widely thought to be financed by the Kremlin. In December 2015, he traveled with his son to Moscow to attend a gala for the network, sitting next to President Vladimir Putin—a trip undertaken with the approval of the Pentagon. Paid a fee through a speakers' bureau, Flynn gave an interview to one of the network's top presenters.[11]

Though he had called Putin a "dictator" and a "thug," Flynn also advocated for a rapprochement with Russia, enabling both countries to focus on fighting Islamic terrorism.[12]

Everything about Flynn was anathema to Obama, Rice, Brennan, and Clapper.[13] Flynn knew the rot that existed within Clapper's and Brennan's intelligence agencies. Who knew what radical changes he might undertake? What he might discover about their spying activities?

In late July 2016, not long after Flynn's book *The Field of Fight: How We Can Win the Global War Against Radical Islam and Its Enemies* was released, John Brennan issued an EC, or electronic communication, to Comey, which became the basis of an FBI enterprise counterintelligence investigation dubbed "Operation Crossfire Hurricane." The EC named four Trump campaign associates as targets: George Papadopoulos, Carter Page, Paul Manafort, and Michael Flynn. FBI counterintelligence agent Peter Strzok signed the document opening the aptly named inquiry—which would rely on intelligence assets from the United States, the United Kingdom, Australia, and Italy—on July 31.[14]

The investigation of Flynn was "based on his relationship with the Russian government," according to Comey and a DOJ official.[15] What relationship? Attending a dinner in Moscow? Giving a speech? Talking to the ambassador? The Flynn investigation was built on sand.

" 'Because of the sensitivity of the matter,' the FBI did not notify congressional leadership about this investigation during the FBI's regular counterintelligence briefings," according to a report by the House Permanent Select Committee on Intelligence (HPSCI).[16] Comey didn't want the Bureau's overseers to know they were targeting a political campaign.

"This was a designed plot to frame General Flynn so they could figure out a way to go after Donald Trump," said diGenova.[17]

Though the probe was opened in late July, by December little had been gleaned that could be used against Flynn. The investigation needed a big push. On December 29, 2016, Obama's intelligence apparatus issued a Joint Analysis Report (JAR) alleging that the Russians had interfered in the 2016 election through hacking and propaganda spread by media outlet RT.

The JAR, in which the administration referred to Russian cyberactivity as "Grizzly Steppe," provided little evidence to tie senior officers of Russian intelligence services to the plan to influence the election.[18] Even so, the Obama administration ejected thirty-five suspected Russian spies, closed several of their compounds, and imposed sanctions on Russia's two leading intelligence services.[19]

The timing of the sanctions—in the waning days of Obama's administration—appeared designed to box in the Trump administration and to provoke a response that could be twisted to prove that Trump was soft on Putin. On the same day, Flynn spoke to Kislyak by phone. That was not unusual. Flynn had made a condolence call to Kislyak on December 19 after a Russian ambassador had been murdered. He'd called on December 28, expressing condolences after a Russian plane had been shot down on its way to Syria. They discussed setting up a Trump-Putin phone call after the inauguration. Possibly a Trump administration official would visit Kazakhstan for a conference in late January.[20]

The next day, to the surprise of the intelligence community, Putin announced that there would be no response to the sanctions; Trump praised his comments on Twitter.[21]

The FBI had transcripts of the calls between Kislyak and Flynn. But those highly classified conversations apparently revealed no crime. If they had, why not just arrest Flynn? The FBI needed the media to gin up a controversy and give cover for monitoring the national security advisor's phone calls. So someone leaked.

On January 12, 2017, David Ignatius of the *Washington Post* published a column on Russian hacking. "According to *a senior U.S. government official*, Flynn phoned Russian Ambassador Sergey Kislyak several times on Dec. 29, the day the Obama administration announced the expulsion of 35 Russian officials as well as other measures in retaliation for the hacking. What did Flynn say, and did it undercut the U.S. sanctions?" [author's italics][22]

That leak of highly classified information was a serious crime under 18 U.S.C. § 798.[23] An investigation of McCabe by the OIG later indicated a high probability that he had been the leaker.[24] Any underlying motivation to serve the public good by disclosing a lie or misrepresentation is of no legal consequence under the statute.

The leak may have also been driven by personal animus. McCabe had clashed with Flynn after the general's intervention in 2014 on behalf of a female counterterrorism agent who had accused top FBI brass of sexual discrimination. Flynn had written a letter supporting supervisory special agent Robyn Gritz on his official Pentagon stationery and offered to testify in her case.

"[Flynn's] offer put him as a hostile witness in a case against McCabe, who was soaring through the bureau's leadership ranks," reported *Circa*'s John Solomon and Sara Carter. "The FBI sought to block Flynn's support for the agent, asking a federal administrative law judge in May 2014 to keep Flynn and others from becoming a witness in her Equal Employment Opportunity Commission (EEOC) case."[25]

The pending case was "serious enough to require McCabe to submit to a sworn statement to investigators" that provided "some of the strongest evidence in the case of possible retaliation."[26]

Several FBI employees were "uncomfortable" due to McCabe's apparent personal antagonism toward Flynn and believed he was driving the investigation into the national security advisor. "Three FBI employees told *Circa* they personally witnessed McCabe make disparaging remarks about Flynn before and during the time the retired Army general emerged as a figure in the Russia case."[27]

The FBI agents' concerns escalated when a description of the intercepted calls between Kislyak and Flynn became public. "The Flynn leaks were nothing short of political," said one FBI official. "The leaks appeared to be targeted to take Flynn out."[28]

Gritz also believed it was political. "McCabe is vicious to anyone who either stands up to him or is a threat to his 'power' and [he] is a screamer," said Gritz. "McCabe wrote false and nasty comments on numerous documents about me when he had not one bit of proof of any lack of performance."[29]

The issue would prompt Senator Charles Grassley to call for McCabe to recuse himself from the Flynn case. In a letter to Deputy Attorney General Rod Rosenstein, Grassley wrote, "[The] evidence and the failure to recuse calls into question whether Mr. McCabe handled the Flynn investigation fairly and objectively, or whether he had any retaliatory motive against Flynn for being an adverse witness to him in a pending proceeding." [30]

The FBI had obviously been monitoring Kislyak's calls. Perhaps Flynn's name was "unmasked"—or his communications were being monitored through the EC. The phone conversation was perfectly legal and in keeping with Flynn's role as incoming NSA. Though a full transcript has never been released, Flynn later said he had asked Kislyak not to "overreact," to wait a few weeks to retaliate until the incoming Trump administration could review the situation, a reasonable approach.

The following day, the *Wall Street Journal* reported the call in more detail, as well as a denial by Trump spokesman Sean Spicer that Flynn had discussed the sanctions. He said that the two men had discussed four other topics, including a phone call between the presidents of the two countries. Exactly why the communications would have been illicit was not made clear.[31]

On January 15, Vice President Mike Pence appeared on the CBS show *Face the Nation* and was asked a question about contacts between members of the Trump incoming team and the Russian government.[32]

"Did any advisor or anybody in the Trump campaign have any contact with the Russians who were trying to meddle in the election?" asked John Dickerson.

"Of course not," Pence said. "And I think to suggest that is to give credence to some of these bizarre rumors that have swirled around the candidacy."

Dickerson conflated the campaign and the transition. On hearing Pence's answer, Comey and crew knew they had an issue to exploit. At the same time, Brennan lashed out at Trump, accusing him of underestimating Putin as the president sought a reset of relations with Russia, stating that he was open to lifting the recently imposed sanctions. Their feud heated up, and Flynn was in the middle of it.[33]

DOJ insiders floated stories speculating that Flynn had violated the Logan Act, an obscure law enacted in 1799 that makes it a felony for a private citizen to interfere in international disputes between the US and foreign governments.

But virtually no one has ever been prosecuted under the Logan Act, principally because most lawyers, legal scholars, and judges agree that it is probably unconstitutional, and in this case, it was irrelevant. Flynn was the incoming NSA, not a private citizen. He was preparing the administration for the foreign policy challenges that lay ahead and establishing the kind of vital contact that assists a new president in formulating effective relationships and policies. In other words, Flynn was doing his job.

The day before Trump's inauguration, the insiders—Comey, Yates, Clapper, and Brennan—met to discuss whether to brief the new president or his staff on the Flynn situation. Comey argued against it, concerned that it would complicate the agency's investigation—meaning they would first have to get Flynn into the trap. He prevailed.[34]

Instead of notifying Trump of any concerns about his new national security advisor, the media was alerted through leaks. A published story would then serve as a convenient excuse for the FBI to ask for an interview of Flynn at the White House before he had even unpacked his boxes in his new office there.

"Michael Flynn is the first person inside the White House under Mr. Trump whose communications are known to have faced scrutiny as part of investigations" by the FBI, CIA, NSA, and Treasury Department, reported the *Wall Street Journal* on January 22.

"The counterintelligence inquiry is aimed to determine the nature of Mr. Flynn's contact with Russian officials, and whether such contacts may have violated laws, *people familiar with the matter said.* [author's italics]"[35] One of the authors was Devlin Barrett, who would come under scrutiny as the recipient of leaks by McCabe. This story surfaced several targets: Paul Manafort, Roger Stone, and Carter Page. But the crosshairs, for now, were trained on Flynn.

The following day, Spicer brushed off the story, saying that Flynn hadn't discussed US sanctions with Kislyak.[36] With the media panting for more, the spadework for the scam had been accomplished.

Comey's call to McCabe on January 24 to "send a couple guys over" laid the predicate for a vicious, malevolent attack on a man who had honorably served his country for thirty-three years that would cost Flynn his hard-won reputation, his home, and more than $5 million in legal fees.[37]

The FBI director's publicly stated reason for sending the agents to talk to Flynn centered around statements he'd made to Pence that contradicted the phone conversation between Flynn and Kislyak.[38] In other words, what was being reported in the press—based on information planted by the FBI—triggered the interview.

Comey was later pressed on his unusual decision to defy protocol by members of the House Oversight and Judicial Committee in a closed-door session. "It is not the FBI's job, unless I'm mistaken, to correct false statements that political figures say to one another," said Representative Trey Gowdy. "So why did you send two Bureau agents to interview Michael Flynn?"[39]

"Because one of the FBI's jobs is to understand the efforts of foreign adversaries to influence, coerce, corrupt the Government of the United States," Comey responded. It was a feeble excuse. If the FBI were to open investigations on all political figures making comments at odds with one another, it would do nothing else.

So at Comey's direction, at 12:35 p.m. on January 24, 2017, the first Tuesday after the inauguration, McCabe picked up the phone and called Flynn at his new office in the West Wing.[40]

Flynn's lawyers later described the conversation this way: "General Flynn had for many years been accustomed to working in cooperation with the FBI on matters of national security. He and Mr. McCabe briefly discussed a security training session the FBI had recently conducted at the White House before McCabe, by his own account, stated that he 'felt that we needed to have two of our agents sit down' with General Flynn to talk about his communications with Russian representatives."[41]

According to McCabe, "I explained that I thought the quickest way to get this done was to have a conversation between [General Flynn] and the agents only. I further stated that if LTG Flynn wished to include anyone else in the meeting, like the White House Counsel, for instance, that I would need to involve the Department of Justice. [General Flynn] stated that this would not be necessary and agreed to meet with the agents without any additional participants." [42]

In other words, McCabe deliberately manipulated Flynn into not having an attorney present. Juxtapose with that the treatment of Hillary Clinton, who was allowed a team of lawyers to accompany her when she was interviewed by FBI agents, including Peter Strzok, in early July 2016.

Less than two hours later, at 2:15 p.m., two agents arrived at Flynn's office: Peter Strzok and Joe Pientka. Unsuspecting, even "jocular," General Flynn offered to give the agents a tour of the area around his office.

"Prior to the FBI's interview of General Flynn, Mr. McCabe and other officials 'decided the agents would not warn Flynn that it was a crime to lie during an FBI interview because they wanted Flynn to be relaxed, and they were concerned that giving the warnings might adversely affect the rapport.'" [43]

Here's the neon sign that flashes "PERJURY TRAP."

The agents had a transcript of the December 29 phone conversation between Flynn and Kislyak but did not inform Flynn of that fact. If they had, the FBI would have admitted that it was spying on the incoming NSA through incidental surveillance collection.

On the way over, the FBI agents decided that if "Flynn said he did not remember something they knew he said, they would use the exact words Flynn used . . . to try to refresh his recollection. If Flynn still would not confirm what he said . . . they would not confront him or talk him through it." [44]

The agents reported that Flynn was "unguarded" during the interview and "clearly saw the FBI agents as allies." [45] Why wouldn't he? He'd done nothing wrong.

But the notorious Strzok-Page text messages later revealed that the FBI agents had not been his allies. It was a setup from the beginning. The day before he interviewed Flynn, Strzok texted his paramour Lisa Page, "I can feel my heart beating harder, I'm so stressed about all the ways THIS has the potential to go fully off the rails."[46]

Page replied, "I know. I just talked with John [unknown last name], we're getting together as soon as I get in to finish that write up for Andy [McCabe] this morning. I reminded John about how I had told Bill [probably Priestap] and the entire group that we should wait 30 to 60 days after the inauguration to change how we're managing this stuff. As it is, he went ahead, and everything is completely falling off the rails."[47]

After the interview, the two FBI agents briefed McCabe, who in turn informed Comey. Strzok later maintained that he and his fellow agent had "both had the impression at the time that Flynn was not lying or did not think he was lying." Flynn "did not parse his words or hesitate in any of his answers."[48] They memorialized their conversation with Flynn in an FD 302. But McCabe had a hand in the summary's final form, according to the Strzok-Page texts. On February 14, 2017, Strzok texted, "Also, is Andy good with F 302?" Page texted back, "Launch on f 302." It was officially entered into the record a day later.[49]

When informed that agents were interviewing Flynn, Yates "was not happy." Nor were others on the FBI's seventh floor. Possibly the Bureau's numerous lawyers were raising red flags about the legality of the entrapment scheme, but Comey and McCabe plowed ahead anyway.[50]

What happened next was reprehensible. Yates received a detailed summary of the interview. The Logan Act and perjury were discussed "at great length" within the DOJ. A holdover from the Obama administration, Yates believed "it was important to get this information to the White House as quickly as possible."[51]

The leaks intensified. As later reported by the *Washington Post*, Yates considered Flynn's comments in the intercepted call to be "highly significant" and "potentially illegal, according to an *official familiar with her thinking.* [author's italics]"[52]

She had the full support of Clapper and Brennan. "They feared that 'Flynn had put himself in a compromising position' and thought that Pence had a right to know that he had been misled, according to one of the officials, who, like others, *spoke on the condition of anonymity.* [author's italics]"[53] Also from the *WaPo* story: "The FBI, Yates, Clapper, and Brennan declined to comment on the issue."

On January 26, Yates and DOJ official Mary McCord met with White House counsel Don McGahn. "The first thing we did was to explain to Mr. McGahn that the underlying conduct that Gen. Flynn had engaged in was problematic in and of itself," Yates later testified before a Senate judiciary subcommittee.[54]

She said that Flynn had lied to Pence about his conversations with Kislyak, leading the VP to give erroneous information to reporters that "we knew not to be the truth."[55]

Because the Russians knew he had lied, Flynn was in danger of being compromised, blackmailed, Yates insisted. That was absurd. The phone conversations were now public; by exposing the lie, how could the Russians extort Flynn?

McGahn asked if Flynn should be fired. Yates responded, "That really wasn't our call." However, "it wouldn't really be fair of us to tell you this and then expect you to sit on your hands," she said.[56]

The next day, McGahn asked Yates to return to his office. "Why does it matter to the Department of Justice that one White House official lied to another?" he asked.[57]

"It was a whole lot more than that," Yates said. "First of all, it was the vice president of the United States and the vice president had then gone out and provided that information to the American people who had then been misled and the Russians knew all of this, making Mike Flynn compromised now."[58]

Days later, Yates was out of a job, fired by Trump on January 30 for refusing to enforce his executive order banning travel to the United States from certain countries.[59] She became a hero to the

#Resistance, and at least one colleague inside the DOJ, attorney An-
drew Weissmann, whose path would soon cross that of Flynn. He
emailed his boss to say he was "proud" and "in awe" of her defiance
of the president.[60]

But Yates's role in the Flynn debacle was appalling.

"Yates represents [former Obama attorney general] Eric Holder's
most enduring legacy—normalization of political law enforcement,"
said J. Christian Adams, president of the Public Interest Legal Foundation and a former DOJ lawyer. "She saw it as her mission to sabotage the incoming administration."[61]

Meanwhile, the White House said nothing about Flynn. So
Comey and McCabe ramped up the pressure.

As the *Wall Street Journal* reported, "After Ms. Yates relayed the
concerns, *some intelligence officials* waited for White House officials
to issue a new statement—to correct the public record in some way
about Mr. Flynn's contacts with the ambassador, according to *people
familiar with the matter.* As time went on, it seemed to Justice Department officials that the White House didn't plan to do so, *these people
said.* [author's italics]"[62]

Flynn, through a spokesman, backed away from his denial that he
had discussed sanctions with Kislyak, saying that "while he had no
recollection of discussing sanctions, he couldn't be certain that the
topic never came up."

Trump administration officials said that "they did not see evidence that Flynn had an intent to convey an explicit promise to take
action after the inauguration." But "*nine current and former officials,*
who were in senior positions at multiple agencies at the time of the
calls, [and who] spoke on the condition of anonymity to discuss intelligence matters [author's italics]," said Flynn's references to the
election-related sanctions were explicit.[63]

Add up the folks on the top floors of the FBI and DOJ, throw in
Clapper and Brennan, and you've got nine who should face charges of
leaking highly classified information.

In an interview with The Daily Caller, Flynn insisted that the conversation with Kislyak "wasn't about sanctions. It was about the 35 guys who were thrown out. . . . It was basically, 'Look, I know this happened. We'll review everything.' I never said anything such as 'We're going to review sanctions,' or anything like that."[64]

But the damage was done. Flynn had lost the confidence of Trump and Pence. "This was an act of trust, whether or not he misled the vice president was the issue," said Spicer, adding that White House officials had determined that Flynn hadn't violated the law.[65]

At Trump's request, Flynn stepped down as NSA on February 13, 2017, after serving only twenty-four days. In his letter of resignation, Flynn apologized, saying that due to the fast pace of events, he had "inadvertently briefed" Pence and others with "incomplete information."[66]

"I have nothing to be ashamed of and everything to be proud of," he told Fox News.[67]

The *New York Times* reported the next day that "Obama advisers grew suspicious that perhaps there had been a secret deal between the incoming [Trump] team and Moscow, which could violate the rarely enforced, two-century-old Logan Act."[68]

All preposterous. But Comey and McCabe had succeeded in ousting Flynn and giving oxygen to the "collusion" investigation. That outrageous illegal maneuver was at the heart of the Trump-Russia hysteria.

The *Wall Street Journal* raised the "troubling" issue that Flynn might have been targeted by intelligence officials who were out to get him; few other media outlets seemed to understand or care about that issue.[69]

If Flynn hoped that his resignation would quell the furor, he was mistaken. Top-ranking Democrats demanded an investigation—all over a phone call that had been perfectly legal and in keeping with Flynn's role in the new administration.[70]

Almost as soon as Flynn resigned, CNN aired a story that the FBI would not pursue any charges against him. "The FBI interviewers believed Flynn was cooperative and provided truthful answers," according to CNN's Evan Perez. "Although Flynn didn't remember all of what he talked about, they don't believe he was intentionally misleading them, the officials say."[71]

TWISTING THE KNIFE

At first, it looked as if Flynn's departure was simply an early administration casualty. But Comey, McCabe, and their cohorts set out to destroy his life, digging deep to find any wrongdoing.

Did he file the right paperwork regarding his speech in Moscow, for which he was paid $33,750, to speak about US foreign policy and intelligence matters? (Far less than Bill Clinton had received for a speech in Moscow.) Before joining the Trump administration, did he register as a foreign agent when he lobbied for the Turkish government, for which his firm was paid $530,000?[72] Did he put it on his security clearance form? Did he disclose a twenty-minute public conversation with a graduate student with Russian and British nationalities at a 2014 UK security conference?[73] Why did he meet with Kislyak and Kushner at Trump Tower in mid-December 2016?[74]

The interaction with a graduate student, Svetlana Lokhova, amounted to a brief introduction to Flynn and a twenty-minute presentation at a dinner organized by Sir Richard Dearlove, the former head of MI6, peddled to the media by FBI informant Stefan Halper as a nefarious conversation. Flynn was most vulnerable on filling out appropriate security clearance forms regarding foreign payments. However, he had registered as a foreign lobbyist for Turkey prior to his appointment as NSA, according to his attorney Robert Kelner.[75]

"Gen. Flynn briefed the Defense Intelligence Agency, a component agency of the DoD, extensively regarding the RT speaking-event trip both before and after the trip, and he answered any questions that were posed by DIA concerning the trip during those briefings," Kelner said.[76]

Even though the two FBI agents who interviewed Flynn felt that he had not been deceptive, Special Counsel Robert Mueller decided to bring charges against him for lying. On December 1, 2017, in the District of Columbia court, Flynn pleaded guilty to one count of making a materially false statement to federal agents under 18 U.S.C. § 1001 and agreed to cooperate with the special counsel in its investigations.[77] The "criminal information" filed in court outlined four lies told by Flynn in two conversations with Kislyak.

Senior prosecutor Brandon Van Grack said that Flynn had "falsely stated" to the FBI that he hadn't asked the Russian ambassador to refrain from escalating a response to the sanctions.

In fact, Flynn had talked to a senior official with the transition team about "what, if anything, to communicate to the Russian ambassador."[78] Immediately after that conversation, Flynn had called Kislyak and asked that Russia respond only "in a reciprocal manner." On December 31, Kislyak called him back to tell Flynn that his country had "chosen not to retaliate in response to Flynn's request."[79]

In addition, he had "falsely stated" that during calls to the governments of Russia and several other countries, he had asked only about their positions on a US resolution submitted by Egypt condemning Israeli settlements in disputed territories. In fact, at the direction of a senior Trump administration official, Jared Kushner, he had asked them to delay the vote or defeat the resolution.[80]

Taken together, the charges sounded serious, but they were not. There was nothing nefarious about Flynn's communications with Moscow. Had Strzok and Pientka never visited his office, had Flynn refused to talk to them, there would had been no "crime." And had Flynn lied? Or forgotten details?

Pleading guilty to such a flimsy process charge was puzzling. If Flynn went to trial, he likely would have prevailed, since the only two witnesses to the alleged lie were on record in their FBI reports as having said he had not appeared to be lying. But Flynn stood up in court and said he had lied. It was clear that the warrior was falling on his sword.

"I recognize that the actions I acknowledged in court today were wrong, and, through my faith in God, I am working to set things right," Flynn said in a written statement. "My guilty plea and agreement to cooperate with the Special Counsel's office reflect a decision I made in the best interests of my family and of our country. I accept full responsibility for my actions."[81]

Flynn was the fourth and most prominent member of the Trump campaign or administration to be publicly charged by Mueller. Former Trump campaign chairman Paul Manafort and his deputy and business associate Richard Gates III were charged with financial improprieties related to earnings from work they'd done in Ukraine unrelated to Trump and were facing prison time. And there was the hapless campaign adviser George Papadopoulos, who had pleaded guilty to lying to the FBI about his contact with an alleged Russian spy.

But Flynn was the prize.

Trump "can't get away with claiming these charges aren't about his inner circle's contacts with Russia, and he can't dismiss Michael Flynn as some low-level aide," said DNC chairman Tom Perez.[82]

"That Mueller would treat Flynn as someone worth flipping, presumably in pursuit of a bigger case, is, to say the least, suggestive," said Amy Davidson Sorkin in *The New Yorker*.[83]

"The deal delivers to Mueller a witness who spent nearly all the 2016 campaign and the presidential transition at Trump's elbow, one who was involved with many conversations with foreigners, including key Russians," wrote Miles Parks for NPR. "Flynn's cooperation resets the clock on the Russian imbroglio as he begins debriefing Mueller and his team."[84]

Judge Emmet G. Sullivan, the presiding judge, probably wondered if Flynn was freely entering a guilty plea, if his lawyers had had access to any and all exculpatory evidence. On December 12, 2017, Sullivan ordered Mueller to produce "any information which is favorable to defendant and material either to defendant's guilt or punishment."

He issued another order in February 2018: "If the government has identified any information which is favorable to the defendant but which the government believes not to be material, the government shall submit such information to the Court for in camera review."[85] The court, not the prosecutors, would thus decide whether the material should be produced. He postponed Flynn's sentencing hearing until May.

That order shocked legal observers and probably Mueller as well. "It certainly appears that Sullivan's order supersedes the plea agreement and imposes on the special counsel the obligation to reveal any and all evidence suggesting that Flynn is innocent of the charge to which he has admitted guilt," wrote *National Review*'s Andrew McCarthy.[86]

Byron York of the *Washington Examiner* reported that according to two sources familiar with his closed-door congressional testimony on March 2, 2017, "Comey told lawmakers that the FBI agents who interviewed Flynn did not believe that Flynn had lied to them, or that any inaccuracies in his answers were intentional. As a result, some of those in attendance came away with the impression that Flynn would not be charged with a crime pertaining to the Jan. 24 interview."[87]

Comey emerged on his book tour to play Grand Weasel. "I don't know where that's coming from," he told George Stephanopoulos of ABC. "That—unless I'm—I said something that people misunderstood, I don't remember even intending to say that. So, my recollection is I never said that to anybody."[88]

In May 2018, a less redacted version of the report on Russia's election meddling by the House Intelligence Committee was released, revealing that Comey's memory was faulty—or he had lied to Congress in May 2017—and that the FBI had tried to hide that information. New details of the probe were revealed.[89]

Comey had testified that the FBI had conducted a counterintelligence investigation of Flynn during 2016, but he had ended it by late December of that year. However, the file had been "kept open due to the public discrepancy surrounding General Flynn's communications with Kislyak"—a discrepancy that didn't arise until mid-January.[90]

A CI investigation of a US citizen proceeds on the suspicion that he or she is an agent of a foreign power. The DOJ and the FBI seemed to be pursuing the theory that Flynn—despite his stated views about Putin and Russia—was an agent of the Kremlin.[91]

The House report quoted Comey's previously redacted statement to the committee on March 2, 2017: "The agents . . . discerned no physical indications of deception. They didn't see any change in posture, in tone, in inflection, in eye contact. They saw nothing that indicated to them that he knew he was lying to them."[92]

Perhaps there were more extensive quotes not in the report to support Comey's revisionism. But Comey, as FBI director, didn't recommend charging Flynn with lying.

Add to that McCabe's testimony to Congress on December 19, 2017: "The two people who interviewed [Flynn] didn't think he was lying, [which] was not [a] great beginning of a false statement case."[93]

All exculpatory. Did Flynn's lawyers have access to this testimony?

If Mueller's team had been hiding crucial documents, Flynn could have withdrawn his guilty plea. However, that wouldn't have ended the prosecution's case. It could go after him on other charges—or target someone he loved.

THE FEDS SQUEEZE FLYNN'S FAMILY

The original authorization by Rod Rosenstein outlining Mueller's investigative mandate as special counsel was filed on May 17, 2017. However, when the Mueller Report was released in April 2019, it revealed two additional scope memos authorizing specific targets.[94]

The first was dated August 2, 2017. Though much of it remains redacted, it included four sets of allegations against Flynn, including conspiracy with a foreign government, lying to the FBI, unregistered lobbying, and making false statements and omissions on government documents regarding his representation of Turkey.

Flynn Intel Group was probed for its work on behalf of the Turkish government involving making a documentary about Fethullah Gülen, a Turkish cleric living in Pennsylvania. President Recep Tayyip Erdoğan had accused Gülen of orchestrating a bungled coup attempt. Gülen had denied the allegations, and the US government had refused Turkey's demands for his extradition.[95] The work had occurred during the campaign but had ended in November 2016.[96]

The second scope memo, dated October 20, 2017, expanded the targets to include Flynn's son Michael G. Flynn, and established the authority to pursue "jointly undertaken activity." Though his name was redacted, it was easy to identify in context.[97]

Perhaps up to that point Flynn had fought off Mueller's bullies. But when they trained their "lawfare" weapons on his son, who had a four-month-old child, the general faced an escalating danger that threatened not only him and his wife, their home and life savings, but his entire family.

The squeeze play authorized Mueller to demand documents, phones, and laptops from Michael G. Flynn and any joint businesses. He could be indicted as a coconspirator.

A month after the second scope memo was filed, Flynn signed the plea agreement. He put his house on the market for $895,000 to pay his legal fees. Friends set up a fund so supporters could donate to his defense.

"I'm not going to sugarcoat it, this has been a trying experience," said Joe Flynn, his brother. "It has been a crucible, and it's not over."[98]

After Sullivan's order that the Mueller team disgorge any hidden evidence, including original FBI documents, Mueller's prosecutors filed a five-page memo offering more information, outlining lies Flynn had told to the media, the transition team, and Pence. None of those actions were crimes, just the excuses for the interview. In a separate memo referenced by Mueller, McCabe admitted that he had pressed Flynn not to have a lawyer present. This memo confirmed that the agents had had no legal basis for interviewing Flynn.[99]

Mueller also referred to a summary of an FBI interview of Strzok on July 19, 2017, undertaken in order "to collect certain information regarding Strzok's involvement in various aspects of what has become the Special Counsel's investigations." The Strzok 302 had been entered into FBI files on August 22, 2017.

The interview was done four days after his lover, Page, left the OSC and days before Strzok would also be removed. Strzok said he and Comey had "at various times" updated Yates about "the entire span of the FBI's Russia election interference/collusion investigations."[100]

Flynn told Strzok that "he had been trying to build relationships with the Russians." That doesn't sound like someone deep in Putin's pocket. Strzok said that Flynn had had a "very sure demeanor and *did not give any indicators of deception*. . . . Flynn struck Strzok as 'bright, but not profoundly sophisticated.' [author's italics]" The most important observation was this one line: "Strzok and [redacted agent] both had the impression at the time that Flynn was not lying or did not think he was lying." Another interesting tidbit: "He then stated that I probably knew what was said" in the phone call.[101] So why lie?

Mueller didn't publicly file even a redacted form of the original Flynn 302, written by agent Pientka. Why not? Senator Charles Grassley, the chairman of the Senate Judiciary Committee, repeatedly demanded that the DOJ present Pientka for testimony but was rebuffed. After Flynn's guilty plea, Grassley asked for a transcript of the Kislyak call, the Flynn 302, and an interview with Pientka, only to be told no.[102]

Pientka could give his impressions on whether Flynn had told the truth and shed light on whether the document had been altered or falsified. What were they hiding? Representative Mark Meadows shed some light on why Pientka was the DOJ's invisible man: he was on Mueller's team.[103]

On December 4, 2018, Mueller filed a sentencing memo in Judge Sullivan's court recommending little or no prison time for Flynn, saying that the retired general had provided "substantial" help to investigators about "several ongoing investigations." But the memo wasn't a "smoking gun" showing that Trump had colluded with the Russians or done anything else illegal. It wasn't even a squirt gun. It praised his military and public service and said that Flynn had been interviewed nineteen times by prosecutors on Mueller's team and other DOJ lawyers. A heavily redacted addendum suggested that he had provided no meaningful information that Trump had conspired or coordinated with Russia to win the election.[104]

The documents made a passing reference to the moribund Logan Act—pointless, except to provide cover to the Obama officials who had used it as an excuse to go after Flynn. The memo did not address how the investigation had begun.

Trump's attorney Rudy Giuliani likened Flynn's offenses to "spitting on the sidewalk, with major repercussions for many." He called the Mueller team "overzealous media inspired prosecutors. They are sick puppies."[105]

On December 11, 2018, Flynn's attorneys filed a lengthy sentencing brief alleging that McCabe had pressed Flynn not to have an attorney present during the interview and that the interviewing agents had not warned him that any false statements he made could constitute a crime.[106] The document described the aggressive tactics the FBI had taken in targeting Flynn, in sharp contrast to the kid-glove strategies used with Clinton and her cronies over her illegal home email server. Investigators had believed that several people, including Clinton, had been untruthful in those interviews, but no one had been charged with making a false statement.

Footnote 23 in Flynn's sentencing brief raised a huge red flag, referencing the date on Flynn's 302 as August 22, 2017—nearly seven months *after* he had been interviewed and about a week after Strzok had been removed from the special counsel's team over his anti-Trump text messages.[107]

Judge Sullivan ordered the Flynn team to turn over documents backing its allegations and gave Mueller forty-eight hours "to file on the docket FORTHWITH the cited Memorandum and FD-302" and any other documents related to the inquiry.[108]

Shockingly, Mueller was still trying to hide the truth from the public. On December 14, he filed redacted versions of several documents—the Strzok 302 and McCabe's memo—urging the judge not to see the "circumstances" surrounding the events as "mitigating."[109]

Mueller wrote, "The defendant chose to make false statements about his communications with the Russian ambassador weeks before the FBI interview, when he lied about that topic to the media, the incoming Vice President, and other members of the Presidential Transition Team. When faced with the FBI's questions on January 24, during an interview that was voluntary and cordial, the defendant repeated the same false statements. The Court should reject the defendant's attempt to minimize the seriousness of those false statements to the FBI."[110]

But what had the agents said at the time? Still no Flynn 302. Under pressure from Judge Sullivan, Mueller's team finally handed over two redacted versions on December 17.[111]

Let's follow the bouncing Flynn 302.

All 302s have three dates on them: the date of interview, the date of drafting, and the date the supervisor approved it. The interview of Flynn by agents Strzok and Pientka had been done on January 24, 2017. Pientka had written a summary the same day.

But McCabe hadn't approved it until February 14.[112]

That was strange, but stranger still was the fact that it had been "re-signed and re-certified" on May 31, 2017, a few weeks after the appointment of Mueller.[113]

In submitting the document, Van Grack, the senior assistant special counsel heading up the Flynn prosecution, explained that the content of both documents was identical, but the original 302 had had a header labeled "DRAFT DOCUMENT/DELIBERATIVE MATERIAL." That had been removed, and the document had been resubmitted for the use of the Mueller team.[114]

A simple header mistake? Were the documents really identical? Why the three-month gap?

Flynn had been under investigation since the EC filed by Brennan in mid-July 2016 for "Russian collusion" that didn't exist, and possibly even earlier. Either his name had been illegally unmasked during the calls to Kislyak or he had been the subject of a FISA warrant, and agents had scooped up all his communications. McCabe or someone else at his level had leaked highly classified information to the press to trigger a furor that had put Flynn under the microscope. Comey had "got[ten] away" with sending a "couple of guys over." He and McCabe had ignored protocol by not contacting White House counsel. McCabe had pressed Flynn not to have a lawyer, ostensibly to make it easier to talk about the "significant media coverage" he had generated as a pretext.

The FBI agents didn't warn Flynn that he was being interviewed because they suspected he had committed a crime. They wanted him "relaxed" and "unguarded." They didn't need to know what he had said; they had the transcripts. They wanted to see if he would give them a weapon to bludgeon the president.

Flynn regarded them as allies; he was ignorant of the rabid anti-Trump malice held by Comey, McCabe, and Strzok—all of whom were fired and remain under criminal investigation. Although Yates "was not happy" about Comey's scheme, she joined in, marching over to the White House and engineering the firing of Flynn.

All those deliberately deceitful machinations, and they nailed him for telling four inconsequential lies. Of course, no one should lie to a federal agent. But the lesson from that episode is that no one should ever talk to an FBI agent without a lawyer present under any circumstances.

Flynn's statements amounted to an anthill compared to the mountain of lies, prevarications, and obfuscations made by Comey in his congressional testimony, also indictable offenses. That elaborate scheme, given a final twist of the knife by Mueller and his crew, destroyed a man's life. Flynn should never have been charged.

Flynn pleaded guilty not because he lied but because Mueller crushed him financially and threatened to take legal action against his son. Mueller's prosecution made it impossible for Flynn to find employment. The conduct of the FBI and the anti-Trump zealots on Mueller's team of prosecutors was egregious.

The release of the Mueller Report revealed more chinks in the carefully constructed artifice of the case against Flynn. The second volume of the Mueller Report, citing possible obstruction of justice violations, quoted a voice mail left by John Dowd, Trump's personal attorney, on November 22, 2017, for Flynn's counsel Robert Kelner, after Flynn had withdrawn from a joint defense agreement with the president.

Dowd said, "I understand your situation but let me see if I can't state it in starker terms. . . . [I]t wouldn't surprise me if you've gone on to make a deal with . . . the government. . . . [I]f . . . there's information that implicates the President, then we've got a national security issue . . . so, you know . . . we need some kind of heads up. Umm, just for the sake of protecting all our interests if we can. . . . [R]emember what we've always said about the President and his feelings toward Flynn and, that still remains."[115]

Flynn's attorneys returned the call the next day, saying, according to Mueller, that "they were no longer in a position to share information under any sort of privilege. According to Flynn's attorneys, the President's personal counsel was indignant and vocal in his disagreement. The President's personal counsel said that he interpreted what they said to him as a reflection of Flynn's hostility towards the President and that he planned to inform his client of that interpretation."[116]

Dowd's comments were portrayed by Mueller—and the press— as an attempt to obstruct justice by urging Flynn not to cooperate. But after Judge Sullivan ordered the government to release the full transcript of the voice mail, it became clear that Mueller's team had deceptively edited his call.

The Mueller version omitted significant words from Dowd's voice mail: "I'm sympathetic. . . . I understand that you can't join the joint defense; so that's one thing. If, on the other hand, we have maybe a national security issue . . . some issue, we got to—we got to deal with, not only for the President, but for the country . . . without you have to give up any confidential information." [117]

By leaving out the context and Dowd's clear intent not to interfere with Flynn's cooperation with the special counsel, Mueller twisted his words to mean the opposite.

A highly esteemed fixture of the DC bar, Dowd was understandably furious, accusing Mueller of seeking to "smear and damage the reputation of counsel and innocent people." [118] Dowd was not attempting to obstruct justice but rightfully doing his job as the president's counsel.

By "taking out half my words, they changed the tenor and the contents of that conversation with Robert Kelner," Dowd said. "Isn't it ironic that this man who kept indicting and prosecuting people for process crimes committed a false statement in his own report?" [119]

A prosecutor who deceptively edits evidence in a court submission, leaving out information that goes against his argument, ends up in deep trouble with the judge and becomes subject to discipline by the bar. There was no excuse for Mueller's dishonesty. The revelation of the Dowd edits reinforced the realistic concern that either McCabe or Mueller's team had altered Flynn's 302.

(So far, Mueller's team has not provided either the transcript or the audio recording of the Flynn/Kislyak call.)

In early June 2019, after the Mueller Report revealed serious discrepancies with the case against him, Flynn surprised everyone by firing his lawyers. Signaling a change in strategy, he hired former US Attorney Sidney Powell, the author of the hard-hitting book *Licensed to Lie: Exposing Corruption in the Department of Justice*, detailing corrupt and abusive tactics used by dishonest federal prosecutors involved in the Enron Task Force, especially Andrew Weissmann, then general counsel and deputy director of the FBI. [120]

Powell, who defended an employee of Merrill Lynch in that massive case, has called Weissmann, Mueller's lead deputy, the "kingpin of prosecutorial misconduct." [121] In 2012, she filed a complaint against Weissmann with the attorney grievance committee of the

New York State Unified Court System over his actions during the Enron prosecutions.[122]

Flynn's decision was excellent news for his supporters. They believed that Powell—who was familiar with the tricks used by Weissmann, such as hiding exculpatory evidence, using false summaries of interviews, and threatening witnesses—would drill down to uncover who had been involved in the plot to take the general down.

As of this writing, Flynn has not been sentenced. After his career was destroyed, we learned that he was the least of the liars.

CHAPTER 9

TARGETED INTIMIDATION

You don't really care about Mr. Manafort's bank fraud. What you really care about is what information Mr. Manafort could give you that would reflect on Mr. Trump or lead to his prosecution or impeachment.

—US District Judge T. S. Ellis III,
rebuke to special counsel lawyers, May 4, 2018

The Mueller team's outrageous and physically abusive treatment of former Trump campaign manager Paul Manafort began in late July 2017, the week he voluntarily met with the Senate Intelligence Committee looking into possible Russian "collusion."

One morning at 6:00 a.m., while the sixty-eight-year-old lobbyist and his wife were in bed, a dozen FBI agents raided his Alexandria, Virginia, condo without warning to execute a sealed search warrant. They stayed ten hours to "mirror" electronic devices and seize records, including "privileged and confidential materials" prepared by his lawyers to aid in his testimony.[1]

This move was designed to intimidate and instill fear. Trump's lawyer John Dowd slammed the tactic in an email to the *Wall Street*

Journal, accusing investigators of committing a "gross abuse of the judicial process" for "shock value." Dowd wrote, "These methods are normally found and employed in Russia, not America."[2]

Before the raid, Manafort had provided documents voluntarily, and Mueller had not requested the records that had been seized. A former Justice Department official described the move as unusual. "I think it sends a very strong message to both Manafort himself and potentially other people who might be targets of this investigation that Mueller is going to pursue this aggressively."[3]

Mueller's prosecutors should have been sanctioned for the seizure of Manafort's privileged communications with his lawyer. The raid was a clear sign the special counsel investigation had quickly gone off the rails, a failure of both Mueller and Rod Rosenstein, who had given the special counsel latitude to investigate "any other matters," contrary to the special counsel statute.

TAKING OUT MANAFORT

A political strategist and lobbyist who had worked for forty years on Republican political campaigns, Paul Manafort was brought in by Trump in March 2016 in advance of the Republican National Convention for his expertise in securing delegates. Two months later, he became campaign chairman.

He resigned in August 2016 after the *New York Times* reported that he had received $12.7 million in undisclosed cash payments from former Ukrainian president Viktor Yanukovych's pro-Russian party between 2007 and 2012.[4]

Manafort had been the subject of an FBI investigation into his business practices that had been dropped in 2014 after being deemed not prosecutable by the Obama DOJ. But Mueller's team, given carte blanche, blew the dust off those files and used escalating intimidation tactics to get Manafort to turn on Trump.[5]

As a lobbyist on behalf of unsavory dictators such as Ferdinand

Marcos of the Philippines and Mobutu Sese Seko of the Democratic Republic of the Congo, Manafort was no stranger to criticism. He defended his political work as "in support of White House foreign policy goals."[6]

But he had been targeted more than a year before the special counsel was appointed by both Christopher Steele and a contractor for the Democratic National Committee, Alexandra Chalupa, a Ukrainian American lawyer and activist. During the Clinton administration, she had served in the White House Office of Public Liaison.[7]

Steele's so-called Source E claimed that Manafort was in charge of a "well-developed conspiracy" between the Trump campaign and the Kremlin to release hacked emails and meddle in the 2016 election.[8] As Mueller would discover, that allegation was fraudulent—completely made up.

On May 3, 2016, Chalupa emailed a top DNC official that a "big Trump component" would hit "the pipe" in the next few weeks.[9] Less than a month later, from the Ukraine there emerged a handwritten "black ledger" identifying large cash payments to Manafort. A story in the *New York Times* about the ledger prompted his resignation as Trump's campaign chairman.[10]

Though instrumental in pushing information on Manafort into official channels, Chalupa's name does not appear in the Mueller Report. And it turned out that the FBI had repeatedly been warned that the "black ledger" might be fake.

In February 2017, soon after Flynn's resignation, Manafort issued a statement that he had "never had any connection to Putin or the Russian government—either directly or indirectly—before, during or after the campaign."[11]

But he became a primary target of Mueller's investigation, which focused on four disparate threads: a possible Ponzi scheme involving real estate in Manhattan, work for Russian oligarch Oleg Deripaska from 2006 to 2009, his having been paid to influence US opinion of Russia on behalf of Ukrainian president Yanukovych, and potential FARA violations.[12]

Though Manafort and Gates had filed FARA paperwork retroactively in June 2017 for their work in Ukraine, Mueller used that as a way to ratchet up the pressure.[13]

Ukraine's top anticorruption prosecutor, Nazar Kholodnytsky, cautioned the US State Department's law enforcement liaison and multiple FBI agents in late summer 2016 that "Ukrainian authorities who recovered the ledger believed it likely was a fraud."[14]

A Ukrainian businessman who worked with Manafort, Konstantin Kilimnik, wrote an email to a senior US official on August 22, 2016, to express his concerns that the document was probably fake. He pointed out that payments to Manafort had always been in the form of wire transfers, not cash.[15]

The FBI used the document anyway but in a devious fashion. In its request for a search warrant on Manafort's house, it didn't quote the ledger; it referenced media reports about it, just as it had used Isikoff's story—sourced by Christopher Steele's phony "dossier"—to buttress its request for a FISA warrant on Carter Page:

"On August 19, 2016, after public reports regarding connections between Manafort, Ukraine and Russia—including an alleged 'black ledger' of off-the-book payments from the Party of Regions to Manafort—Manafort left his post as chairman of the Trump campaign," an FBI agent's affidavit read. It cited an AP story as a footnote to the affidavit.[16]

Weissmann and his crew ginned up "collusion" hysteria, but the charges against Manafort boiled down to tax evasion and bank fraud. Manafort made it clear that he was not interested in becoming a cooperating witness. To raise the stakes, the FBI went after Manafort's son-in-law, Jeffrey Yohai, with whom he had partnered in business deals, to "get into Manafort's head," using a technique in white-collar criminal probes called "climbing the ladder."[17]

Within a month of the heavy-handed FBI raid, Manafort changed law firms, retaining attorneys who specialized in international tax issues as Mueller's team began issuing subpoenas to global financial institutions for his personal and business bank records.[18]

"It's obvious that it has morphed into an open-ended investigation that is way beyond Russian collusion," said David Rivkin, an attorney who worked in the DOJ during the Reagan and George H. W. Bush administrations, "and the only unifying principle seems to be that it covers people who are close to Trump or worked with Trump. And that is the classical definition of a fishing expedition."[19]

After the raid, leaks of classified information escalated the pressure. CNN reported that Manafort had been under FBI surveillance beginning in 2014. The surveillance, reportedly authorized by a FISA warrant, had included wiretapping, searches, and other types of observation, triggered by work done by a group of Washington consulting firms on behalf of Yanukovych's party in Ukraine.

"The intelligence collected by the FBI led to officials' concerns that Manafort encouraged Russian interference in the election," reported Hallie Detrick in *Fortune*. "A second warrant obtained in 2016 required the FBI to provide evidence for the suspicion that Manafort was acting as an agent of a foreign power. That warrant was directly related to the FBI's investigation into ties between the Trump campaign and people suspected of operating on behalf of the Russian government."[20]

Leakers liberally sprinkled "collusion" bread crumbs around news outlets, portraying Manafort in a negative light.

But the stories were often untrue. For example, four "current and former American officials" told the *New York Times* that phone records and intercepted phone calls showed repeated contact between Trump officials and senior Russian intelligence officers in the year before the election. The story was illustrated with a picture of Manafort. "The officials said that one of the advisers picked up on the calls was Paul Manafort." However, the OSC later told Manafort's lawyers that there had been no such intercepts.[21]

"Before signing up with Donald Trump, former campaign manager Paul Manafort secretly worked for a Russian billionaire with a plan to 'greatly benefit the Putin Government,'" reported the Associated Press on March 21, 2017.[22]

That was also untrue.

After yet another leak, Manafort's lawyers were alerted to a pre–special counsel meeting in April 2017 involving three FBI agents, Weissmann, then the chief of the DOJ's fraud division, and four AP reporters. "The meeting raises serious issues about whether a violation of grand jury secrecy occurred," they wrote in a court filing. Notes by an FBI agent revealed that reporters had even offered investigators a "code" allowing access to a storage facility used by Manafort.[23]

The following day, the AP reported an exclusive story: "Manafort Firm Received Ukraine Ledger Payout." Two reporters who had been at the meeting shared the byline with a third journalist.[24]

On October 30, 2017, Manafort and Gates surrendered to the FBI after being indicted by a federal grand jury on charges that included conspiracy against the United States, conspiracy to launder money, being unregistered agents of the Ukrainian government, making false and misleading statements, and failure to file reports of foreign banks and financial accounts.[25]

The indictment made much of Manafort's "lavish lifestyle," saying he had failed to pay taxes, "laundered" more than $18 million, and conspired to defraud the United States by "impeding, impairing, obstructing and defeating the lawful governmental functions" of the DOJ and Treasury Department by routing money through "scores of United States and foreign corporations, partnership and bank accounts" between 2006 and "at least 2016."

In essence, the two men were accused of hiding their income from lobbying work for Ukraine to avoid paying taxes, then lying about it. The thirty-one-page indictment made no mention of Trump, Russia, or "collusion." The media seemed as dejected as a kid who wakes up on Christmas morning only to find there are no presents under the tree. The alleged crimes all predated Manafort's and Gates's involvement with Trump's campaign.[26]

The Manafort and Gates indictments were unsealed the same day Papadopoulos pleaded guilty to lying. Now Flynn, Papadopoulos, Manafort, and Gates were being squeezed to flip on Trump. None of them did. And that made the special counsel team very angry.

It was clear that Papadopoulos was a dud. The team had nothing on Flynn besides lying, a potential FARA charge, and the useless Logan Act. But it had a paper trail of financial improprieties committed by Manafort and Gates, which could result in substantial prison time.

The charges were politically motivated. On the same day Manafort was indicted, Democrat power lobbyist Tony Podesta, the brother of Clinton adviser John Podesta, stepped down from his post at the Podesta Group. The special counsel had opened an investigation into Tony Podesta because he had done work on a Ukrainian public relations project organized by Manafort. He had filed a retroactive FARA declaration in April and has never been charged.[27]

Indeed, Washington is home to many political consultants who flocked to post–Soviet Union countries to ply their trade. Manafort worked for Yanukovych and his Party of Regions. Obama-connected consultants lobbied for his rival, Yulia Tymoshenko. The Clinton-istas took the side of Viktor Yushchenko, a Putin antagonist. Like Manafort, everyone else got rich.[28]

Yes, such activities could be unsavory, but they could also be politically useful to the United States. For example, though Yanukovych was a "thug," as former federal prosecutor Andrew McCarthy pointed out, Manafort remade him "from the ground up: Learn English, warm to Europe, embrace integration in the European Union, endorse competitive democracy."

But as his various businesses soured, Manafort owed some tycoons money, particularly Oleg Deripaska, a Russian aluminum baron. Manafort was accused of leveraging his position with the Trump campaign by sharing polling data with Deripaska, as if to prove that his candidate did indeed have a shot at the presidency.[29]

Manafort's associate Kilimnik supposedly passed the polling data to two Ukrainian oligarchs after an August 2016 meeting at a Manhattan cigar club. According to the FBI, Kilimnik had ties to Russian intelligence.[30] A federal judge described that allegation as an "undisputed core" of Mueller's investigation.[31]

The Kilimnik connection was accidentally revealed when

Manafort's attorneys failed to redact a footnote in a filing. "The document provided the clearest evidence to date that the Trump campaign may have tried to coordinate with Russians during the 2016 presidential race," reported the *New York Times*.[32]

"This is the closest thing we've seen to collusion," said Clint Watts, a senior fellow with the Foreign Policy Research Institute.[33]

The sharing of polling data was not a "collusion" conspiracy. But Kilimnik, often described as a "shadowy" character, was the Russian that Mueller desperately needed to connect Trump to Putin.

In fact, Kilimnik had stronger ties to Senator John McCain than to Putin.

A graduate and former language instructor at an elite Russian military academy, Kilimnik began working in 1995 as a translator in the Moscow office of the International Republican Institute (IRI), a nonprofit pro-democracy think tank funded in part by the US government.[34] The IRI was headed by McCain for twenty-five years until he stepped down in August 2018.[35]

Kilimnik, who is five foot three, was nicknamed "the midget" by his fellow staffers and kidded about being a spy because of his background, according to political consultant Michael Caputo, an American who worked in Moscow and later on the Trump campaign. Kilimnik became the deputy director for IRI's Moscow office. In 2004, he did work on the side as a translator for Manafort's then partner Rick Davis, who was in Ukraine consulting for billionaire Deripaska. Fired by IRI for moonlighting, he began translating for Manafort. That began a long and mutually beneficial association as both built valuable contacts among Ukrainian government officials and the country's political elite.[36]

Kilimnik, who lives in Moscow, told the *Wall Street Journal* he met Deripaska only once, when interpreting for Davis. "I have never had anything to do with any intelligence service of Russia," he wrote by email, saying the accusations against him had been "intentionally pushed out to make me a 'missing link' in a story that is built of flawed foundation."[37]

In fact, according to hundreds of government documents, Kilimnik was a "sensitive" intelligence source for the US State Department—information deliberately omitted from the Mueller Report.[38]

Kilimnik's communications with the State Department went back to 2013. He often met with the chief political officer at the US Embassy in Kiev, relaying messages to Ukraine's leaders and delivering written reports to US officials. They believed he was impartial, with no particular allegiance to Moscow.[39]

Three government officials confirmed to reporter John Solomon that Mueller's team had had all the interviews the FBI had done with State Department officials, in addition to Kilimnik's written reports, "well before they portrayed him as a Russia sympathizer with ties to Moscow intelligence."[40]

The Mueller Report also painted a proposed Ukraine-Russia peace plan that Kilimnik had presented to the Trump campaign in August 2016 as nefarious. But Mueller neglected to point out that Kilimnik had proposed a similar peace plan to the Obama State Department in May 2016.

"That's what many in the intelligence world would call 'deception by omission,'" Solomon wrote.[41] That exclusion by Mueller not only was unethical but raised more questions about what other crucial information was missing from his report.

However, once the fishing expedition began, fish were hooked. Both Manafort and Gates had reaped the benefits of lucrative consulting work and—instead of paying Uncle Sam his share—parked large sums of money overseas.

Gates pleaded guilty in February 2018 to two charges: conspiracy to defraud the United States involving undeclared income and making a false statement. He became the fifth person targeted by Mueller to admit to criminal misconduct and the third Trump associate to state his willingness to cooperate with the special counsel. Since Gates had continued his work for Trump long after Manafort's resignation, he presumably had dirt to dish.[42]

As of March 2019, Gates had not been sentenced and was still providing help in "several ongoing investigations," perhaps activities surrounding Trump's inauguration.[43]

A superseding indictment, obtained from a federal grand jury in Virginia, alleged that the two men had engaged in a complex scheme to defraud lenders and obtain more than $20 million in loans, doctoring financial statements to show millions of dollars in income they didn't have.[44]

Manafort said he would continue to defend himself "against the untrue piled up charges. . . . I had hoped and expected my business colleague would have had the strength to continue the battle to prove our innocence. For reasons yet to surface, he chose to do otherwise."[45]

Manafort faced charges in two jurisdictions, the District of Columbia and the Eastern District of Virginia. One new charge included the allegation that Manafort "secretly retained" a group of "former senior European politicians" in 2012 to support his Ukrainian client's positions while appearing to be independent.[46]

US District Judge T. S. Ellis III in Washington, DC, cast a gimlet eye on the charges against Manafort, chastising the Mueller team for using an earlier investigation to pressure him to testify against Trump.

"You're now using [that material] . . . as a means of persuading Mr. Manafort to provide information," Ellis said during a heated hearing in May 2018.[47] "You don't really care about Mr. Manafort's bank fraud. What you really care about is what information Mr. Manafort could give you that would reflect on Mr. Trump or lead to his prosecution or impeachment. . . . The vernacular to 'sing' is what prosecutors use. What you have to be careful of is that they may not only sing, they may compose."[48]

Even former Clinton attorney Lanny Davis said that Manafort's attorneys had a point in trying to limit what the special counsel could pursue. "I don't know whether that line has been crossed," Davis said. "I certainly think there's a question to be raised about what bank fraud has to do with Russian collusion, and that's something that needs to be addressed by the court."[49]

Mueller's team found a tenuous connection. In June 2018, they indicted Manafort and Kilimnik on two counts of obstruction of justice by attempting to contact potential witnesses, believed to be high-ranking European political leaders, via phone calls, text messages, and encrypted apps.[50]

Kilimnik denied the allegations. Based on Manafort's communications with Kilimnik, prosecutors asked the judge to revoke Manafort's $10 million bail.[51] US District Court judge Amy Berman Jackson complied, remanding him to jail in Virginia on June 16, 2018.[52] Though held as a "VIP" in a self-contained unit for his "safety," Manafort was effectively kept in solitary confinement twenty-three hours a day, a policy designed to punish troublemakers. The facility was two hours south of Washington, making it "effectively impossible" for him to prepare for two trials, his attorneys argued.[53]

After prosecutors complained that Manafort's "VIP" treatment was too luxurious, he was moved to another jail in mid-July.[54] By the time Manafort went to trial on eighteen counts of tax evasion and bank fraud in the Virginia courtroom of Judge Ellis in the summer of 2018, his health had visibly deteriorated.

Mueller's prosecutors painted Manafort as a profligate spender, going into elaborate detail about Manafort's spending on seven homes, a $15,000 ostrich-skin jacket, cars, and other luxuries. Judge Ellis told them to move along, that it wasn't a crime to be a big spender.[55] He also instructed them not to use the pejorative term *oligarchs*, which implied that Manafort consorted with "despicable people."[56]

The jury convicted him on eight counts of tax and bank fraud.[57] To avoid a second trial in Washington, DC, Manafort agreed in September 2018 to plead guilty to conspiring against the United States and obstruction of justice and cooperate with the special counsel. He was interrogated more than a dozen times by the FBI and Mueller's attorneys; nine of those times came after his plea deal.[58]

Investigators repeatedly brought up the June 9, 2016, meeting at

Trump Tower that Manafort had attended with Donald Trump, Jr., Jared Kushner, and Russian lawyer Natalia Veselnitskaya, pressing him on whether the presidential candidate had known about it in advance.[59]

It would not have been a crime if he had. But Manafort repeatedly denied that Trump had had prior knowledge of the meeting; he was accused in court of lying. The piling on of those "process" crimes was just another weapon in Mueller's arsenal.

In a heavily redacted December 2018 court filing, Mueller's team listed five topics Manafort had lied about: his interactions with Kilimnik, the role Kilimnik had played in trying to influence witness testimony; a business payment; his Trump administration contacts; and information for another investigation. Though Manafort denied having intentionally provided false information, the Mueller team insisted that he had breached his agreement, erasing any prospect for leniency.[60]

In February 2019, the Mueller team filed a highly anticipated eight-hundred-page sentencing brief. The media pundits were disappointed to find that it contained no evidence that Manafort, Gates, Kilimnik, Trump, or anyone else on the candidate's campaign staff had conspired with Russia.

As Andrew McCarthy said, perhaps Manafort was a big-spending scoundrel. But did that justify the predawn raid in which his wife had been held at gunpoint? The "months of solitary confinement that have left him a shell of his former self"?

Manafort didn't collude with Russia, McCarthy pointed out. "He worked for Ukraine, not Putin. Indeed, for much of his time in Ukraine, he pushed his clients *against* Putin's interests."[61]

But even more outrage was yet in store. Within minutes of Manafort's final sentencing, Manhattan district attorney Cyrus Vance, Jr., indicted him for mortgage fraud and more than a dozen other New York state felonies. The core allegation was that Manafort had lied about buying a Manhattan condo to use as a home, as opposed to a rental. Some of the charges overlapped with the charges of which Manafort had already been found guilty, putting him into double jeopardy.

That "nakedly political prosecution," said legal scholar Jonathan Turley, was intended to punish Manafort in case President Trump pardoned his former campaign manager, by convicting him on state charges out of the reach of executive power. It was lauded in the out-for-blood media as a clever "insurance policy."[62]

In her zeal to get Manafort, New York attorney general Letitia James campaigned for office on changing the state constitution to reduce double-jeopardy protections, which would "remove constitutional protections from all citizens," said Turley. "What is clear is that New York should not reduce protections for everyone in the state just to punish one man."[63]

The effort to change the state constitution succeeded, but it may not survive judicial review. Though the indictments were unethical and constitutionally flawed, DA Vance brought them anyway.

"The state prosecutors have brought a case they otherwise would never waste time on—not because the case should be done, but to try to block a pardon," McCarthy said. "This raw politicization of prosecutorial power ought to frighten everyone. . . . The New York district attorney did not indict Manafort because he committed mortgage fraud. The DA indicted Manafort because he worked on the Trump campaign and could be pardoned during Trump's presidency. That's disgraceful."[64]

If there was any doubt that the prosecution by Vance was political, the announcement that while awaiting trial in Manhattan, Manafort would be transferred from a penitentiary in Loretto, Pennsylvania, to Rikers Island dispelled that notion. In a letter to the Loretto penitentiary warden, Manafort's attorney objected and asked that the move be blocked because New York prosecutors were "insisting that Mr. Manafort remain on Rikers Island, likely in solitary confinement, pending trial."[65]

Reserved for hard-core criminals and plagued by violence, Rikers is no place for an ailing elderly man convicted of white-collar offenses. "The decision to move Paul Manafort . . . from the decent federal prison to which he was sentenced to solitary confinement to the dangerous hell hole that is New York City's Rikers Island seems abusive and possibly illegal," wrote former Harvard Law School professor Alan Dershowitz. "I know Rikers well having spent time there visiting numerous defendants accused of murder and other violent crimes. . . . Mass murderers and torturers are among those incarcerated at Rikers Island."[66]

Former New York City police commissioner Bernard Kerik wrote that solitary confinement is "basically a deathtrap." Even Representative Alexandria Ocasio-Cortez (D-NY) agreed, tweeting "A prison sentence is not a license for gov torture and human rights violations. That's what solitary confinement is."[67]

Attorney Sidney Powell, who had taken over Flynn's case, also weighed in: "When a witness or defendant from whom prosecutors want 'cooperation' does not do as they demand, they put him in solitary confinement. And it works. It literally breaks people."[68]

That was Vance's goal: to break Manafort to get Trump. However, Jeffrey A. Rosen, Barr's new deputy attorney general, blocked the move to Rikers—for now.[69]

Kilimnik lives in Moscow. He will never be extradited to face trial, but Mueller's team smeared him as a spy, a central player in the Russian collusion narrative that didn't exist.

BUSTING ROGER STONE

Just after 6:00 a.m. on January 25, 2019, seventeen SUVs and two armored vehicles, sirens blaring and lights flashing, screamed into a quiet residential neighborhood in Fort Lauderdale where many houses back up to a canal. A helicopter hovered overhead, and two police boats roared up to the back yard of a home.

Twenty-nine FBI agents, wearing tactical gear and wielding M4 rifles, swept across the lawn. Four agents approached the front door, two with a battering ram. One pounded on the front entrance, shouting "FBI! Open the door!" When a white-haired man wearing shorts and a T-shirt, clearly just awakened from sleep, complied, he saw two rifle barrels pointed at his head.[70]

Sixty-six-year-old Roger Stone was told he was under arrest. The feds scoured his home for documents, computers, phones, and, apparently, a meth lab or illegal gun-running operation, perhaps a terrorist hiding under a bed.

The bust was shown live on CNN, which conveniently happened to have a camera van parked down the street. The commentator described the scene as an "extraordinary arrest" involving "many lights, heavy weaponry."[71]

The jackbooted tactics were to bust not an armed and dangerous criminal but a writer, self-promoter, and longtime friend and political adviser of Donald Trump, indicted the previous day on seven counts of lying to federal agents. The feds knew that Stone had no criminal record, owned no firearms, and had an expired passport and thus was not dangerous or a flight risk.

The raid was similar to that on Manafort's condo, but on steroids.

"I opened the door to pointed automatic weapons," Stone said later. "I was handcuffed. There were 17 vehicles in the street with their lights on, they terrorized my wife and my dogs."[72]

Former FBI agent Kenneth Strange said the video shot by CNN showed all the markings of a predawn "knock and announce" arrest warrant, usually reserved to nab a cartel member.[73]

"The manner, timing, resources, and manpower used to effect Stone's arrest bordered on the surreal," said Strange. "And while Mueller and his minions may have been within their right to execute an arrest warrant in such a manner, their indifference to the optics of this 'reality show arrest' is baffling."[74]

The extraordinary presence of the TV cameras made it clear that the heavy show of force was a message from Mueller to Stone and Trump. As former federal prosecutor Paul Butler described it, "This is personal."[75]

Another former FBI agent, Danny Coulson, who rose to become FBI deputy assistant director, was appalled at the excessive use of force and the way Mueller had exploited the FBI to frighten and intimidate. He had made more than a thousand arrests during his career and had never used that kind of force for a white-collar suspect.[76] Before the Bureau was "hijacked" by Mueller's office, a defendant such as Stone—who had stated publicly that he expected to be indicted and vowed not to cooperate with the OSC—would have been told to self-surrender at an appointed date and time, especially because prosecutors knew that he was represented by counsel.

The bizarre scene raised questions about the judgment of not only Mueller but also FBI director Christopher Wray. And why had the CNN reporters shown up an hour before the raid? Had they been tipped off? The network attributed their fortuitous presence to "reporter's instinct."[77]

Was Mueller trying to get Stone to "sing . . . or compose" as Judge Ellis said in the Manafort case? Was the special counsel trying to goad Trump into firing him? Then Democrats could impeach Trump for obstruction of justice. Or maybe they were trying to terrify Stone into submission as it became clear that, with the investigation winding down, they had no evidence of "collusion" and needed another indictment for the body count.

The raid pointed up selective and unequal treatment by the FBI and DOJ. Hillary Clinton's top aides, Huma Abedin and Cheryl Mills, made false statements, and their emails proved it. Yet they were not charged; instead, they were given immunity in exchange for nothing. So if you were a friend of Hillary's, you got a free pass. If you were Trump's friend, you got guns in your face at dawn.

A dandy who writes a fashion blog, a former adviser to Richard Nixon and self-described dirty trickster, Stone's peculiar personality and lurid speculations about Hillary and Bill Clinton had earned him a special loathing in Democrat circles.[78]

A coauthor of *The Clintons' War on Women*, he portrayed the former president as a serial sex offender and Hillary as an enabler, both willing to commit diabolical deeds for power.[79] He's an equal-opportunity basher, however: as a coauthor of *Jeb! And the Bush Crime Family*, he earned the enmity of Republicans as well.[80]

But for all his flamboyance and preening, Stone is no different from many political hacks, pundits, and writers on both sides of the aisle who view American culture and politics through a conspiratorial lens.

More than any other single event, the abusive wielding of the OSC's power in the arrest of Roger Stone revealed the rot at the heart of the entire investigation and the personal animosity of those driving it. Americans took notice when Trump tweeted: "Greatest Witch Hunt in the History of our Country! NO COLLUSION! Border Coyotes, Drug Dealers, and Human Traffickers are treated better. Who alerted CNN to be there?"[81]

Despite his enduring friendship with Trump, Stone had been fired from the campaign by the candidate in August 2015—though Stone said he had resigned voluntarily.[82]

The twenty-four-page indictment against Stone was filed on January 24, 2019, by three prosecutors from the District of Columbia and three from the Special Counsel's Office, including Jeannie Rhee, who had represented Hillary on her email scandal and never should have been in the same room as they discussed this case.[83]

The gaseous windbag of a document told a tantalizing story about Trump, WikiLeaks, and Julian Assange, suggesting that Stone might have had some advance knowledge or inside information about the content of hacked Clinton campaign emails that were released by WikiLeaks in the summer of 2016. "A senior Trump campaign official was directed to contact Stone about any additional releases and what other damaging information Organization 1 [WikiLeaks] had regarding the Clinton campaign."[84]

And there was the reference to the movie *The Godfather Part II*: to keep "Person 2" from ratting him out, Stone had allegedly told him to do a "Frank Pentangeli," a reference to a person testifying before Congress who "claims not to know critical information that he does in fact know."[85]

But the froth boiled down to process crimes: obstruction of an official proceeding by making false statements to the House Intelligence Committee, denying he had records, making false statements, and persuading a witness to provide false testimony. Together they exposed Stone to a sentence of up to fifty years in the slammer.[86]

I don't want to minimize "process crimes." No person should ever lie, mislead, or obstruct a legitimate law enforcement investigation. But none of the charges had anything to do with Trump-Russia "collusion." It was not alleged that Stone had conspired with Russians to hack or steal documents.

But Stone was accused of reaching out to WikiLeaks and asking others to do so. As did hundreds of journalists, including myself. That's not a crime. An examination of Stone's emails shows that he provided little more than the same information that WikiLeaks had already stated publicly.

He speculated that the Clinton emails would be damaging. But that was stating the obvious. Trying to insert himself into the action, Stone created the appearance that he knew more than he did—a frequent habit of his.

He tried to contact Trump adviser Steve Bannon with his prediction about WikiLeaks but whined to an editor for Breitbart, "I'd tell Bannon, but he doesn't call me back." After the editor prodded Bannon, Bannon replied, "I've got important stuff to worry about." Bannon finally did email Stone, who told him nothing that Assange hadn't already said publicly. Then Stone insisted that Bannon get campaign surrogates to push his allegation that Bill Clinton had a black love child.[87] It was as if, sidelined after decades of being in the middle of hot campaigns, Stone was desperate to be relevant.

Mueller's job was to uncover crimes that had occurred before he was appointed. But his investigation generated or created Stone's alleged offenses. Five of the charges against Stone were for making false statements to federal authorities. If he goes to trial, it will be exceedingly difficult for prosecutors to prove his guilt because the statute governing those offenses (18 U.S.C. § 1001) requires proof that the statements were "knowingly and willfully" false.

A faulty memory or diminished recollection is a complete defense. If Stone recalled events differently from the way Mueller interpreted them, it's not a crime. Moreover, Stone amended some of his testimony with corrected statements, which will be introduced as evidence in his defense.

As Comey dismissively told the House Judiciary Committee regarding Abedin and Mills, "There's always conflicting recollections of facts."[88]

That from the same guy who twisted the facts and contorted the law to clear Clinton from the felony statutes she had so flagrantly violated. Comey's buddy Mueller was now applying the same double standard of justice.

Stone did not retreat into silence after his arraignment but instead went on a media tour. "I think the American people need to hear about it," he told the *New York Times*.[89]

Even before the raid, he insisted he had done nothing more than "posture, bluff, hype," based on WikiLeaks' Twitter feed and tips from others who were following Assange. "I didn't need any inside knowledge to do that. They keep looking for some direct communications with WikiLeaks that doesn't exist."[90]

And despite Clapper's opining that Stone's indictment revealed "connection, coordination, synchronization, whatever you want to call it," the opposite was true.[91] If the Trump campaign conspired with Russia to hack and release DNC emails through WikiLeaks, why did they need Stone to contact Assange?

After Stone's arrest, Trump said he had never spoken with his friend about WikiLeaks and the stolen DNC emails, nor had he directed anyone to do so.[92]

Mueller filed a motion in February 2019 that he had evidence that Stone had communicated with WikiLeaks. His source was convicted perjurer Michael Cohen. He claimed to have been in Trump's office during a call when Stone was on the speaker. "Mr. Stone told Mr. Trump that he had just gotten off the phone with Julian Assange and that Mr. Assange told Mr. Stone that, within a couple of days, there would be a massive dump of emails that would damage Hillary Clinton's campaign. Mr. Trump responded by stating to the effect of 'wouldn't that be great.'" [93]

Stone denied it, and WikiLeaks responded via Twitter, saying "WikiLeaks publisher Julian Assange has never had a telephone call with Roger Stone." If Cohen is the only source, good luck trying to prove that allegation in court. [94]

The allegation of witness tampering involved communications Stone had exchanged with Randy Credico, a radio host who had interviewed Assange. Stone's emails implied that Credico had been his "back channel" to WikiLeaks but this was later discounted as an exaggeration. [95]

Mueller alleged that Stone had obstructed justice by threatening Credico if he talked. "If you testify you're a fool," Stone said in one message to Credico. "I guarantee you you [sic] are the one who gets indicted for perjury if you're stupid enough to testify." [96]

Stone insisted that the exchanges had been humorous, that texts before and after the ones in question were jocular, undercutting the notion that Credico might have felt threatened. "They're taking things out of context to present them in a light that mischaracterizes their significance. I never told Mr. Credico to lie."

He said he had forgotten about some exchanges with Credico when he told House investigators that they did not exist. "I am human and I did make some errors, but they're errors that would be inconsequential in the scope of this investigation."

Though once friends, the two men fell out over the investigation into collusion. "You are an inveterate liar everybody knows that," Credico texted to Stone in May 2018. Stone texted back, "You ain't exactly George Washington yourself." [97]

Insisting he would "fight to the bitter end," facing what he predicted would be $2 million in legal fees, Stone put out an appeal for
donations to his legal defense fund. He and his wife moved to a smaller
condo and canceled their medical insurance to economize.[98] Stone
has pleaded not guilty and asked the court to dismiss the charges.[99]

TARGETING JEROME CORSI

In their effort to build a case against Stone, Mueller's team targeted
conservative radio host Jerome Corsi, a correspondent for WorldNet-
Daily and Infowars. Author of many controversial books, Corsi is
most famous for claiming Obama was born in Kenya, not Hawaii,
as detailed in his book *Where's the Birth Certificate?: The Case That
Barack Obama Is Not Eligible to Be President.*[100]

Corsi, who had known Trump for years, talked to Stone for the
first time on February 22, 2016, for a piece on Stone's book about
the Bush family. The Trump campaign was heating up. Stone was on
the outside but claimed he had still talked to Trump by phone "every
day," offering political advice and strategy.

The men met for dinner and talked about how to support Trump's
presidential campaign, later exchanging emails and texts. "I crossed
over from the reporter's role to work behind the scenes as a political
operative, working secretly with Roger Stone to engineer events that
would affect the news cycle favorably for the Trump campaign during
the 2016 presidential election," Corsi wrote in *Silent No More: How I
Became a Political Prisoner of Mueller's "Witch Hunt,"* a book about his
experience with Team Mueller.[101]

On August 28, 2018, two FBI agents rang the doorbell of his
home in northern New Jersey. He was shaken but not surprised. Corsi
knew the Mueller team was interviewing Stone's contacts about his
interactions with Julian Assange and WikiLeaks. They presented him
with a subpoena to appear before a grand jury in ten days.[102]

Corsi said he would contact his attorney and get back in touch with them. After conferring with his lawyer, David Gray, who explained that the FBI would have a harder time of establishing that he had intended to lie if he cooperated voluntarily, Corsi agreed. However, he was wary. "Even with the advice of David Gray, I was not sure I could avoid a perjury trap, even though I fully intended to tell the truth," he wrote.[103]

Gray notified President Trump's attorney Jay Sekulow that Corsi had been subpoenaed. They entered into a verbal "mutual defense" agreement, which allowed them to share information privately.

On September 6, 2018, Corsi and Gray went to Washington, DC, and met with prosecutors Aaron Zelinsky, Jeannie Rhee, and Andrew Goldstein and a phalanx of FBI agents. Corsi voluntarily turned over his phone, two laptops, a Time Machine application with a backup hard drive, and access to his email accounts. He assumed they had all his communications anyway.

The conversation turned to Julian Assange. Corsi said that although Stone had wanted him to contact Assange, he had declined, not willing to involve himself in the inevitable intelligence investigations that would follow.[104]

The meeting "blew up," with the prosecutors and FBI agents walking out.

Corsi was left alone for an hour and a half to wonder what crime he'd committed. Rhee and Zelinksy returned, "visibly agitated."

"We have demonstrable proof that what you said was false," Zelinksy said.

Rhee said she was glad he had not made the statement about Assange to the grand jury. "It would be extremely difficult to expunge that testimony from the record." Meaning that if Corsi lied, he'd be a poor witness against Stone. Zelinsky said he had a week to review his emails and come back. They returned his laptops and Time Machine external hard drive backup.

Corsi had not reviewed his emails prior to the interrogation. At home, he reloaded 60,000 old emails on a new computer and examined them. Only then did he realized that he had forgotten a lot of them.

After the first big Clinton-related email dump by WikiLeaks in March 2016, it was no secret that there was more to come. In June 2016, Assange told *The Guardian* that WikiLeaks had been planning to publish emails sent and received when Clinton was secretary of state.[105]

Corsi found an email from Stone on July 25 with the subject line "Get to Assange." The text read, "At the Ecuadorian Embassy in London and get the pending WikiLeaks emails . . . they deal with [Clinton] Foundation, allegedly." [106]

WikiLeaks had been dropping DNC emails over the previous few days to coincide with the beginning of the Democrat National Convention. The furor over the DNC's treatment of Bernie Sanders would result in the resignation of DNC chair Debbie Wasserman Schultz.

Corsi had forwarded the email to Ted Malloch in London, whom he had met while researching the Clinton Foundation, with a simple message: "Ted. From Roger Stone. Jerry." [107]

Stone and Malloch, a professor and senior fellow at the Saïd Business School at Oxford University, had met at a dinner set up by Corsi. A strong Trump supporter, Malloch, who had served on the executive board of the World Economic Forum at Davos, was no gumshoe. As Corsi had anticipated, Malloch did not contact Assange.

Corsi found other emails from Stone and other people he had forgotten. While in Italy to celebrate his wedding anniversary, his WorldNetDaily editor, Joseph Farah, emailed him, upset that no one had contacted Assange. Corsi replied, "We can reach Assange, but someone may have to go to London. I'm sure if we went to the embassy, he would talk to us." [108]

But Farah didn't assign the story to Corsi, so nothing ever came of it. And if he had, it would not have been a crime for Corsi, a journalist, to talk to Assange.

Corsi published a story on August 15, 2016, about Stone, who claimed that his computer and personal bank accounts had been hacked for declaring that WikiLeaks had obtained Clinton's 30,000 scrubbed emails and was planning to release them.

From open-source research on WikiLeaks, his knowledge about computer infrastructure, and his personal study of John Podesta, Corsi theorized that Assange didn't have Clinton's emails. He had Podesta's emails, and he'd drop them in October, in a *drip-drip-drip* style for maximum effect. Corsi explained his theory to Stone. On August 21, 2016, Stone posted on Twitter: "Trust me, it will soon [be] Podesta's time in the barrel." [109]

At the time, that was interpreted by many as proof that Stone had "advance knowledge" that the emails Assange had yet to publish were Podesta's.

"This is the central email that had become the focus of the Special Counsel's criminal investigation," Corsi wrote. "Stone was the link between Julian Assange and Donald Trump that was essential if the Trump campaign's alleged collusion with Russia" were to be proved. Mueller wanted to use Corsi to prove the link between Stone and Assange. [110]

After Stone took heat for his tweet regarding Podesta, Corsi sent him a lengthy background memo on Podesta to be used as a "cover story." Corsi then wrote a piece for Infowars titled " 'Blame Me!' Corsi says. 'Not Assange.' " [111]

Corsi didn't consider that fake story immoral or illegal. It was just politics. [112]

In his next interview with Team Mueller, Corsi explained that he'd found three emails he'd exchanged in 2016 with Stone that he had not remembered. With Mueller's consent, he amended his earlier statements. [113]

"What astonished Zelinksy and the Special Counsel's prosecutorial team was how I had obtained information this accurate in July and August, months before Assange began dropping Podesta's emails on October 7, 2016," Corsi wrote.

They believed not only that Corsi had a direct connection to Assange but that he'd played a role in timing the WikiLeaks October release after Stone had tipped Corsi off about the *Access Hollywood* tape, which would soon be leaked, and asked him to get word to Assange. However, with access to all Corsi's communications, they couldn't find his connection.

Zelinsky pushed Corsi to remember by using a regression technique. He urged Corsi to imagine himself back in Italy on his vacation when his editor had emailed him about Assange. Whom had he talked to? Corsi tried to cooperate but failed to conjure up a person giving him secret details on Assange. "I think if we continue this, I will be telling you next that I was Alexander the Great in a former life," he said.[114]

Corsi testified before a grand jury on September 21, 2018. It took about thirty minutes. The prosecutors shook his hand and congratulated him. But it wasn't over. In preparation for a second appearance before the grand jury, he met again with the Mueller team on October 31.

Zelinsky and Rhee again peppered him with questions. But his memory was poor. They finally showed Corsi an email he had written to Stone while in Italy in August 2016. They wouldn't let him have a copy, but it said something such as "Word is . . . Assange will make two drops, one soon on Clinton foundation, second one in October on Podesta."[115] Corsi had no recollection of sending that email, but it seemed to him that he had merely been placating the excitable Stone.

That triggered another barrage of questions, hours and hours of questions over every email and phone call he'd made in 2016.

Corsi underwent forty hours of interrogation. But the prosecutors refused to accept that his memory was poor.[116] They pressured him to plead guilty to making three false statements under the 18 U.S.C. § 1001 statute. He accused the OSC of setting him up for a perjury trap.[117]

"If I was obstructing justice, why would I keep the external hard drive?" Corsi said. "I would have thrown all of that away." On November 21, his lawyer sent a letter to the special counsel: "The issue is that the statements that Dr. Corsi made were, in fact, the best he could recall at the time. From the beginning, Dr. Corsi immediately provided all of his computers, emails, phones, social media accounts, etc., and his intent was always to tell you the truth to the best of his recollection, which he admitted to you, was not very good as these events took place years ago." [118]

Amending testimony is not uncommon: after delivering demonstrably false testimony before Congress, Comey and Clapper had amended their statements. Neither had been charged with a crime.

But in the twisted world of Special Counsel Mueller, a person could be charged with a crime for forgetting about three emails sent two years earlier. You'd be accused of lying because you're human. And then, under threat of imprisonment, you'd be forced to confess that your imperfect recollection had been an intentional deception, even though it hadn't been. In other words, you'd be told to lie about the truth.

That happened to Flynn and Papadopoulos, who yielded to the pressure. Corsi dug in his heels. After weeks of coercion, intimidation, and legal bullying by Rhee and her cohorts, he refused to acquiesce to their demands.

"I will not sign a plea agreement that is a lie," he wrote in a memo to his attorney on November 25, 2018. "I never 'willfully and knowingly' gave false information to the FBI or the Special Counsel." [119]

Pleading guilty could have cost Corsi his securities license in the banking and securities business where he had been active. [120] Corsi claimed that Rhee had advised him not to tell federal regulatory authorities of the plea agreement because it would be sealed. "Rhee just advised us to commit a felony." [121] He refused the plea deal.

Mueller's team went after Corsi's stepson, Andrew Stettner, subpoenaing him to testify before a grand jury. [122] Notice a pattern?

Corsi went public in November 2018, saying he expected to be indicted any day. "This has been one of the most frightening experi-

ences of my life," he said, speaking of the pressure from the OSC. "At
the end of the two months, my mind was mush." [123]

He claimed that the special counsel had wanted him to support
a Trump-Russia "collusion" narrative devised by Mueller's team. Insisting that his only crime was supporting Trump, Corsi said he was
afraid he'd go to jail "for the rest of my life because I dared to oppose
the deep state." [124]

At roughly the same time, Trump's lawyers became aware of the
unconscionable pressure being applied on Corsi by the special counsel when documents involved in his case were delivered to them by
an anonymous source. They included a proposed indictment and
plea agreement, as well as a statement or allocution for Corsi to read
during a guilty plea that would falsely accuse the president of having
received some collusive information. Rudy Giuliani said they took
action late that same night:

> We immediately reported it to the FBI and Department of Jus-
> tice telling them that Mueller's team was attempting to suborn
> perjury which is a felony. They were dangling an offer that
> he'd get probation and never have to go to jail if only he agreed
> to say things under oath that were false. They were threatening
> him to get him to lie. [125]

Trump's lawyers met with prosecutors at the Justice Department.
However, with Rod Rosenstein in charge, no action appears to have
been taken over those complaints of gross prosecutorial misconduct,
if not crimes.

After Stone's over-the-top raid by the FBI, Corsi wondered if
agents wearing body armor would haul him out of bed one morning
at gunpoint. But perhaps Mueller's team realized that their dramatic
bust of Stone had backfired. Charging Corsi with such thin gruel
would have made them look petty and vengeful. Mueller issued his report in March 2019 with the announcement that no new indictments
would be brought.

THE FBI EMPLOYED SCORCHED-EARTH TACTICS

People got caught in the crosshairs of a zealous FBI and a prosecution team eager to run up its count as it became clear that there was no real reason for the special counsel's existence. Process crimes would do. Among those convicted were:[126]

- Alex van der Zwaan, a Dutch attorney based in London but arrested in the United States, who pleaded guilty to lying to the FBI about his contacts with Rick Gates and an unnamed person in Ukraine. He served thirty days in jail and was deported.
- Sam Patten, a "Never Trumper" and lobbyist who pleaded guilty to a violation of the Foreign Agents Registration Act (FARA), for failing to register as a foreign agent for Ukrainian clients. After spending more than $140,000 in legal fees, he was sentenced to three years of probation and a $5,000 fine. "It's proof positive of selective enforcement [of FARA]," he said. "It's only used when the government has nothing else." [127]
- Richard Pinedo, who pleaded guilty to identity theft in relation to the Internet Research Agency Russian troll farm operation (see below).

Charges remain pending against:

- Konstantin Kilimnik, a Ukrainian colleague of Manafort, charged with obstruction of justice.
- Iranian American businessman Bijan Kian and Turkish businessman Ekim Alptekin, business associates of Mike Flynn, charged with conspiring to violate foreign lobbying laws. Kian was tried by a federal jury and convicted.
- Gregory Craig, a former White House counsel for Barack Obama, indicted for lying and concealing information

from federal authorities regarding foreign lobbying. He is the only Democrat charged to date in connection with Mueller's investigation; however, criminal referrals were made on lobbyists Tony Podesta and Vin Weber.

Literally hundreds of people were caught up in the investigation as witnesses or potential suspects, particularly conservative journalists, Republican congressional staffers, and conservative think tank analysts.

Journalist Paul Sperry tracked down a dozen or so witnesses—ignored by most media outlets—and found strikingly similar tactics. Agents and prosecutors had pressured them to admit crimes, threatened their wives and girlfriends, and intimated that they'd be going to jail. Their communications—emails, texts, call records, travel records—had been scooped up and scrutinized, even though they had little evidence to provide. In some cases, they believed that information they had presented to a grand jury had been leaked, a federal crime.[128]

Compelled to hire lawyers, many suffered emotional and mental distress. Some have filed legal complaints or formally complained to the DOJ. They have decried Mueller's "scorched earth" tactics, and, given their description of events, it's hard to disagree with them.[129]

FBI informants had even targeted some others.

Michael Caputo, a former aide to Trump's campaign who had lived in Russia during the 1990s, was approached by a Florida-based Russian named Henry Oknyansky, aka Henry Greenberg, offering to sell derogatory financial information on Clinton, according to the Mueller Report. Caputo and Stone met with Greenberg in May 2016. Stone refused the offer, "stating that Trump would not pay for opposition research." [130]

Caputo ran up more than $125,000 in legal bills as a result of interrogations by Mueller's prosecutors and preparation for his testimony before two congressional committees. He was forced to remortgage his home to pay his legal bills.

"Forget about all the death threats against my family," he said in a statement after a Senate committee interviewed him. "I want to know who cost us so much money, who crushed our kids, who forced us out of our home, all because you lost an election. . . . I want to know because God damn you to hell." [131]

It turned out that Greenberg had another name as well: Gennady Vostretsov. He had a long string of arrests in the United States and Russia. In 2015, he told an immigration judge he had worked for the FBI for seventeen years throughout the world, including in the United States, Iran, and North Korea. [132]

Caputo hired a private investigator to look into Greenberg's background, posting the eye-opening details online on Democrat Dossier. "When he tried to sell us dirt on Clinton, it was an FBI operation. . . . The FBI sent a violent, illegal-alien criminal Russian to meet with us." [133]

Greenberg's criminal record and history as an FBI informant are not mentioned in the Mueller Report. Caputo called for Mueller's team to be investigated for prosecutorial misconduct, saying "Ruining lives was blood sport for them." [134]

Two FBI agents knocked on the door of the journalist Art Moore, an editor for the news site WorldNetDaily, demanding to talk about Corsi and WikiLeaks. "They were clearly on a fishing expedition," said Moore, who lives in Washington state. "They seemed desperate to find something to hang onto the narrative" of Russian collusion. He believed that the Mueller team had secretly obtained his emails, phone records, and text messages. [135]

Joseph Farah, the founder of and an editor at WorldNetDaily, got the same treatment. Not long after agents grilled him about Corsi, Farah suffered a stroke; his nationally syndicated column has been suspended. [136]

Trying to connect the dots among WikiLeaks, Stone, and Corsi, FBI agents went after Jason Fishbein, a Florida lawyer who had done legal work for WikiLeaks. Interrogated for six hours over two days, he gave them more than five hundred pages of documents, and then a grand jury subpoenaed him for all records of his communications with WikiLeaks, which should have been protected by attorney-client privilege. Though a small cog in the wheel, he spent $20,000 for legal fees to handle the demands.[137]

"That doesn't nearly account for all the time I have had to spend occupied with this," he said. Though not charged with a crime, he came to believe the investigation was nothing more than a "politically manufactured criminal investigation and coordinated media outrage."[138]

In March 2018, Ted Malloch, the author and political analyst whom Corsi had asked to talk to Assange, was detained and interrogated by FBI agents at Logan Airport about Roger Stone and WikiLeaks and interrogated about having visited the Ecuadorian Embassy in London.[139]

Malloch told the agents he had not been to the embassy or communicated with Assange. Issued a subpoena to testify before a grand jury, Malloch complied but argued he had no information pertinent to the investigation. "I am not an operative, have no Russian contacts, and—aside from appearing on air and in print often to defend and congratulate our President—have done nothing wrong. What message does this send?" He published a book arguing that the covert intelligence activity surrounding the Trump campaign had been not Russian but Western.[140]

CNN reporters showed up in April 2018 on the Bethesda doorstep of Joseph Schmitz, a former Pentagon inspector general who had worked on the Trump campaign as a foreign policy adviser. They waylaid him in his front yard with "16 questions as salacious as the stuff in the [Steele] dossier," he said. Schmitz declined to answer their questions. CNN's resulting "exclusive" story suggested that Mueller had been investigating Schmitz about his efforts "to expose damaging information about Clinton" by looking for her deleted emails on the dark web. Though Mueller's team and the FBI refused to comment for the story, Schmitz believed they had been coordinating with CNN, a suspicion reinforced when network cameras magically appeared at dawn to record the raid on Roger Stone's home.[141]

WHY DID THE OBAMA ADMINISTRATION NOT TARGET RUSSIAN TROLLS AND HACKERS?

The introduction to the Mueller Report stated as undisputed fact that "the Russian government interfered in the 2016 presidential election in sweeping and systematic fashion." The interference included a "social media campaign that favored presidential candidate Donald J. Trump and disparaged presidential candidate Hillary Clinton. Russian military intelligence conducted computer-intrusion operations against entities, employees, and volunteers working on the Clinton Campaign and then released stolen documents." [142]

Attorney General Barr has questioned why more was not done by the Obama administration to counter Russian efforts. "That's one of the things I'm interested in looking at as part of my review of the Russia collusion investigation," he said. As the Obama administration had warnings of such efforts as early as April 2016, Barr wondered "what, exactly, was the response to it if they were alarmed. Surely the response should have been more than just, you know, dangling a confidential informant in front of a peripheral player in the Trump campaign." And why had the FBI not given the Trump campaign a defense briefing regarding Flynn, Papadopoulos, and Carter Page, who had been under investigation well before the election?[143]

Obama's national security advisor, Susan Rice, told the White House cybersecurity coordinator to "stand down" in 2016 regarding the meddling by Russia. Michael Daniel, special assistant to the president and cybersecurity coordinator for the National Security Council, told the SSCI that he and his staff had responded in "disbelief." [144]

Even when Mueller charged actual Russians, he could not make a connection between them and Trump.

On February 16, 2018, Mueller indicted three Russia-based companies and thirteen Russian individuals. He alleged that they had
conspired to defraud the United States by "impairing, obstructing,
and defeating the lawful functions" of the Federal Election Commission, the DOJ, and the State Department. Additional charges included conspiracy to commit wire fraud and bank fraud, as well as
aggravated identity theft.[145]

In a multimillion-dollar operation, the Internet Research Agency
(IRA) and two related companies were accused of conspiring to interfere in the 2016 election by hiring people to pose as Americans and
post political comments and memes on the internet. In other words,
"to troll" or provoke reactions on Facebook, Twitter, and Instagram.
They also staged rallies with the "strategic goal to sow discord in the
U.S. political system."

Announcing the charges, Deputy Attorney General Rod Rosenstein made it clear that there had been no alleged wrongdoing by the
Trump campaign, nor had the activity affected the outcome of the
election.[146]

Based in Saint Petersburg, the IRA began its effort in 2014—
well before Trump announced his run for office. Russians entered
the United States, hiding their true identities, and began operating
through shell companies. With a budget in the millions of dollars, the
IRA had more than eighty employees working on "the translator project," an effort to manipulate social media. They created fake personas
and pages and communicated with "unwitting members, volunteers,
and supporters of the Trump campaign involved in local community
outreach, as well as grass-roots groups that supported then-candidate
Trump."

The operatives purchased ads on Facebook and Instagram to promote Trump rallies, paid one Floridian to dress up as Hillary in a
prison uniform, and paid another to build a cage.[147]

Their operations included "supporting the presidential campaign
of then-candidate Trump . . . and disparaging Hillary Clinton." So it
was proof that Russia wanted Trump to win!

"It's hard to read the indictment and not see the 'troll factory' as conducting a Kremlin-sponsored covert action aimed at the U.S. political system," said Andrew Weiss, a former Clinton White House national security aide. He cited their use of cameras, SIM cards, and "drop phones" to communicate secretly as "intelligence-style trade-craft," suggesting a sophisticated Kremlin-type operation. Or maybe a garden-variety adulterer cheating on his wife.[148]

John Brennan posted this tweet: "DOJ statement and indictments reveal the extent and motivations of Russian interference in 2016 election. Claims of a 'hoax' in tatters. My take: Implausible that Russian actions did not influence the views and votes of at least some Americans."[149]

It's doubtful that Instagram accounts such as "Woke Black" posting this had much effect: "A particular hype and hatred for Trump is misleading the people and forcing Blacks to vote Killary. We cannot resort to the lesser of two devils. Then we'd surely be better off without voting AT ALL." Or "Blacktivist" posting this comment: "Choose peace and vote for Jill Stein. Trust me, it's not a wasted vote." Then there was the "United Muslims for America" posting "American Muslims [are] boycotting elections today, most of the American Muslim voters refuse to vote for Hillary Clinton because she wants to continue the war on Muslims in the middle east and voted yes for invading Iraq."[150]

In the social media chaos of the election, those Russian-backed accounts' efforts amounted to children's popguns in a shooting war, as did their ads on Facebook: "Hillary Clinton Doesn't Deserve the Black Vote!" "Donald wants to defeat terrorism . . . Hillary wants to sponsor it."

And did the organization really support Trump? Before the election, it sponsored rallies for Trump; a few days after he won, it sponsored rallies *against* him: "Trump is NOT my President," held in New York, and "Charlotte Against Trump," in North Carolina.

The IRA and its related companies also encouraged business owners to buy into marketing campaigns, getting them to turn over private information such as log-in credentials and IP addresses, under the auspices of a Russian called "Yan Big Davis," who claimed to be involved with several black activist groups.

It used those businesses "to make its network for fake accounts seem legitimate," according to an investigation by the *Wall Street Journal*. "Working with real Americans allowed the Russian trolls to 'eliminate the detection and exposure risk of inauthentic personas.'" [151]

The Russian operatives ran a Los Angeles–based start-up called Your Digital Face, which provided social media programs for small-business owners. It operated in the United States, Russia, Iran, China, Vietnam, the United Arab Emirates, and Cuba. "The group took control of social-media accounts, posted advertisements, and deployed software designed to add followers." Several accounts attracted more than 100,000 followers on Instagram. Twitter revealed that it had identified about 3,100 accounts linked to the "troll farm." [152]

A former intelligence officer told the *Wall Street Journal* that such marketing schemes had been used to map out business networks. In other words, it was a scam, like many other internet-based businesses.

No matter; with all the named defendants safely in Russia, which does not extradite accused individuals to the United States, the indictment would stand as proof that the Russians had been behind Trump's election. One American, Richard Pinedo, of Santa Paula, California, pleaded guilty to taking part in identity fraud to aid the IRA in setting up PayPal accounts to purchase online ads. His was the third known guilty plea, after Papadopoulos's and Flynn's. [153]

The indictments of the Russians "very specifically didn't make a connection between the Internet Research Agency and the Russian government," as Matt Taibbi of *Rolling Stone* pointed out. "So that piece of it was not really reported all that well. It quickly became, 'The Russians attacked us.' Well, what does 'The Russians' mean? Is it anybody in Russia? Is it necessarily a Russian government operation?" [154]

CNN seized upon the indictments to ambush an elderly pro-Trump woman in Florida who had unwittingly endorsed an event promoted by the Russian trolls. It published her full name, inciting an avalanche of hate on social media. The network didn't acknowledge that it had enthusiastically covered anti-Trump rallies staged by the IRA.[155]

Then, for Mueller, the unthinkable happened: one of the three named Russian entities, Concord Management and Consulting, sent two American lawyers to court to challenge allegations that it had funded the IRA's efforts.

Concord was controlled by a Russian businessman, Yevgeny Prigozhin, a close Putin ally sanctioned by the State Department in 2017 for ties to senior Russian government officials and Russia's defense ministry.[156] Named as a defendant in the indictment, Prigozhin operates some of Saint Petersburg's most prestigious restaurants, and his companies have contracts to feed Russian soldiers.[157]

On April 11, 2018, attorneys Eric Dubelier and Katherine Seikaly entered appearances on behalf of Concord, and made discovery demands seeking full disclosure of the evidence. Insisting that their client was not guilty, they sought "information about more than 70 years of American foreign policy," including each time the United States had attempted to interfere in a foreign election.[158]

When asked if he represented a third company, Concord Catering, Dubelier said, "I think we're dealing with the government having indicted the proverbial ham sandwich. That company didn't exist as an entity during the period alleged by the government."[159]

Mueller's team backpedaled, first seeking to postpone the arraignment on the grounds that Concord had not been properly served. It then filed a protective order opposing a proposal by Concord's attorneys that they be allowed to share information with the officers or employees of the company, including Prigozhin.[160]

"To anyone with an IQ above that of a celery stalk, such a fundamental and entirely proper move should have been anticipated," said former prosecutor George Parry. "Nevertheless, this seems to have caught Team Mueller by surprise."[161]

At a hearing in May 2018, Dubelier accused Mueller of filing charges against Concord "to justify his own existence," because he "has to indict a Russian—any Russian." He pointed out that the DOJ had never brought any case of "an alleged conspiracy by a foreign corporation to 'interfere' in a Presidential election by allegedly funding free speech."

He wrote, "The obvious reason for that is that no such crime exists in the federal criminal code."[162] If successfully prosecuted, the DOJ's strategy had the potential to make a criminal of anyone who posted political speech on the internet under a pseudonym.

At a hearing in autumn 2018, the judge acknowledged that Dubelier had an argument. "I'll give it to you, Mr. Dubelier," he said, "this is an unprecedented case," adding that the government would have a "heavy burden at trial."[163]

Dubelier's insistence that Mueller turn over documents proving his case was a major ace up his sleeve. In March 2019, OSC prosecutors asked the judge to withhold 3.2 million "sensitive" documents because of national security and law enforcement concerns. Even though none of the material was deemed classified, the government had provided only five hundred documents to that point. DOJ attorney Jonathan Kravis told the judge that the accused individuals could view some of the documents at Dubelier's Washington law firm, but the entire set could reveal how the government had acquired them.[164] Since Prigozhin would be arrested if he entered the United States, Dubelier called that idea a "nonstarter."

The rest of the hearing was sealed and the public removed from the courtroom. It is doubtful that the case will ever go to trial, but it could prove entertaining, based on Dubelier's cheeky filings. "Could the manner in which [the special counsel] collected a nude selfie really threaten the national security of the United States?" he asked in one court brief.[165]

"Mueller risked exactly what has happened: one of the businesses showing up to contest the case at no risk, in effect forcing Mueller to show this Kremlin-connected defendant what he's got," wrote Andrew McCarthy, "even though he has no chance of getting the Kremlin-connected defendant convicted and sentenced to prison." [166]

In other words, the IRA indictment was a Mueller public relations stunt.

The second effort by Russians to influence the election, according to the Mueller Report, involved Russian "government actors." On July 13, 2018, the special counsel charged a dozen Russian intelligence agents with hacking the computers of several Democratic Party organizations, including the DNC. Though the detailed twenty-nine-page indictment outlined the technological trickery used to allegedly subvert the 2016 presidential election, his charges did not allege collusion or conspiracy by the Trump campaign with the Russians. Nor did he allege that any votes had been affected. [167]

But the convenient timing of the indictments—just before the Trump-Putin summit—provided fodder for critics to attack Trump. Senator John McCain urged Trump to cancel the summit unless he was "prepared to hold Putin accountable." [168]

Mueller was controlling and shaping the narrative. Never mind the fact that because the FBI and the DOJ had never physically examined the DNC servers, the indictment would be extremely difficult to prove in court. And why hadn't Assange been charged with criminal collusion along with those Russian operatives? After all, the indictment said that Assange and WikiLeaks had urged Russian hackers to send them "new material" to raise the chances of Senator Bernie Sanders becoming the Democrat nominee. The only US charge against Assange to that point was a single cybertheft count alleging a conspiracy between him and Bradley (now Chelsea) Manning. [169]

"Why leave obvious, serious charges on the cutting room floor?" asked Andrew McCarthy. "Mueller brought a dozen felony charges against the Russian operatives with whom, we've been told, Assange conspired. . . . If I were a cynic (perish the thought!), I'd suspect that the government does not want Special Counsel Mueller's Russian-hacking indictment to be challenged." [170]

He's not the only cynic. The indictment of the intelligence agents was a PR gambit to put to rest the nagging question of how Americans could believe that the Russians were guilty. If Assange were to be tried, that question might be answered.

THE RECKONING

Most guys would've crawled in a corner with their thumb in their mouth. Okay? Saying "Mommy, take me home." I found it to be an incredible challenge. What shocked me was the level of corruption. They really did try to take away an election.

What went on in this country with our intelligence agencies and the FBI is unprecedented in American history. It should never, ever happen to another president again.

—Author's interview with President Donald J. Trump,
Oval Office, White House, June 25, 2019

I n civil society, power to act for the benefit of the many is vested in the few. We instill our public *trust* that this power is dispensed for the common good without passion or prejudice. Such faith is an indispensable necessity to sound governance. Democracy fails when that trust is breached for personal gain or political design. It failed here.

Witch Hunt is the story of ambitious and unscrupulous people in high positions of government who abused their authority. They sought to subvert the rule of law and undermine the democratic process. They weaponized their powers to influence a presidential election, undo the result they did not like, and extrude the elected president from office. Our intelligence community and the FBI were at the heart of this illicit and unprecedented scheme.

Their cudgel was "collusion." It was nothing more than a cleverly conjured-up lie. There was never any evidence of a treasonous conspiracy hatched by Trump and Putin in the bowels of the Kremlin. It was all a hoax, contrived to masquerade as the truth. Yet the hoax gained popular currency because Democrats and the media drove the fictive narrative with ferocity. They were convinced that Trump was an illegitimate president. To them, it was inconceivable that he had won the highest office in the land, absent some sinister plot to steal the election. Many people across the United States believed it. They accepted without question or challenge that Trump was a poseur—a Russian agent disguised as a US president. The carefully crafted hoax begot the witch hunt, a series of investigations designed to sully and destroy him.

There are many villains in this morass of misbegotten deeds. Clinton ally John Brennan, who was CIA director in the summer of 2016, instigated and fueled the hoax by collecting foreign source information on Trump that proved to be utterly bogus. Undeterred, he exploited those phony tips to create an "interagency task force," which was fundamentally a spying operation, despite US laws prohibiting his agency from targeting US citizens domestically.[1] That was why some members of the Trump campaign were lured overseas, where Brennan's cabal could spy on them more freely and where undercover informants, such as Stefan Halper, could be deployed without limits. James Comey, Peter Strzok, Andrew McCabe, and others at the FBI were in on the task force and eagerly used it as the foundation to open the Bureau's Trump-Russia "collusion" probe, dubbed "Operation Crossfire Hurricane." Intelligence from the United Kingdom, Australia, and Italy was mined, with timely contributions by James Clapper, the director of national intelligence, who was Brennan's favorite collaborator. They were all virulently opposed to Trump and strong supporters of Clinton.

Of course, the centerpiece of the hoax was the anti-Trump "dossier" composed by ex–British spy Christopher Steele and commissioned by the Clinton campaign and Democrats but cloaked behind a double firewall of payments that allowed them to smear Trump and influence the election. Based on either Russian disinformation or fabricated hearsay from anonymous Moscow sources (or both), those garbage documents were "unverifiable," as Steele admitted.[2] But that minor detail did not dissuade the FBI, which zealously appropriated the material to launch its counterintelligence investigation of Trump as a covert Russian asset. Try as it might, the FBI could corroborate none of the "collusion" allegations contained therein. Its reported "spreadsheet" of findings was barren of proof.[3] The agency was repeatedly warned of the "dossier's" "credibility issues."[4] Nevertheless, Comey signed off on a Foreign Intelligence Surveillance Act (FISA) warrant application to surveil former Trump campaign adviser Carter Page, while concealing vital evidence from the FISA court and deceiving the judges. Comey represented that his information was "verified" when he knew it was not. He vouched that his source was "reliable" when he knew he was not.[5] The artifice also allowed the FBI to gain retroactive access to all of Page's electronic communications with the campaign. No evidence of a criminal conspiracy with Russia was ever found.

As all of those Machiavellian machinations were unfolding, Steele and the man who had hired him, Glenn Simpson of Fusion GPS, were furiously disseminating the "dossier" to any reporter who would listen in the hope that media stories would scuttle Trump's bid for the White House. The FBI, which knew that Steele was both biased against Trump and pushing dubious material, set up an "information-laundering" system to conceal the identity of its source.[6] Simpson and Steele would furtively feed each new "dossier" memo to Bruce Ohr at the Justice Department. Ohr would be debriefed by his "handler,"

FBI agent Joe Pientka, who would then pass the material on to his cohort, Peter Strzok, who would deliver it to Andrew McCabe, FBI director Comey's chief deputy. By transferring it through that complex sequence, its origins could be obscured and the fact that it was paid political propaganda funded by Trump's opponent could be shrouded. Similar to money laundering, Steele and his dirty information were sufficiently cleansed, and his product would thereafter be referenced only as emanating from a reliable and neutral foreign intelligence source. In truth, the Trump-hating Steele was the antithesis of neutral. Although he was fired by the FBI for leaking to the media and lying about it, the FBI continued to use him as its source while pretending to disassociate from him to avoid raising suspicions.[7] Even after Trump won the presidency, Steele was composing and circulating his specious material that the new president was in league with Putin. His clandestine contacts with the FBI and DOJ continued well into Trump's first year in office.[8] FBI documents uncovered in early August 2019 confirmed that the bureau and the Justice Department were well aware of Steele's anti-Trump bias, but persisted in their reliance on his "dossier" to gain permission from the FISA court to wiretap Page.[9] Calling the warrant application a "fraud," Sen. Lindsey Graham (R-SC), chairman of the Senate Judiciary Committee added, "Here's what we're looking at: Systemic corruption at the highest level of the Department of Justice and the FBI against President Trump and in favor of Hillary Clinton."[10]

Comey, Brennan, and Clapper were determined to sabotage Trump before he took the oath of office. Their ploy to selectively brief him about the "dossier" as a pretext for leaking it to the media was the definition of devious.[11] Trump was not told that he and his campaign were under investigation or that the FBI had obtained a warrant to surveil a former campaign associate.[12] This information was also withheld from congressional leadership. The unmasking and leaking of National Security Advisor Michael Flynn's conversations with the Russian ambassador were both unconscionable and felonious.[13] Comey and McCabe manipulated him into an FBI interrogation under the guise of the moribund Logan Act.[14] Flynn was fired and later charged with lying after agents concluded that he had *not* appeared to be lying.[15] Comey was also busy compiling confidential presidential memos that he later stole and leaked to effectuate the appointment of his friend and mentor Robert Mueller as special counsel.[16] By his own admission, Comey's intent was to influence the investigation of the man who had fired him. The witch hunt assumed an odious dimension as Mueller hired a "hit squad" of partisans, while Deputy Attorney General Rod Rosenstein connived to depose the president under the Twenty-fifth Amendment.[17]

None of those corrupt acts seemed to interest Mueller in the least. Tasked to find proof of a nonexistent "collusion" conspiracy by Trump to filch the election from Clinton, no plausible evidence surfaced over the course of the twenty-two-month probe. Instead of determining who had engineered the hoax and exposing its manipulation to destroy a presidency, the special counsel was determined to *imagine* an obstruction of justice offense that was unsupportable under the law. Mueller and his confederates surely knew that. His explanation for why he decided *not* to decide the issue was stunningly "unintelligible."[18] Attorney General William Barr admitted he had been baffled by the special counsel's "strange statement," forcing him and other top lawyers at the Justice Department to correct Mueller's tortured interpretation of the law.[19] The evidence did not sustain an obstruction offense.[20]

Beyond the special counsel's incomprehensible reasoning and his shameless derogation of Trump, what was extraordinary about the Mueller Report was what was missing. Nowhere in the exhaustive 448 pages was there an earnest examination of the surreptitious actions of the Clinton campaign and Democrats to hustle and hype their false Russian information to damage Trump and impact the outcome of the 2016 election.[21] Trump didn't "collude" with Russians, but his opponent did. Mueller conspicuously ignored that fact. There were a few passing references to Steele and only an indirect reference to his infamous "dossier." How is that possible? The special counsel reportedly interviewed Steele twice.[22] Was Mueller looking only for incriminating evidence against Trump, while ignoring the role of his opponent in successfully paying for Russian information to influence the election? It was a remarkable abdication of duty for a man whose primary responsibility was to determine *any* Russian interference in the United States' democratic process. That glaring omission reinforced how the special counsel's investigation was infected with bias against Trump.

Mueller waxed at length about Trump adviser Carter Page's trip to Moscow before concluding that it had not constituted a "collusion" conspiracy.[23] Yet the special counsel never addressed the question of how his friend Comey and the DOJ appeared to have defrauded the FISA court by using unverified information from Steele and his "dossier" to spy on Page. In his report, Mueller dismissed one of his own conflicts of interest in a short footnote but left his other conflicts unchallenged.[24] He never sought to dispel Rosenstein's disqualifying conflict as both a key witness and supervisor of the investigation. He did not bother to defend the conflicts of his chief prosecutor, Andrew Weissmann, whose favoritism toward Clinton was well documented and who had been given the "dossier" material in the summer of 2016.[25] Weissmann knew then that the "collusion" allegations were based on biased and defective information.

Mueller's decision to assemble a team of partisans ruined the integrity and credibility of his work. He fired Strzok only when his profane and politically charged anti-Trump text messages surfaced. Instead of confiscating the FBI agent's cell phone to preserve the evidence contained therein, Strzok's iPhone was recycled and wiped clean. Mueller failed to question the agent about whether his bias had already contaminated the investigation. It likely had, although there were plenty of prejudiced prosecutors on the special counsel team to carry the torch. The tactics they employed were reprehensible, pressuring witnesses and defendants to lie. People were threatened with prosecution unless they signed statements falsely accusing Trump of some malefic act of "collusion."[26] Those unprincipled prosecutors were committing the equivalent of attempts to suborn perjury, extortion, and bribery. If you or I did the same, we'd be behind bars. They should be investigated. It is doubtful that they ever will be.

In the end, Mueller's confederates "did not establish that members of the Trump Campaign conspired or coordinated with the Russia government in its election interference activities."[27] That should have come as no surprise to any serious person paying attention. There were never any witches.

The sad coda to Mueller's work occurred during an embarrassing spectacle on July 24, 2019, when he appeared before two separate congressional committees in televised hearings. The abstruse arguments and analysis in his report were laid bare. As he stumbled and stammered his way through several hours of testimony, Mueller seemed lost and confused. He struggled to understand basic questions. His answers were slow, halting, and uncertain. It was obvious that Mueller had not written the report that bore his name. Nor had he authored the letter sent to Attorney General William Barr that complained about the AG's summary of the report's conclusions.[28] Nor had he penned the script he labored to read at his puzzling nine-minute news conference on May 29, 2019.[29] In his testimony, Mueller admitted attending "very few" of the witness interviews.[30] He had a feeble grasp of fundamental facts in the report and seemed oblivious to the law that supposedly supported the evidence cited therein.

Representative John Ratcliffe (R-TX) asked Mueller to explain how he came up with the unprecedented legal standard of a prosecutor announcing that he was not exonerating the subject of an investigation.

RATCLIFFE: Can you give me an example other than Donald Trump, where the Justice Department determined that an investigated person was not exonerated . . .
MUELLER: I—I . . .
RATCLIFFE: . . . because their innocence was not conclusively determined?
MUELLER: I cannot, but this is a unique situation.

RATCLIFFE: OK. Well, I— you can't—time is short. I've got five minutes. Let's just leave it at, you can't find it because—I'll tell you why: It doesn't exist.[31]

Armed with unchecked power, Mueller and his team dreamed up a novel legal standard that robbed Trump of the presumption of innocence, a bedrock principle of law. His "does not exonerate" chimera was an unconscionable maneuver that bastardized the burden of proof, shifting it away from prosecutors and onto the shoulders of the presumptively innocent target. But in Mueller's sinuous universe the president was "unique" and deserved only the presumption of guilt. Why? Mueller did not explain. How could he? It was inexplicable. As Ratcliffe observed, the president was treated by the special counsel as if he were "below the law."[32] He was accused of obstructing justice when he proclaimed his innocence and dared to criticize the special counsel or contemplate his removal over legitimate conflicts of interest. Representative Louie Gohmert (R-TX) reasoned that the president was not obstructing justice, he was vociferously protesting injustice.[33] Trump had every right to do so. If the president's words or actions constituted an obstruction offense, the special counsel would have said so in no uncertain terms. They knew the law did not support it, so they chose to otherwise smear him.

Mueller was repeatedly asked whether he ever investigated the origins of the Russia "collusion" allegations and the FBI's counterintelligence operation against Trump and his campaign. Did he examine the contents of the "dossier" or Steele's purported Russian sources, all of which were central to the conspiracy case? Apparently not. Did he investigate the role of the Clinton campaign and Democrats who paid for it to influence the election? Absolutely not. In one disturbing exchange, Mueller did not seem to know what Fusion GPS is or how the firm and its founder, Glenn Simpson, had peddled the false information that drove the hoax.[34] To all of these vital questions, Mueller replied that it was "beyond my purview" and "predated" his appointment as special counsel.[35] This was mind-splitting to even the casual observer. In truth, the entirety of Mueller's "collusion" investigation predated his tenure. It was a nonsensical rationale for failing to investigate the real "collusion" conspiracy instead of the imagined one.

If nothing else, the debacle of his agonizing testimony drew back the curtain on the great and powerful Mueller. Like the Wizard of Oz, he was not what his vaunted reputation promised. His diminished mental acuity made plain that he had been little more than a detached figurehead, ceding command and control to his team of partisans. His top two lieutenants, Andrew Weissmann and Aaron Zebley, had likely assumed authority over the investigation and the report's composition. They should never have been allowed to join the special counsel. Weissmann harbored pro-Clinton bias, having attended her election night event. Zebley, along with Jeannie Rhee, offered legal representation in Clinton-related lawsuits.[36] This accounts for a great deal of the anti-Trump slant that shaped their probe and its final report. Although Mueller insisted "it is not a witch hunt," his inability to convincingly explain and justify the special counsel's work left many to conclude otherwise.[37] Even some Democrats and members of the liberal media grudgingly confessed that his disoriented discourse had been a "disaster" politically and a calamity personally for Mueller.[38] It brought into sharper focus how Trump had been framed with a non-existent "collusion" conspiracy and re-victimized by a one-sided squad of partisan zealots who inhabited the special counsel team. The actions taken and decisions made were neither fair nor impartial. It was a "witch hunt," exactly as Trump so often described it. Ken Starr, who once served as Independent Counsel in an investigation that led to the impeachment of President Bill Clinton, criticized Mueller for failing to ensure that his special counsel staff would be fair and unbiased. "I love Bob Mueller as a human being, as a patriot—but I think he's done a grave disservice to our country in the way he conducted this investigation," observed Starr just moments after Mueller concluded his testimony.[39]

In *The Russia Hoax*, I argued that the case against Trump was neither warranted by facts nor supported by law. *Witch Hunt* chronicles the wealth of new evidence that proves the original thesis. The fallacy of Trump's being a collusive agent of the Kremlin was blindly accepted by many people without skepticism or scrutiny. For some, it was an emotional response to his unexpected election. They countenanced a suspicion and accepted it as truth. Their dislike of the man, his policies, or both, clouded their analysis. Still others embraced the lie for malevolent purposes. The liberal media, in concert with Democrats, endorsed false accusations despite readily available evidence to the contrary. They betrayed their responsibility to be objective and fair. They allowed their enmity and bias to obscure reasoned judgment.

I have no doubt that if the same facts and circumstances had surrounded someone *other* than Donald Trump, there would have been no FBI investigation initiated, no warrants to spy issued, and no special counsel appointed to perpetuate the witch hunt. But Trump has always been a lightning rod. His outspoken nature attracts both criticism and praise. He was not a conventional politician. He advocated turning Washington upside down. Voters liked what they heard. The entrenched establishment did not. They feared him. The dramatic change he envisioned threatened their very existence and their hold on power. He had to be stopped. The "collusion" hoax was the means to an end.

I am frequently asked whether those who bent the rules or broke the law to achieve their end will ever be held accountable. That, of course, is up to Attorney General William Barr. I have no crystal ball. However, Barr seems well aware that the explanations he has received so far are inconsistent with the known facts. He is on record as stating that "the counter-intelligence activities directed at the Trump Campaign were not done in the normal course and not through the normal procedures."[40] Indeed they were not. He cautioned that government abuse of power is equally as dangerous as foreign interference. He promised to investigate the investigators.

In May 2019, it became known that Barr had assigned John H.

Durham, a top prosecutor in Connecticut, to examine the origins of the Russia investigation initiated by Comey, McCabe, Strzok, and others at the FBI.[41] His assignment encompasses an investigation into the actions of intelligence agencies under the direction of Brennan and Clapper. As Barr noted in his congressional testimony, "I think spying did occur. The question is whether it was adequately predicated. I need to explore that."[42] Durham, as a US attorney, is well versed in FBI corruption and CIA abuses.

There is a separate investigation under way by DOJ inspector general Michael Horowitz.[43] He is determining whether officials at the FBI and DOJ deceived or otherwise misled the FISA court to gain approval for four successive warrants to spy on Carter Page. Horowitz is also probing the FBI's deployment of undercover informants on the Trump campaign, as well as the use of Christopher Steele and his "dossier." The ex-spy was interviewed in Britain by the IG investigators in early June 2019.[44] Upon completion of Horowitz's report, it is possible that criminal referrals may be lodged with the Justice Department if evidence of wrongdoing justifies it.

Barr and Durham have their work cut out for them. Comey will likely plead ignorance or acute amnesia, as he did some 245 times when questioned by Congress in late 2018.[45] Other former FBI officials may come up with an incurable case of laryngitis, invoking their Fifth Amendment right against self-incrimination. Piercing the thick veil of secrecy that people such as Brennan and Clapper routinely hide behind will pose an exceptional challenge. The intelligence community and the FBI have vowed to cooperate. But this is a clever feint. They will resist and obstruct at every turn. However, Barr has been given unilateral authority to declassify a plethora of heretofore undisclosed documents that may prove to be a rich trove of corruption evidence.[46] He should do so aggressively.

Witch Hunt presents compelling evidence that high officials in government whom we entrusted to uphold the law instead breached that trust and violated the law. They are the ones who meddled in the election to help Clinton. Thereafter, they aspired and conspired to evict Trump from office. Evidence against him was invented or embellished. Laws were perverted or ignored. Trump was framed for a "collusion" conspiracy that had never existed. They *knew* it was untrue.

Inventing the lie was easy. Spreading it was even easier. Uncovering the full truth will be hard. It will take time to unravel the layers of venal acts intended to take down Trump and his presidency. But the only cure for a lie is the truth. The only remedy for lawlessness is justice.

The reckoning awaits.

LIST OF MAJOR CHARACTERS

FEDERAL BUREAU OF INVESTIGATION

James Baker: Top FBI attorney involved in decision making on both Clinton email investigation and Trump-Russia "collusion." Prepared and approved the initial application for a FISA warrant to spy on Carter Page.

James Comey: Became Obama's director of the FBI in September 2013. Wrote exoneration of Hillary Clinton for her illicit email server before she was interviewed. Initiated investigation of Trump's campaign. Signed three of four FISA warrant applications to spy on Carter Page. Was fired by Trump on May 9, 2017. Leaked his memos to trigger special counsel investigation.

Michael Gaeta: FBI agent and assistant legal attaché based in Rome. Interviewed Steele in Rome and London, passed his early memos to Victoria Nuland at the State Department.

Andrew McCabe: Deputy director of the FBI. Wife ran for state senate in Virginia after receiving $700,000 from Hillary Clinton as-

sociates. After James Comey firing, McCabe initiated a secret investigation of the president. Discussed removing Trump by recruiting
cabinet members to invoke the Twenty-fifth Amendment. Signed
Carter Page FISA warrant renewal request in June 2017. Testified that
without the Christopher Steele "dossier," no warrant from the FISC
would have been sought. Fired from FBI for lying.

Lisa Page: FBI attorney infamous for exchanging anti-Trump
texts with her paramour, Peter Strzok. General counsel to Andrew
McCabe and on special counsel team. Investigated for leaking classified information.

Joe Pientka: FBI agent who interviewed Michael Flynn with Peter
Strzok. Interviewed DOJ attorney Bruce Ohr as a back channel to
Steele after he was terminated.

E. W. "Bill" Priestap: Assistant director of FBI Counterintelligence
Division. Testified that verification of the Christopher Steele "dossier"
was in its "infancy" when the FBI and DOJ filed for Carter Page FISA
warrant.

Peter Strzok: Veteran FBI counterintelligence agent. Involved in
both "Midyear Exam" and "Crossfire Hurricane." Tapped by Robert
Mueller to lead special counsel probe. Was removed in August 2017
after IG discovered thousands of anti-Trump texts he had traded with
Lisa Page. Was fired from the FBI.

DEPARTMENT OF JUSTICE

William Barr: Appointed attorney general by Trump. Said he believes there was spying on the Trump campaign.

Dana Boente: US attorney for the Eastern District of Virginia. After Trump fired Sally Yates, became acting attorney general until Jeff Sessions was confirmed as attorney general. Then served as acting deputy attorney general until confirmation of Rod Rosenstein. Signed Carter Page FISA warrant application in April 2017.

Michael Horowitz: DOJ inspector general. Uncovered the Peter Strzok–Lisa Page texts in July 2017. In June 2018, released a scathing report on FBI and DOJ actions during the election. Is currently preparing a report on FISA abuse.

Loretta Lynch: Attorney general under President Barack Obama. On June 27, 2016, met secretly with Bill Clinton on the tarmac at Phoenix airport days before Hillary Clinton was interviewed by the FBI.

Bruce Ohr: Top DOJ official married to Russia expert Nellie Ohr. Served as a secret conduit to feed Christopher Steele "dossier" memos to FBI in an information-laundering scheme. Warned the FBI that Steele was desperate that Trump not be elected and the information needed verifying.

Rod Rosenstein: Deputy attorney general from April 2017 to May 2019. Wrote memo recommending that James Comey be fired. Appointed Robert Mueller special counsel on May 17, 2017. Discussed with Andrew McCabe the possibility of wearing a wire to record Trump and recruiting cabinet members to remove the president under the Twenty-fifth Amendment. Refused to recuse himself from oversight of the special counsel investigation even though he was a key witness in the case. Signed final Carter Page FISA warrant application in June 2017.

Jeff Sessions: Attorney general from February 2017 to November 2018. Head of Trump campaign's foreign policy advisory team. Recused himself from Russia investigation. Resigned after 2018 mid-term election.

Sally Yates: Acting attorney general from January 20 to January 30, 2017. Signed two Carter Page FISA applications in October 2016 and January 2017. Engineered the firing of Michael Flynn. Sacked by Trump for refusing to implement travel ban.

STATE DEPARTMENT

Elizabeth Dibble: Deputy chief of mission at US Embassy in London from 2013 to July 2016. Contacted regarding Alexander Downer's conversation with George Papadopoulos in late July 2016.

Kathleen Kavalec: Top State Department official. Interviewed Christopher Steele on October 11, 2016, and warned the FBI of his political motives, contacts with the press, and errors in his "dossier."

David J. Kramer: Former State Department official. Gave copy of Christopher Steele "dossier" to Senator John McCain, leaked it to BuzzFeed.

Victoria Nuland: Top Obama State Department official. Gave permission for FBI agent Michael Gaeta to meet with Christopher Steele in London on July 5, 2016, when Gaeta first received the "dossier."

Jonathan Winer: Top State Department official. Passed on a two-page summary of Christopher Steele "dossier" to Victoria Nuland and Secretary of State John Kerry. Received salacious Cody Shearer memo from Sidney Blumenthal, passed it to Steele.

INTELLIGENCE COMMUNITY AND "ASSETS"

Julian Assange: Founded WikiLeaks. Claims that the Russians were not involved in the hacking or disclosure of DNC and John Podesta emails published by WikiLeaks during the 2016 election. Now fighting extradition to the United States.

John Brennan: Director of the CIA under Barack Obama, MSNBC pundit. The epicenter of the collusion narrative. Triggered "Crossfire Hurricane."

James Clapper: Director of national intelligence. Directed James Comey to brief Trump on the salacious "pee dossier," then leaked that news to CNN and lied about it. Testified that the Christopher Steele "dossier" had not been verified by December 2016.

Sir Richard Dearlove: Ex-chief of MI6, close associate of Stefan Halper and Christopher Steele. Vouched for Steele's reputation.

Alexander Downer: Australian ambassador to the United Kingdom. Arranged a $25 million donation to the Clinton Foundation in 2006. Met with George Papadopoulos on May 10, 2016. Reported to US Embassy that Papadopoulos had told him the Russians had damaging information on Clinton, the alleged basis of "Crossfire Hurricane."

Henry Greenberg: Russian businessman and FBI informant, also known as Henry Oknyansky and Gennady Vostretov. Tried to peddle opposition research on Hillary Clinton to Roger Stone.

Stefan Halper: Professor at Cambridge University, US spy for more than thirty years. At direction of the FBI, he targeted Carter Page, George Papadopoulos, and others in the Trump campaign.

Robert Hannigan: Director of Government Communications Headquarters (GCHQ), the British equivalent of the NSA. In the summer of 2016, flew to the United States to brief Brennan. Abruptly resigned in January 2017.

Nagi Khalid Idris: Director of the London Centre of International Law Practice (LCILP) who hired George Papadopoulos. Insisted that Papadopoulos accompany him to Rome to meet Joseph Mifsud.

Joseph Mifsud: Professor from Malta who told George Papadopoulos that Russia had thousands of Hillary Clinton's emails. Has links to Western intelligence.

Sergei Millian: Belarusan-born American citizen, real name Siarhei Kukuts. Offered George Papadopoulos $30,000 a month to work as a consultant, provided he also work for the Trump administration. Reportedly a source of material in the Christopher Steele "dossier."

Susan Rice: National security advisor under Barack Obama. Involved in unmaskings of Trump associates.

Michael Rogers: Director of the National Security Agency. Rogers traveled to Trump Tower on November 17, 2016, without informing his boss, James Clapper. The next day, the transition team moved to a Trump property in New Jersey.

Christopher Steele: Former MI6 spy, hired by Fusion GPS to research Trump; his seventeen memos constituted the Steele "dossier," used by the DOJ and the FBI to target Trump associates Carter Page, Paul Manafort, Michael Cohen, and Michael Flynn.

Erika Thompson: Australian diplomat. Introduced George Papadopoulos to Alexander Downer.

Azra Turk: Pseudonym of informant sent to London by the FBI to meet with George Papadopoulos. Real identity remains unknown.

Olga Vinogradova: "Vladimir Putin's niece," who met George Papadopoulos with Joseph Mifsud, identified as someone who could broker meetings between the Trump campaign and the Putin regime. Unrelated to Putin. Real name Olga Polonskaya.

Sir Andrew Wood: Former British ambassador to Russia. Met with David Kramer and Senator John McCain about the Steele "dossier." Vouched for Christopher Steele's credibility.

OFFICE OF THE SPECIAL COUNSEL

Robert Mueller III: Appointed special counsel by Deputy Attorney General Rod Rosenstein on May 17, 2017. Director of the FBI from 2001 to 2013. Close friend of James Comey. Hired a team of partisan Democratic attorneys to investigate Trump. His final 448-page report found no collusion by the Trump campaign.

Jeannie Rhee: Defended the Clinton Foundation in a civil racketeering case, represented Hillary Clinton in litigation regarding her emails. Involved in the prosecutions of George Papadopoulos, Roger Stone, and Paul Manafort.

Andrew Weissmann: Chief of DOJ Criminal Fraud Division and Robert Mueller's chief deputy. Served as general counsel of the FBI under Mueller. Noted for wrongful prosecutions that earned both reversals and rebukes. Donor to DNC, Hillary Clinton, and Barack Obama. Attended Clinton's election-night "victory party."

Aaron Zebley: Former FBI agent, rose to become Robert Mueller's chief of staff. While at Mueller's law firm, represented Justin Cooper, who set up Hillary Clinton's private email server.

THE TRUMP CAMPAIGN AND ADMINISTRATION

Michael Caputo: Trump campaign adviser targeted by Russian FBI informant Henry Greenberg.

Michael Cohen: Former vice president of Trump Organization and personal attorney of Trump. Approached by real estate developer Felix Sater, an FBI informant, to build Moscow Trump tower. Pleaded guilty to nine charges.

Jerome Corsi: Author targeted by Robert Mueller for communications with Roger Stone regarding WikiLeaks.

Michael Flynn: Director of the Defense Intelligence Agency from 2012 to 2014. Trump's national security advisor for twenty-four days. Targeted by "Crossfire Hurricane." Pleaded guilty to lying about communications with Russian ambassador. As yet has not been sentenced.

Richard Gates III: Paul Manafort's partner and deputy campaign manager for Trump. Pleaded guilty to conspiring to defraud the United States, as well as making false statements.

Paul Manafort, Jr.: Longtime Washington political strategist and lobbyist for foreign governments. Adviser to former Ukrainian president. Worked on Trump campaign from March through August 2016. Convicted of tax evasion and bank fraud.

Don McGahn: White House Counsel. Met with Sally Yates regarding Michael Flynn's conversations with Sergey Kislyak.

Carter Page: FBI cooperating witness who helped convict a Russian spy. On Trump's foreign policy advisory team. Christopher Steele "dossier" accused him of being a Russian agent. FBI obtained a FISA warrant on Page in October 2016, renewed in January 2017, April 2017, and June 2017.

George Papadopoulos: Foreign policy adviser to Trump campaign. Met with Mifsud, who told him the Russians had thousands of Hillary's emails. His comments to Australian diplomat Downer were reported to have triggered Crossfire Hurricane. Pleaded guilty to one count of lying to federal agents.

Roger Stone: Political strategist, longtime adviser to Donald Trump. Fort Lauderdale home raided at 6:00 a.m. by two dozen FBI agents. Indicted for lying to Congress, witness tampering, and obstruction of justice. Has pleaded not guilty.

Donald Trump, Jr.: President Trump's son. Attended the June 2016 meeting with Russian lawyer Natalia Veselnitskaya.

THE CLINTON CAMPAIGN AND THE DEMOCRATIC PARTY

Sidney Blumenthal: Notorious political operator for Bill and Hillary Clinton. Passed second "dossier" written by Cody Shearer to the State Department.

Alexandra Chalupa: A lawyer, Ukrainian activist, and former DNC contractor who targeted lobbyist Paul Manafort.

Hillary Clinton: Democratic nominee for president. Cleared by James Comey of violating the Espionage Act and other criminal statutes. Paid for Christopher Steele "dossier" through law firm Perkins Coie. Blamed James Comey and Russian meddling for her election loss.

Lanny Davis: Attorney, lobbyist, and longtime political adviser to Clintons. Represented Michael Cohen in Robert Mueller investigation.

Marc E. Elias: Attorney at Perkins Coie, general counsel for Hillary Clinton's 2016 presidential campaign, hired Fusion GPS on behalf of Clinton campaign.

Mary Jacoby: Married to Glenn Simpson and cofounder of Fusion GPS. Wrote Facebook post crediting her husband as the architect of "Russiagate," then deleted it.

Terry McAuliffe: Former governor of Virginia, longtime Hillary Clinton associate. Approached Andrew McCabe's wife to run for the Virginia state senate, funneled about $700,000 to her campaign.

Cheryl Mills: Longtime attorney for Hillary Clinton. Given limited immunity in email investigation and allowed to sit in on Peter Strzok's interrogation of Clinton, despite being a fact witness. Involved in sorting and deleting Clinton's emails.

Robby Mook: Top Hillary Clinton campaign official, first to publicly say that Russia wanted Trump to win.

Barack Obama: Described by Lisa Page in a text to Peter Strzok as wanting to "know everything" they were doing. Used a pseudonym to communicate with Hillary Clinton on her unauthorized email server. Gave TV interview on April 10, 2016, saying Clinton had merely been "careless."

Nellie Ohr: Russia expert who worked for CIA Open Source Center, married to Bruce Ohr, approached Fusion GPS in late 2015. Applied for a ham radio license on May 23, 2016. Invoked "spousal privilege" during testimony.

John Podesta: Top Clinton campaign adviser. After he clicked a phishing link, hundreds of his emails were stolen, then published by WikiLeaks.

Samantha Power: US ambassador to the United Nations under Barack Obama. Accused of requesting hundreds of unmaskings of US citizens related to the Trump campaign. Claims that others used her name.

Heather Samuelson: Attorney for Hillary Clinton. Reviewed Clinton's emails and sorted out more than 30,000 that were then deleted from the server.

Cody Shearer: Dirty trickster and longtime associate of Clintons. Gave second "dossier" to Sidney Blumenthal, saying the Russians had compromising information on Trump.

Glenn Simpson: Cofounder of Fusion GPS. Hired by Perkins Coie to do opposition research on Trump. Hired Christopher Steele and Nellie Ohr. Created Trump-Russia "collusion" narrative and fed it to the FBI, the State Department, and the media. Invoked Fifth Amendment privilege when subpoenaed by Congress.

Anthony Weiner: Former New York Democratic congressman who went to prison for sexting a teenager. Married to Clinton aide Huma Abedin. The FBI seized a laptop in his home containing emails sent or received by Hillary Clinton and Huma, triggering a hasty second Clinton email investigation shut down by James Comey days before the election.

FOREIGN INTELLIGENCE
SURVEILLANCE COURT (FISC)

Rosemary Collyer: Chief judge of FISC, wrote scathing report on rampant abuses of the FISA process by the FBI and the DOJ released April 26, 2017. Approved first Carter Page FISA warrant on October 21, 2016, before she became aware of abuses.

Rudolph "Rudy" Contreras: Mentioned as "Rudy" in Peter Strzok and Lisa Page texts. Appointed to FISC on May 19, 2016. Removed from presiding over Flynn case in December 2017 without explanation after the IG discovered the texts.

UKRAINIANS AND RUSSIANS

Oleg Deripaska: Russian billionaire approached by FBI in September 2016 for confirmation of Trump-Russia "collusion." Deripaska told them that there had been no collusion.

Igor Divyekin: According to Carter Page FISA application, a Vladimir Putin associate who met secretly with Page, promising *kompromat* that the Kremlin possessed on Hillary Clinton. No such meeting took place.

Konstantin Kilimnik: Russian/Ukrainian political consultant and associate of Paul Manafort. Indicted by Robert Mueller.

Sergey Kislyak: Former Russian ambassador to the United States.

Yevgeny Prigozhin: Head of Russian companies Concord Catering and Concord Management and Consulting. Indicted for financing the Internet Research Agency.

Felix Sater: Russian-born real estate developer who worked with Michael Cohen to pursue Moscow Trump Tower project. Longtime FBI informant.

Igor Sechin: Chief executive officer of Rosneft, a Russian state-controlled oil company. According to Christopher Steele, offered Carter Page a 19 percent stake in the company in exchange for the lifting of US sanctions.

Natalia Veselnitskaya: Russian lawyer who met with Donald Trump, Jr., at Trump Tower in New York City. Also worked with Fusion GPS.

TIMELINE

2016

April 12   Fusion GPS hired by lawyers for Clinton campaign and DNC to do opposition research on Donald Trump.

April 26   Joseph Mifsud meets George Papadopoulos in London hotel and allegedly says that Russians have dirt on Hillary Clinton; "they have thousands of her emails."

May 2   James Comey drafts statement that Hillary Clinton was "grossly negligent" under the Espionage Act but later tells staff he wants to absolve her anyway.

May 10   George Papadopoulos meets with Alexander Downer and allegedly repeats rumor that the Russians have damaging material on Clinton.

June 6   Peter Strzok changes James Comey's statement about Hillary Clinton from "grossly negligent" to "extremely careless."

June (date is unclear)   Christopher Steele is hired by Hillary Clinton's campaign and the DNC through Fusion GPS.

June 20 — Steele files his first "dossier" memo with Fusion GPS claiming Trump-Russia "collusion."

June 27 — Attorney General Loretta Lynch meets with Bill Clinton on an airport tarmac, five days before Hillary Clinton's FBI interview.

July 1 — Victoria Nuland at the State Department grants permission for FBI agent Mike Gaeta to meet with Christopher Steele in London.

July 2 — Peter Strzok and others interview Hillary Clinton for 3.5 hours. Not under oath, she answers "I do not recall" thirty-nine times.

July 5 — James Comey exonerates Hillary Clinton. FBI agent Mike Gaeta meets Christopher Steele and is shown the "dossier."

July 7 — Carter Page visits Moscow to give a speech.

July 11 — Page attends a Cambridge symposium, meets undercover informant Stefan Halper.

July 22 — WikiLeaks releases stolen Democratic National Committee emails.

July 27 — Donald Trump says, "Russia, if you're listening, I hope you're able to find the 30,000 emails that are missing."

July 30 — Christopher Steele meets with Bruce and Nellie Ohr and gives them "dossier" information; Bruce Ohr immediately passes the information to top FBI official and DOJ prosecutors, including Andrew Weissmann.

July 31 — FBI opens "Crossfire Hurricane."

Aug. 6 — Clinton campaign suggests Trump-Russia "collusion."

Aug. 8 — Peter Strzok texts about Trump presidency, "We'll stop it."

Aug. 15 — Infamous Peter Strzok–Lisa Page text on "insurance policy."

Sept. 2 — Stefan Halper invites George Papadopoulos to London.

Sept. 15 — Papadopoulos meets FBI undercover informant "Azra Turk."

Sept. 23 — Michael Isikoff publishes story about Carter Page in Moscow.

Oct. 3   Christopher Steele meets FBI in Rome.

Oct. 7   WikiLeaks begins releasing John Podesta emails.

Oct. 11   Christopher Steele meets with Kathleen Kavalec at State Department; she notes errors in his "dossier."

Oct. 21   FBI obtains approval for FISA warrant against Carter Page.

Oct. 26   Michael Rogers gives FISC an audit showing abuse of NSA database.

Oct. 31   *Mother Jones* publishes story referencing Christopher Steele.

Nov. 1   Steele terminated as an FBI source.

Nov. 9   Donald Trump wins the election.

Nov. 17   Michael Rogers meets Donald Trump and transition team. Trump moves all transition activity to New Jersey.

Nov. 18   Trump names Mike Flynn national security advisor.

Nov. 19   Sir Andrew Wood and David Kramer tell Senator John McCain about the Steele "dossier."

Nov. 22   FBI agent Joe Pientka begins interviewing Bruce Ohr as back channel to Christopher Steele.

Nov. 28   David Kramer flies to London to meet with Steele.

Nov. 29   Kramer gets dossier from Glenn Simpson, gives it to John McCain.

Dec. 9   McCain gives James Comey the "dossier."

Dec. 22   Michael Flynn speaks to Russian ambassador Sergey Kislyak; their conversation is recorded.

Dec. 29   Obama administration expels thirty-five Russian diplomats and issues sanctions against Russia.

Dec. 30   Vladimir Putin announces that he will not retaliate.

2017

Jan. 5   Barack Obama holds meeting with Joe Biden, James Comey, Michael Rogers, James Clapper, John Brennan, Sally Yates, and Susan Rice.

Jan. 6   Comey briefs Donald Trump on the "salacious and un-
          verified" portion of the Steele "dossier." Clapper leaks
          to CNN.

Jan. 10   CNN reports on Comey briefing. BuzzFeed releases the
          Steele "dossier."

Jan. 12   Michael Flynn's call to Sergey Kislyak on December 29 is
          leaked.

Jan. 20   Donald Trump's inauguration.

Jan. 22   Michael Flynn sworn in as national security advisor.

Jan. 24   James Comey directs Andrew McCabe to send Peter
          Strzok and Joe Pientka to interview Flynn.

Jan. 26   Acting Attorney General Sally Yates discusses Flynn with
          White House counsel Don McGahn.

Jan. 27   FBI agents interview George Papadopoulos.

Jan. 30   Donald Trump fires Yates for refusing to enforce travel
          ban.

Feb. 8   Senator Jeff Sessions confirmed as attorney general.

Feb. 13   Michael Flynn resigns.

Feb. 14   James Comey claims Trump asked if he could see fit to
          "letting Flynn go."

Feb. 16   George Papadopoulos interviewed again by FBI agents.

March 2   Jeff Sessions recuses himself from Russia inquiry.

March 4   Trump tweets that Obama had him "wiretapped."

April 19 DOJ/FBI obtain second renewal on Carter Page FISA
          warrant.

April 25 Christopher Steele admits that the "dossier" is "unverified."

April 26 Rod Rosenstein is sworn in as deputy attorney general.

May 9   Trump fires James Comey.

May 15-16 Comey gives the memos to a friend, who leaks to force
          appointment of a special counsel.

May 16   Trump meets with Robert Mueller III. Andrew McCabe
          and Rod Rosenstein discuss invoking the Twenty-fifth
          Amendment to remove Trump from office.

May 17   Rosenstein appoints Mueller as special counsel.

July 18   Carter Page FISA warrant is renewed for the final time.

July 26   FBI raids home of Paul Manafort, Jr.

July 27   George Papadopoulos is arrested. Peter Strzok leaves Robert Mueller team.

Aug. 2   Rod Rosenstein issues a revised scope memo to Mueller.

Oct. 5   George Papadopoulos pleads guilty to lying to FBI.

Oct. 20   Rod Rosenstein issues a second scope memo.

Oct. 24   The *Washington Post* reveals that the Clinton campaign and DNC funded Fusion GPS and the Steele "dossier."

Oct. 30   Robert Mueller indicts Paul Manafort and Richard Gates III. Papadopoulos's guilty plea is made public.

Dec. 1   Michael Flynn pleads guilty. Judge Rudolph "Rudy" Contreras is recused from his case.

Dec. 2   The *Washington Post* reveals anti-Trump texts between Peter Strzok and Lisa Page.

Dec. 4   Robert Mueller recommends little or no prison time for Michael Flynn.

Dec. 12   Judge Emmet G. Sullivan, now on the Flynn case, orders Mueller to produce exculpatory evidence.

Dec. 30   The *New York Times* publishes the narrative that the FBI investigation was due to George Papadopoulos's comments to Alexander Downer.

2018

Feb. 2   Devin Nunes memo is declassified. Andrew McCabe testifies that "no surveillance warrant would have been sought from the FISC without the Steele dossier information."

March 12   House of Representatives releases report finding "no evidence of collusion, coordination or conspiracy."

March 16   Andrew McCabe is fired from the FBI.

April 9   Federal agents raid Michael Cohen's home and office.

May 19   Stefan Halper is revealed to be the FBI's "inside source" within the Trump campaign.

June 14   Inspector General Michael Horowitz releases report on Hillary Clinton email investigation, showing FBI bias, media leaks, and insubordination by James Comey.

July 13   Robert Mueller indicts Russian intelligence agents on charges of hacking DNC.

Aug. 10   Peter Strzok is fired.

Aug. 22   Michael Cohen pleads guilty to eight counts.

Aug. 28   Bruce Ohr testifies he warned the FBI that Christopher Steele lacked credibility.

Nov. 7   Attorney General Jeff Sessions resigns.

Dec. 8   Trump nominates William Barr as attorney general.

2019

Jan. 25   FBI agents stage dawn raid on Roger Stone's home.

Feb. 14   William Barr is confirmed as attorney general.

Feb. 17   Andrew McCabe admits that he and Rod Rosenstein discussed removing Trump by invoking the Twenty-fifth Amendment.

March 12   Sentenced to seven years in prison, Paul Manafort is indicted by the Manhattan DA on similar charges.

March 22   Robert Mueller delivers report on Russian interference: no indictments, arrests, or convictions for "collusion."

March 24   William Barr states that there was no obstruction of justice.

April 18   Barr releases the full Mueller Report with minimal redactions.

May 6   Michael Cohen reports to federal prison.

May 29   Robert Mueller says he "could not exonerate" the president of obstruction of justice.

ACKNOWLEDGMENTS

———

Writing a book is a horrible, exhausting struggle, like a long bout of some painful illness. One would never undertake such a thing if one were not driven on by some demon whom one can neither resist nor understand.

—GEORGE ORWELL, "WHY I WRITE," 1946

The late George Orwell was a remarkably eclectic author who penned many incisive books, both fiction and nonfiction. In his seminal novel *Nineteen Eighty-Four* he warned that "Big Brother is watching you." Published in 1949, his prophetic words became synonymous with abuse of power and the ubiquity of government surveillance. The Orwellian vision was a cautionary nightmare that came to life. It still haunts us today. More than we know. Evidence of this can be found in the pages of *Witch Hunt*.

Orwell was controversial, to say the least. But, for me, his assessment of the writing process is well stated. It is not a satisfying or even rapturous endeavor, as some might imagine. Like Orwell, I find it to be grueling, wearying, and debilitating, without a speck of pleasure in it. A daily root canal would be preferable. Skip the Novocain.

Orwell confided that he often wrote because "there is some lie that I want to expose." That, too, is the reason I wrote this book and the one that preceded it. The compulsion to expose a lie and unravel the truth is what drove me. I felt an urge to right a wrong. In a way, it became my demon that needed exorcising.

I am certain that my literary agent, David Vigliano, grew tired of the missives I sent his way blaming him for my misery. He had convinced me to write this second book when I was dead set against it. In retrospect, I am most grateful that he did. He was tolerant and a good sport about it. David was also a constant source of sage advice. With his unflagging humor, he lifted my spirits when they sagged. This occurred rather frequently, which is a poor reflection on me. To David, I wish to express my utmost gratitude.

Rare is the editor who can satisfy a prickly author during the seemingly endless editing process. Eric Nelson, executive editor at Harper-Collins, managed to achieve this. I submitted a manuscript that was twice the length he wanted, yet he managed to apply judicious cuts with a scalpel when a chainsaw was probably merited. As before, Eric was instrumental in shaping the content of the book. He contributed meaningful insights that helped frame the narrative. Many thanks to Eric and his associate, Hannah Long. Theresa Dooley at Harper has been a wonderful publicist.

Unlike with *The Russia Hoax*, I sought assistance for *Witch Hunt*. I enlisted the help of Glenna Whitley, who is an extraordinary writer and investigative journalist in Dallas, Texas. She supplemented my notes research and columns with her own considerable research and crafted the last three chapters. She also updated and revised what I had originally written in my critical analysis of the media. She assembled the Timeline and List of Major Characters that can be found in the Appendix. Her work was a valuable piece of the puzzle. This book could not have been completed on time without Glenna's excellent contributions. I am indebted to her.

Here at Fox News, I thank especially my friends and colleagues at the network's website. Lynne Jordal Martin, Greg Wilson, and Morgan Debelle Duplan gave prominence to my opinion columns. Gregg Re, an editor and lawyer, lent an extra set of eyes to the original manuscript and recommended clarifications that proved helpful.

Kimberley Sialiano solved my computer woes whenever they arose. Outside of Fox, I am most appreciative to several former FBI officials and federal prosecutors who gave freely of their time to answer my many questions. Victoria Toensing, former deputy assistant attorney general in the Criminal Division at the Justice Department, was notably helpful.

My sincere thanks to Sean Hannity for his continuous support and friendship, along with Porter Berry, Tiffany Fazio, Robert Samuel, Alyssa Carey, Christen Bloom, Andrew Luton, Stephanie Woloshin, Haley Caronia, and Lynda McLaughlin. Grateful am I, too, for the kind words and encouragement from Rush Limbaugh and David Limbaugh. Thanks to Jon Sale, former assistant Watergate prosecutor, as well as Joe diGenova, former independent counsel, for reviewing the book.

Finally, but most importantly, I wish to thank my wife, Cate, and our two daughters, Grace and Liv. They are an indispensable part of my life and career. Their love and devotion are an inspiration to me every single day.

NOTES

PREFACE: A MALIGNANT FORCE

1. Adam Entous, Devlin Barrett, and Rosalind S. Helderman, "Clinton's Campaign, DNC Paid for Research That Led to Russia Dossier," *Washington Post*, October 24, 2017.

2. Kimberley A. Strassel, "Brennan and the 2016 Spy Scandal," *Wall Street Journal*, July 19, 2018.

3. Ken Bensinger, Miriam Elder, and Mark Schoofs, "These Reports Allege Trump Has Deep Ties to Russia," BuzzFeed, January 10, 2017; "Dossier" document published by BuzzFeed, https://www.documentcloud.org/documents /3259984-Trump-Intelligence-Allegations.html.

4. Mike Memoli and Alex Moe, "Republican Congressman Releases Full Transcript of Bruce Ohr Hearing," NBC News, March 8, 2019; "Interview of: Bruce Ohr," Executive Session, Committee on the Judiciary, Joint with the Committee on Government Reform and Oversight, August 28, 2018, https://doug collins.house.gov/sites/dougcollins.house.gov/files/Ohr%20Interview%20 Transcript%208.28.18.pdf.

5. John Solomon, "State Department's Red Flag on Steele Went to a Senior FBI Man Well Before FISA Warrant," The Hill, May 14, 2019; Gregg Re and Catherine Herridge, "State Department Official Cited Steele in Emails with Ohr After Flagging Credibility Issues to FBI, Docs Reveal," Fox News, May 15, 2019.

6. Federal Bureau of Investigation, "FBI Domestic Investigations and Operations Guide (DIOG)," https://vault.fbi.gov/FBI%20Domestic%20Investiga tions%20and%20Operations%20Guide%20%28DIOG%29; Department of Justice, "the Attorney General's Guidelines for Domestic FBI Operations"; also found at United States Code, 28 U.S.C. §§ 509, 510, 533, 534; Executive Order 12333, https://www.justice.gov/archive/opa/docs/guidelines.pdf; Andrew McCarthy, "FBI Russia Investigation Was Always About Trump," Fox News, January 13, 2019.

7. Michael Isikoff and David Corn, *Russian Roulette: The Inside Story of Putin's War on America and the Election of Donald Trump* (New York: Twelve, 2018), 153.

8. Brooke Singman, "IG Confirms He Is Reviewing Whether Strzok's Anti-Trump Bias Impacted Launch of Russia Probe," Fox News, June 19, 2018.

9. Author's interview with President Donald J. Trump, June 25, 2019, Oval Office.

10. Virginia Heffernan, "A Close Reading of Glenn Simpson's Trump-Russia Testimony," *Los Angeles Times*, January 14, 2018.

11. Chuck Ross, "Here's How the Steele Dossier Spread Through the Media and Government," Daily Caller, March 18, 2019.

12. John Solomon, "Comey's Confession: Dossier Not Verified Before, or After, FISA Warrant," The Hill, December 8, 2018.

13. Abigail Tracy, "Is Donald Trump a Manchurian Candidate?," *Vanity Fair*, November 1, 2016; Aiko Stevenson, "President Trump: The Manchurian Candidate?," Huffington Post, January 18, 2017; Ross Douthat, "The 'Manchurian' President?," *New York Times*, May 31, 2017; Franklin Foer, "Putin's Puppet: If the Russian President Could Design a Candidate to Undermine American Interests—and Advance His Own—He'd Look a Lot like Donald Trump," Slate, July 4, 2016; Jeffrey Goldberg, "It's Official: Hillary Clinton Is Running Against Vladimir Putin," *The Atlantic*, July 21, 2016; David Remnick, "Trump and Putin: A Love Story," *The New Yorker*, August 3, 2016.

14. Bret Baier and Catherine Herridge, "Samantha Power Sought to Unmask Americans on Almost Daily Basis, Sources Say," Fox News, September 21, 2017; John Solomon, "'Unmasker in Chief' Samantha Power Spewed Anti-Trump Bias in Government Emails," The Hill, June 26, 2019.

15. Tim Hains, "Sessions: Illegal Leak Against General Mike Flynn Is Being Investigated 'Aggressively,'" RealClearPolitics, February 18, 2018.

16. Josh Gerstein and Kyle Cheney, "FBI Releases Part of Russia Dossier Summary Used to Brief Trump, Obama," Politico, December 14, 2018; Alex Swoyer, *Washington Times*, "James Comey: Trump Briefed on Dossier Salacious Parts Only," *Washington Times*, December 18, 2018.

17. Scott Pelley, "Andrew McCabe: The Full 60 Minutes Interview," CBS News, February 17, 2019, https://www.cbsnews.com/news/andrew-mccabe-interview-former-acting-fbi-director-president-trump-investigation-james-comey-during-russia-investigation-60-minutes/.

18. Devlin Barrett, Sari Horwitz, and Matt Zapotosky, "Deputy Attorney General Appoints Special Counsel to Oversee Probe of Russian Interference in Election," *Washington Post*, May 18, 2017.

19. Jonathan Turley, "The Special Counsel Investigation Needs Attorneys Without Conflicts," The Hill, December 8, 2017; Eric Felten, "Does Robert Mueller Have a Conflict of Interest?," *Weekly Standard*, July 5, 2018; Carrie Johnson, "Special Counsel Robert Mueller Had Been on White House Shortlist to Run FBI," NPR, June 9, 2017; Dan Merica, "Trump Interviewed Mueller for FBI Job Day Before He Was Tapped for Special Counsel," CNN, June 13, 2017.

20. Author's interview with John Dowd, lawyer for President Donald Trump, June 13, 2019.

21. Gregg Re, "Comey Reveals He Concealed Trump Meeting Memo from DOJ Leaders," Fox News, December 9, 2018.

22. Ibid.

23. Michael B. Mukasey, "The Memo and the Mueller Probe," *Wall Street Journal*, February 4, 2018; Andrew C. McCarthy, "Rosenstein Fails to Defend His Failure to Limit Mueller's Investigation," *National Review*, August 7, 2017.

24. Adam Goldman and Michael S. Schmidt, "Rod Rosenstein Suggested Secretly Recording Trump and Discussed 25th Amendment," *New York Times*, September 21, 2018.

25. Author's interview with President Donald J. Trump, Oval Office, White House, June 25, 2019.

26. Samuel Chamberlain, "McCabe Says Rosenstein Was 'Absolutely Serious' About Secretly Recording Trump," Fox News, February 17, 2019.

27. Author's interview with John Dowd, June 13, 2019; Byron York, "When Did Mueller Know There Was No Collusion?," *Washington Examiner*, April 26, 2019; Andrew McCarthy, "How Long Has Mueller Known There Was No Trump-Russia Collusion?," Fox News, March 26, 2019.

28. Robert S. Mueller, *The Mueller Report: The Final Report of the Special Counsel into Donald Trump, Russia, and Collusion as Issued by the Department of Justice* (New York: Skyhorse Publishing, 2019), 39.

29. Ibid., 194.

30. Ibid., 195.

31. Tim Hains, "Full Replay: AG William Barr Senate Judiciary Committee Testimony on Mueller Report," RealClearPolitics, May 1, 2019.

32. William P. Barr, Attorney General, letter to the Chairmen and Ranking Members of the House and Senate Judiciary Committees, March 24, 2019, https://assets.documentcloud.org/documents/5779688/AG-March-24-2019-Letter-to-House-and-Senate.pdf.

33. Jan Crawford, "William Barr Interview: Read the Full Transcript," CBS News, May 31, 2019.

34. Ibid., 2.

35. Tim Hains, "Full Replay."

36. Jan Crawford, "William Barr Interview," 9.

CHAPTER 1: A TALE OF TWO CASES

1. The alliance of the mainstream media and corrupt law enforcement officials was a perfect marriage of liberal ideology. Both abhorred the conservative orthodoxy espoused by the Republican nominee. Both ridiculed his foreign policy pronouncements as reckless and demeaned his proposed outreach to Putin as dangerous. This despite the fact that Trump's advocacy on the campaign trail of rapprochement with Russia was not materially different from the promises and practices of his immediate predecessor, President Barack Obama. In fact, they were, to all intents and purposes, identical.

It was Obama who acceded to Russian demands that a missile defense shield in Europe be abandoned amid intense domestic opposition. Then, with

Obama's critical endorsement, Russia became a member of the World Trade Organization. The Trump administration would later call it a mistake. But at the time, Clinton, who was secretary of state, praised the controversial decision. Of course, who can forget the red "reset button" presented by Clinton to Russia's foreign minister, Sergey Lavrov, which was intended to usher in a new era of warmer US-Russian relations? The media cheered. There was little public criticism when Clinton proposed a plan to "collaborate" with Russia on the construction of a new Silicon Valley–like complex of advanced technology in Moscow, despite dire warnings that the project would only serve to strengthen our adversary's cyberabilities and military proficiency. Collusion? Not if you are Obama or Clinton.

The FBI did not launch an investigation of Clinton or accuse her of being a Russian asset, even as they monitored a Russian spy's efforts to gain access to the secretary of state or when a Kremlin-linked bank cut a check to Bill Clinton in the amount of $500,000. The mainstream media and federal prosecutors all but ignored Clinton's highly suspicious role in the sale of the United States' uranium deposits to Russia while the coffers of the Clinton Foundation were enriched by some $145 million from Russian sources connected to that deal. Corruption? Collusion? Not if you're a Democrat or your last name is Clinton.

No one accused Obama of being in league with Vladimir Putin when the president whispered on a hot microphone to his counterpart, Dmitry Medvedev, "After my election I have more flexibility" to negotiate US plans for a missile defense system in Europe. Medvedev faithfully replied, "I will transmit this information to Vladimir." Imagine if Trump had held such a conversation? He would instantly have been branded as a de facto agent of the Kremlin, followed by immediate and sustained calls for his impeachment on the grounds of treason. The double standard is conspicuous. When Obama engaged with the Russians, it was extolled as sage diplomacy. When Trump attempted to do the same, it was roundly condemned as treachery and sedition.

2. David Drucker, "Romney Was Right About Russia," CNN, July 31, 2017.

3. Anne Gearan, Paul Sonne, and Carol Morello, "U.S. to Withdraw from Nuclear Arms Control Treaty with Russia, Raising Fears of a New Arms Race," *Washington Post*, February 1, 2019.

4. Eli Lake, "Reckoning Time: Security Officials Played Politics on Russiagate," *New York Post*, March 27, 2019.

5. United States Code, 18 U.S.C. § 793, "Gathering, Transmitting or Losing Defense Information."

6. Meghan Keneally, Liz Kreutz, and Shushannah Walshe, "Hillary Clinton Email Mystery Man: What We Know About Eric Hoteham," ABC News, March 5, 2015; Josh Gerstein, "The Mystery Man Behind Hillary's Email Controversy," Politico, March 4, 2015.

7. Per FBI interviews with Clinton IT support staff, Clinton had a Sensitive Compartmented Information Facility (SCIF) in her Washington, DC, residence and her Chappaqua, New York, residence. However, the FBI was informed that

both of those SCIFs were unsecured. See Gregg Re, "Attempt to Hack Email Server Stunned Clinton Aide, FBI Files Show," Fox News, May 7, 2019.

8. Clinton received a briefing on how to handle classified information on or about January 22, 2009, upon being confirmed as secretary of state. However, she reportedly skipped a "refresher" briefing two years later, in 2011. Separately, she told the FBI she had received no training at all on the topic. See Richard Pollock, "Exclusive: Clinton Received Training on Classified Docs Just Once in Three Years at State," Daily Caller, March 24, 2016; Catherine Herridge and Pamela Browne, "Clinton Skipped Special Cyber Briefing in 2011, Documents Show," Fox News, March 25, 2016; Kevin Johnson and David Jackson, "Clinton Told FBI She Had No Training on How to Handle Classified Documents," *USA Today*, September 2, 2016.

9. United States Code, 18 U.S.C. § 793 (d) and (e), "Gathering, Transmitting or Losing Defense Information":

> Whoever, lawfully [or unlawfully] having possession of, access to, control over, or being entrusted with any document . . . relating to the national defense . . . has *reason to believe* could be used to the injury of the United States or to the advantage of any foreign nation, *willfully* communicates, delivers, transmits or causes to be communicated . . . to any person not entitled to receive it . . . or willfully retains the same and fails to deliver it on demand to the officer or employee of the United States entitled to receive it . . . shall be fined under this title or imprisoned not more than ten years, or both. [Author's italics.]

10. United States Code, 18 U.S.C. § 1924(a), "Unauthorized Retention of Removal of Classified Documents or Material."

11. "Statement by FBI Director James B. Comey on the Investigation of Secretary Hillary Clinton's Use of a Personal E-Mail System," FBI National Press Office, July 5, 2016, 3, available at https://www.fbi.gov/news/pressrel/press-releases /statement-by-fbi-director-james-b-comey-on-the-investigation-of-secretary -hillary-clinton2019s-use-of-a-personal-e-mail-system.

12. Ibid.

13. United States Code, 18 U.S.C. § 371, "Conspiracy to Commit Offense or to Defraud United States"; United States Code, 18 U.S.C. § 286, "Conspiracy to Defraud the Government with Respect to Claims."

14. Andrew McCarthy, "Restoring the Rule of Law to the Protection of Classified Information," *National Review*, January 6, 2018.

15. United States Code, 18 U.S.C. § 793(f), "Gathering, Transmitting or Losing Defense Information."

16. "Statement by FBI Director James B. Comey on the Investigation of Secretary Hillary Clinton's Use of a Personal E-Mail System."

17. Office of the Inspector General, U.S. Department of Justice, "A Review of Various Actions by the Federal Bureau of Investigation and Department of Justice in Advance of the 2016 Election," https://www.justice.gov/file/1071991/down load, iv.

18. "FBI Records: The Vault," Federal Bureau of Investigation, https://vault.fbi

.gov/hillary-r.-clinton/hillary-r.-clinton-part-33-of-33/view. See also Daniel Chaitin and Jerry Dunleavy, "FBI Docs: Study Found Clinton Email Server Hacked, Info Found on Dark Web," *Washington Examiner*, June 7, 2019; Chris Enloe, "FBI Releases Damning New Hillary Clinton Email Docs That Discuss 'Smoking Gun Document,'" The Blaze, June 7, 2019.

19. Ibid.
20. Ibid.
21. "Statement by FBI Director James B. Comey on the Investigation of Secretary Hillary Clinton's Use of a Personal E-Mail System."
22. Gregg Jarrett, *The Russia Hoax: The Illicit Scheme to Clear Hillary Clinton and Frame Donald Trump* (New York: Broadside Books, 2018), 24–27.
23. "Statement by FBI Director James B. Comey on the Investigation of Secretary Hillary Clinton's Use of a Personal E-Mail System."
24. Office of the Inspector General, U.S. Department of Justice, "A Review of Various Actions by the Federal Bureau of Investigation and Department of Justice in Advance of the 2016 Election," v, vi.
25. Memorandum for the Attorney General, "Restoring Public Confidence in the FBI," Rod Rosenstein, deputy attorney general, May 9, 2017, available at https://www.documentcloud.org/documents/3711188-Rosenstein-letter-on-Comey-firing.html.
26. Ibid., 497.
27. Chris Wallace, interview with President Barack Obama, *Fox News Sunday*, Fox News, April 10, 2016.
28. Office of the Inspector General, U.S. Department of Justice, "A Review of Various Actions by the Federal Bureau of Investigation and Department of Justice in Advance of the 2016 Election," 186.
29. Politico Staff, "Full Text: James Comey Testimony Transcript on Trump and Russia," Politico, June 8, 2017; Michael S. Schmidt, "Comey Memo Says Trump Asked Him to End Flynn Investigation," *New York Times*, May 16, 2017.
30. Author's interview with President Donald J. Trump, Oval Office, White House, June 25, 2019.
31. Office of the Inspector General, U.S. Department of Justice, "A Review of Various Actions by the Federal Bureau of Investigation and Department of Justice in Advance of the 2016 Election," Exhibit C.
32. Ibid., 188, 193.
33. Ibid., 189.
34. Ibid., 191, 192.
35. Ibid., Exhibit C.
36. Pamela K. Browne and Catherine Herridge, "Hillary Clinton Signed Non-Disclosure Agreement to Protect Classified Information While Secretary of State," Fox News, November 7, 2015; Brendon Bordelon, "Clinton Acknowledged Penalties for 'Negligent Handling' of Classified Information in State Department Contract," *National Review*, November 6, 2016; Chuck Ross, "Document Completely Undermines Hillary's Classified Email Defense," Daily Caller, November 6, 2016; Jeryl Bier, "Hillary Signed She Received Briefing on

Classified Information, but Told FBI She Hadn't," *Weekly Standard*, September 2, 2016.

37. Office of the Inspector General, U.S. Department of Justice, "A Review of Various Actions by the Federal Bureau of Investigation and Department of Justice in Advance of the 2016 Election," Exhibit C.

38. Ibid., 190.

39. Ibid., 193.

40. Some lawyers and legal commentators have argued that prosecutors must still prove "intent" under the "gross negligence" provision of the amended Espionage Act, Section (f). Think about that for just a moment. How exactly does a person *intend* to do something in a *grossly negligent* manner? Either people intend to commit an act, or they perform an act in a grossly negligent manner. Can both be done simultaneously? They cannot. The former reflects a specific state of mind, while the latter tends to reflect the absence of it.

Those who conflate "intent" with "gross negligence" have relied mistakenly on a 1941 Supreme Court case called *Gorin v. United States*. In *Gorin*, the high court was interpreting the explicit language of the *original* Espionage Act of 1917 when it held that "intent" must be proven. However, seven years after the *Gorin* decision, the act was amended and supplanted by several sections of 18 U.S.C. 793, including the "gross negligence" provision (f). Thus, *Gorin* has no application or relevance to Section (f) because that provision of the law did not exist at the time the court rendered its ruling.

In the decades since the *Gorin* case, courts have explained that "intent" is required for other sections of the statute, but not Section (f), which established a lower threshold of "gross negligence." For example, in *United States v. McGuinness*, the court stated that "it is clear that Congress intended to create a hierarchy of offenses against national security, ranging from 'classic spying' to mere losing classified materials through gross negligence."

In the famous "Pentagon Papers" case, Justice Byron White wrote that the district court erred in relying on *Gorin* when it determined that the government needed to prove "only willful and knowing conduct." White noted that *Gorin* "arose under other parts of the predecessor [law] . . . parts that imposed different intent standards not repeated in" the successor statute.

41. "Committee on the Judiciary and Committee on Oversight and Government Reform Joint Hearing on 'Oversight of FBI and DOJ Actions Surrounding the 2016 Election: Testimony by FBI Deputy Assistant Director Peter Strzok,'" July 12, 2018, https://judiciary.house.gov/legislation/hearings/committee-judiciary-and-committee-oversight-and-government-reform-joint-0.

42. John Solomon, "FBI Gave Clinton Email Investigation 'Special' Status, Deputy Director Email Shows," The Hill, November 15, 2017; "Rep. Gaetz: Time for Mueller to Show Collusion Evidence or End Investigation," Fox News, December 20, 2017; Mary Kay Linge, "Trump Slams FBI's McCabe over Planned Retirement," *New York Post*, December 23, 2017.

43. Testimony of FBI Director James Comey, Senate Judiciary Committee, May 3, 2017, available at https://www.washingtonpost.com/news/post-politics/wp

/2017/05/03/read-the-full-testimony-of-fbi-director-james-comey-in-which -he-discusses-clinton-email-investigation/?utm_term=.313b13f20553; Peter Baker, "Comey Raises Concerns About Loretta Lynch's Independence," *New York Times*, June 8, 2017; Tal Kopan, "Comey: Lynch Asked for Clinton Investigation to Be Called a Matter," CNN, June 8, 2017; Ed O'Keefe, "Comey Repeats That Lynch Asked Him to Describe Clinton Investigations as a 'Matter,'" *Washington Post*, June 8, 2017; Kelly Cohen, "James Comey: Loretta Lynch Told Me Not to Call Clinton Email Probe an 'Investigation,'" *Washington Examiner*, June 8, 2017.

44. "Ex–Mueller Aides' Texts Revealed: Read Them Here," Fox News, December 13, 2017.

45. Brooke Singman, "New Texts Show 'Fix Was In' for Clinton Email Probe, GOP Lawmakers Say," Fox News, January 26, 2018.

46. "Ex–Mueller Aides' Texts Revealed: Read Them Here"; Samuel Chamberlain, "Newly Released Texts Between Ex–Mueller Team Members Suggest They Knew Outcome of Clinton Email Probe in Advance," Fox News, January 21, 2018.

47. Committee on the Judiciary, U.S. House of Representatives, "Interview of: Peter Strzok," June 27, 2018, https://dougcollins.house.gov/sites/dougcol lins.house.gov/files/06.27.18%20Interview%20Of%20Peter%20Strzok.pdf ?utm_source=Collins+Judiciary+Press+List&utm_campaign=96979c2884 -EMAIL_CAMPAIGN_2019_03_14_11_47&utm_medium=email&utm _term=0_ff92df788e-96979c2884-169043429, 122.

48. Office of the Inspector General, U.S. Department of Justice, "A Review of Various Actions by the Federal Bureau of Investigation and Department of Justice in Advance of the 2016 Election," pages xi, 147, 329, and 420.

49. Committee on the Judiciary, U.S. House of Representatives, "Interview of: Peter Strzok," 193.

50. Committee on the Judiciary, U.S. House of Representatives, "Interview of: Lisa Page," July 13, 2018, https://dougcollins.house.gov/sites/dougcollins.house.gov /files/Lisa%20Page%20interview%20Day%201.pdf.

51. Brooke Singman, Alex Pappas, and Jake Gibson, "More than 50,000 Texts Exchanged Between FBI Officials Strzok and Page, Sessions Says," Fox News, January 22, 2018.

52. "Statement by FBI Director James B. Comey on the Investigation of Secretary Hillary Clinton's Use of a Personal Email System."

53. Committee on the Judiciary, U.S. House of Representatives, "Interview of: Lisa Page," 20.

54. Stephen Loiaconi, "'I'm with Her': Timeline of Texts the OIG Said 'Cast a Cloud' over Clinton Case," WJLA-TV, June 15, 2018.

55. Christopher Sign, "U.S. Attorney General Loretta Lynch, Bill Clinton Meeting Privately in Phoenix Before Benghazi Report," KNXV-TV, ABC15, June 29, 2016 (updated July 4, 2016).

56. Matt Zapotosky, "Attorney General Declines to Provide Any Details on Clinton Email Investigation," *Washington Post*, July 12, 2017.

57. Code of Federal Regulations, 28 § C.F.R. 45.2, "Disqualification Arising from Personal or Political Relationship." This regulation states that "no employee shall participate in a criminal investigation or prosecution he [or she] has a personal or political relationship with any person or organization substantially involved in the conduct that is the subject of the investigation or prosecution." Lynch had *both* a personal and a professional relationship with Bill and Hillary Clinton. The private tarmac meeting underscored their close personal connection, and Bill's selection of Lynch for a high-ranking position at the Justice Department is evidence of their professional association. The two individuals likely knew their meeting was improper because the FBI on the scene instructed everyone that "no photos, no pictures, no cell phones" could be used to capture what took place. (See Douglas Ernst, "FBI Ordered 'No Cellphones' to Airport Witnesses of Lynch-Clinton Meeting, Reporter Says," *Washington Times*, July 1, 2016.) Under a Freedom of Information Act request, the FBI replied that "No records responsive to your requests were located." (See Jordan Seculow, "DOJ Documents Dump to ACLJ on Clinton Lynch Meeting: Comey, FBI Lied, Media Collusion, Spin and Illegality," American Center for Law and Justice, August 2017.) A year later, the DOJ produced more than four hundred pages of emails, although the contents of what Lynch and Clinton discussed were heavily redacted.

58. Zeke J. Miller, "Transcript: Everything Hillary Clinton Said on the Email Controversy," *Time*, March 10, 2015.

59. Eugene Scott, "Hillary Clinton on Emails: 'The Facts Are Pretty Clear,'" CNN, July 28, 2015.

60. Josh Feldman, "Hillary: Use of Personal Email 'Clearly Wasn't the Best Choice,'" Mediaite, August 26, 2015; Glenn Kessler, "Clinton's Claims About Receiving or Sending 'Classified Material' on Her Private Email System," *Washington Post*, August 27, 2015.

61. "Statement by James B. Comey, FBI Director, House Committee on Oversight and Government Reform," July 7, 2016.

62. Harper Neidig, "Clinton to FBI: Didn't Know Parenthetical 'C' Stood for Confidential," The Hill, September 2, 2016.

63. "CNN Exclusive: Hillary Clinton's First National Interview of 2016 Race," CNN, July 7, 2015, http://cnnpressroom.blogs.cnn.com/2015/07/07/cnn -exclusive-hillary-clintons-first-national-interview-of-2016-race/.

64. Chuck Grassley, "Grassley on Investigation: Let's Be Transparent," April 8, 2019, https://www.grassley.senate.gov/news/news-releases/grassley -investigation-transparency-lets-be-consistent.

65. Brooke Singman, "IG Confirms He Is Reviewing Whether Strzok's Anti-Trump Bias Impacted Launch of Russia Probe," Fox News, June 19, 2018.

66. Brooke Singman, "Lisa Page Transcripts Reveal Details of Anti-Trump 'Insurance Policy,' Concerns over Full-Blown Probe," Fox News, March 12, 2019.

67. Catherine Herridge and Cyd Upson, "Lisa Page Testimony: Collusion Still Unproven by Time of Mueller's Special Counsel Appointment," Fox News, September 16, 2018; Chuck Ross, "Lisa Page Testified the FBI Hadn't Seen

Evidence of Collusion by the Time Mueller Was Appointed," Daily Caller, September 17, 2018; John Solomon, "Lisa Page Bombshell: FBI Couldn't Prove Trump-Russia Collusion Before Mueller Appointment," The Hill, September 16, 2018.

68. Gregg Re, "Comey Reveals He Concealed Trump Meeting Memo from DOJ Leaders," Fox News, December 9, 2018.

69. Inspector General Report, "A Review of Various Actions By the Federal Bureau of Investigation and Department of Justice in Advance of the 2016 Election."

70. Ibid.; Brooke Singman, "IG Confirms He Is Reviewing Whether Strzok's Anti-Trump Bias Impacted Launch of Russia Probe."

71. Inspector General Report, "A Review of Various Actions By the Federal Bureau of Investigation and Department of Justice in Advance of the 2016 Election."

72. John Bowden, "FBI Agents in Texts: 'We'll Stop' Trump from Becoming President," The Hill, June 14, 2018; Michael S. Schmidt, "Top Agent Said FBI Would Stop Trump from Becoming President," New York Times, June 14, 2018.

73. "Committee on the Judiciary and Committee on Oversight and Government Reform Joint Hearing on 'Oversight of FBI and DOJ Actions Surrounding the 2016 Election: Testimony by FBI Deputy Assistant Director Peter Strzok.'"

74. Ibid.

75. Byron York, "After Mysterious 'Insurance Policy' Test, Will Justice Department Reveal More on FBI Agent Bounced from Mueller Probe?," Washington Examiner, December 13, 2017.

76. Joseph Wulfsohn, "McCabe Says He Doesn't Recall Discussing Infamous 'Insurance Policy' with Strzok, Page in 2016," Fox News, February 19, 2019; Terence P. Jeffrey, "IG: McCabe 'Does Not Recall' Meeting in His Office Where Page Argued There's 'No Way' Trump's Elected," CNS News, June 18, 2018.

77. Committee on the Judiciary, U.S. House of Representatives, "Interview of: Lisa Page," 38. See also Brooke Singman, "Lisa Page Transcripts Reveal Details of Anti-Trump 'Insurance Policy,' Concerns over Full-Blown Probe."

78. Ibid.

79. Federal Bureau of Investigation, "FBI Domestic Investigations and Operations Guide (DIOG)," https://vault.fbi.gov/FBI%20Domestic%20Investigations%20and%20Operations%20Guide%20%28DIOG%29; "The Attorney General's Guidelines for Domestic FBI Operations," Department of Justice, https://www.justice.gov/archive/opa/docs/guidelines.pdf.

80. "Committee on the Judiciary and Committee on Oversight and Government Reform Joint Hearing on 'Oversight of FBI and DOJ Actions Surrounding the 2016 Election: Testimony by FBI Deputy Assistant Director Peter Strzok.'"

81. Philip Bump, "How The Two Rogue FBI Officials Explain Their Text Messages About Trump," The Washington Post, June 14, 2018; John Solomon, "Opinion: One FBI Text Message In Russia Probe That Should Alarm Every American," The Hill, July 19, 2018.

82. Jan Crawford, "William Barr Interview: Read the Full Transcript," CBS News, May 31, 2019.

83. "Ex-Mueller Aides' Texts Revealed: Read Them Here."

84. Ibid.

85. Andrew G. McCabe, *The Threat: How the FBI Protects America in the Age of Terror and Trump* (New York: St. Martin's Press, 2019), 192.

86. Inspector General Report, "A Review of Various Actions By the Federal Bureau of Investigation and Department of Justice in Advance of the 2016 Election."

87. Ibid.

88. Paul Sperry, "Despite Comey Assurances, Vast Bulk of Weiner Laptop Emails Were Never Examined," RealClearInvestigations, August 23, 2018.

89. James B. Comey, "Letter from FBI Related to Clinton Email Case," *New York Times*, November 6, 2016, https://www.nytimes.com/interactive/2016/11/06/us/politics/fbi-letter-emails.html.

90. Sperry, "Despite Comey Assurances, Vast Bulk of Weiner Laptop Emails Were Never Examined."

91. Office of Inspector General, "A Review of Various Actions By the Federal Bureau of Investigation and Department of Justice in Advance of the 2016 Election," 497.

92. Matt Apuzzo, Adam Goldman, and Nicholas Fandos, "Code Name Crossfire Hurricane: The Secret Origins of the Trump Investigation," *New York Times*, May 16, 2018.

93. Malia Zimmerman and Adam Housley, "FBI, DOJ Roiled by Comey, Lynch Decision to Let Clinton Slide by on Emails, Says Insider," Fox News, October 13, 2016.

94. Jarrett, *The Russia Hoax*, 43–45.

95. Tal Kopan and Evan Perez, "FBI Releases Hillary Clinton Email Report," CNN, September 2, 2016; Aaron Blake, "Hillary Clinton Told the FBI She Couldn't Recall Something More than Three Dozen Times," *Washington Post*, September 2, 2016.

96. Chuck Ross, "Top Clinton Aides Face No Charges After Making False Statements to FBI," Daily Caller, December 4, 2017.

97. Ibid.

98. "Judicial Watch Releases Testimony of Clinton Email Administrator—Clinton Lawyer Cheryl Mills Communicated with Him a Week Prior to Testimony," Judicial Watch, June 18, 2019.

99. "FBI Docs Show Notes About Meeting with Intelligence Community Inspector General About Clinton Emails Are 'Missing' and CD Containing Notes Is Likely 'Damaged' Irreparably," Judicial Watch, June 7, 2019.

100. Gregg Re, "Attempt to Hack Email Server Stunning Clinton Aide, FBI Files Show," Fox News, May 7, 2019.

101. Nolan Hicks and Bruce Golding, "Slow Team Hill Response to Server Hack," *New York Post*, May 8, 2019.

102. "Judicial Watch Releases Testimony of Clinton Email Administrator." See also "JW v State Benghazi Talking Points Transcript 01242," October 16, 2018, https://www.judicialwatch.org/document-archive/jw-v-state-benghazi-talking-points-transcript-01242-2/.

103. Mark Tapscott, "Abedin's Key Clinton Email Claim Contradicted by Former Aide," *The Epoch Times*, June 18, 2019.

104. Ross, "Top Clinton Aides Face No Charges After Making False Statements to FBI."

105. Grassley, "Grassley on Investigation: Let's Be Transparent."

106. Federal Bureau of Investigation, "FBI Domestic Investigations and Operations Guide" (DIOG); "The Attorney General's Guidelines for Domestic FBI Operations."

107. John Solomon, "FBI's Top Lawyer Believed Hillary Clinton Should Face Charges, but Was Talked Out of It," The Hill, February 20, 2019.

108. Ibid.

109. Ibid.

110. Ginni Thomas, "The Obama Administration's 'Brazen Plot to Exonerate Hillary Clinton' Starting to Leak Out, According to Former Fed Prosecutor," Daily Caller, January 20, 2018.

111. United States Code, 44 U.S.C. § 3101 et al., "Records Management by Agency Heads, General Duties"; Foreign Affairs Manual, 5 F.A.M. 441(h)(2), et al.

112. Patrick F. Kennedy, memorandum re "Senior Officials' Records Management Responsibilities," August 28, 2014, https://www.archives.gov/files/press/press-releases/2015/pdf/attachment1-memo-to-department-leadership.pdf. (Kennedy: "All records generated by Senior Officials belong to the Department of State.")

113. United States Code, 18 U.S.C. § 641, "Public Money, Property or Records."

114. "Clinton Email Investigation," Federal Bureau of Investigation, September 2, 2016, 19, available at https://vault.fbi.gov/hillary-r.-clinton.

115. Adam Goldman and Michael S. Schmidt, "6 Things We Learned in the FBI Clinton Email Investigation," New York Times, September 2, 2016.

116. United States Code, 18 U.S.C. § 2071(b), "Concealment, Removal or Mutilation Generally."

117. "Clinton Email Investigation," 18; Byron York, "From FBI Fragments, A Question: Did Team Clinton Destroy Evidence Under Subpoena?," Washington Examiner, September 3, 2016; DeRoy Murdock, "Obstruction of Justice Haunts Hillary's Future," National Review, September 8, 2016.

118. David E. Kendall, letter to Trey Gowdy, Chairman of the House Select Committee on Benghazi, March 27, 2015, available at http://wallstreetonparade.com/wp-content/uploads/2016/07/David-Kendall-Letter-to-Trey-Gowdy-Chair-House-Select-Committee-on-Benghazi-March-27-2015.pdf.; Lauren French, "Gowdy: Clinton Wiped Her Server Clean," Politico, March 27, 2015.

119. United States Code, 18 U.S.C. § 1505, "Obstruction of Justice Proceedings Before Departments, Agencies and Committees"; United States Code, 18 U.S.C. § 1515(b), "Definitions for Certain Provisions, General Provision." See also United States Code, 18 U.S.C. §§ 1503, 1512.

120. McCabe, The Threat, 175–76.

121. "Statement by James B. Comey, Director of the FBI, before the House Committee on Oversight and Government Reform," July 8, 2016, available at https://www.politico.com/story/2017/06/08/full-text-james-comey-trump-russia-testimony-239295.

122. Ibid.

123. Gregg Re and Catherine Herridge, "State Department Identifies 23 Violations, 'Multiple Security Incidents' Concerning Clinton Emails," Fox News, June 17, 2019; Chuck Ross, "State Department Identifies 30 Security Incidents Related to Hillary Clinton's Email Server," Daily Caller, June 17, 2019.

124. Mary Elizabeth Taylor, assistant secretary, Bureau of Legislative Affairs, letter to The Honorable Charles Grassley, chairman, Committee on Finance, June 5, 2019, https://www.grassley.senate.gov/sites/default/files/2019-06-05%20State%20to%20CEG%20%28Security%20Investigation%20Follow-Up%29.pdf.

125. Re and Herridge, "State Department Identifies 23 Violations"; Re, "Attempt to Hack Email Server Stunned Clinton Aide, FBI Files Show."

126. Adam Shaw, "Chinese Company Reportedly Hacked Clinton's Server, Got Copy Of Every Email In Real-Time," Fox News, August 29, 2018; Gregg Re, "Attempt To Hack Email Server Stunned Clinton Aide, FBI Files Show," Fox News, May 7, 2019; Richard Pollock, "Sources: China Hacked Hillary Clinton's Private Email Server," Daily Caller, August 27, 2018.

127. Adam Shaw, "Chinese Company Reportedly Hacked Clinton's Server, Got Copy of Every Email in Real-Time," Fox News, August 29, 2018.

128. Maria Bartiromo, interview with Representative Doug Collins, *Sunday Morning Futures*, Fox News, March 17, 2019. See also transcript posted by Tim Hains, "Rep. Doug Collins on Page, McCabe and Strzok Testimony: 'Loretta Lynch Has Some Explaining to Do,'" RealClearPolitics, March 17, 2019.

129. Hains, "Rep. Doug Collins on Page, McCabe and Strzok Testimony."

130. John Solomon, "Forgetting Hanssen Scandal's Failures: FBI Saw Agent's Affair as Security Risk but Took Little Action," The Hill, March 3, 2019.

131. Jeff Mordock, "Former FBI Official Feared Strzok-Page Affair Could Compromise Them," *Washington Times*, April 2, 2019.

132. Solomon, "Forgetting Hanssen Scandal's Failures."

133. Committee on the Judiciary, U.S. House of Representatives, "Interview of: Peter Strzok." See also Jerry Dunleavy and Daniel Chaitin, "Peter Strzok Said Mueller Never Asked if Anti-Trump Bias Influenced Russia Investigation Decisions," *Washington Examiner*, March 14, 2019.

134. Committee on the Judiciary, U.S. House of Representatives, "Interview of: Peter Strzok."

135. Ibid.

136. Aaron Blake, "Nobody Did More Damage to Robert Mueller Than Peter Strzok," *Washington Post*, August 13, 2018.

137. "Report of Investigation: Recovery of Text Messages From Certain FBI Mobile Devices," Office of the Inspector General, U.S. Department of Justice, December 2018, available at https://oig.justice.gov/reports/2018/i-2018-003523.pdf.

138. Byron York, "New Justice Department Report Asks: In Anti-Trump Text Probe, What Happened To Strzok, Page iPhones?," *Washington Examiner*, December 17, 2018.

139. Crawford, "William Barr Interview."

CHAPTER 2: CLINTON COLLUSION

1. "Hillary Clinton Remarks on Counterterrorism," March 23, 2016, C-SPAN.
2. Katie Reilly, "Read Hillary Clinton's Speech on Donald Trump and National Security," *Time*, June 2, 2016.
3. Ibid.
4. Michael Doran, "The Real Collusion Story," *National Review*, March 13, 2018.
5. Ibid.
6. Gregory Krieg and Joshua Berlinger, "Hillary Clinton: Donald Trump Would Be Putin's 'Puppet,'" CNN, October 20, 2016.
7. Adam Entous, Devlin Barrett, and Rosalind S. Helderman, "Clinton Campaign, DNC Paid for Research That Led to Russia Dossier," *Washington Post*, October 24, 2017; Kenneth P. Vogel, "Clinton Campaign and Democratic Party Helped Pay for Russia Trump Dossier," *New York Times*, October 24, 2017.
8. Kenneth Vogel and Maggie Haberman, "Conservative Website First Funded Anti-Trump Research by Firm That Later Produced Dossier," *New York Times*, October 27, 2017.
9. Senate Judiciary Committee, U.S. Senate, "Interview of: Glenn Simpson," August 22, 2017, https://www.feinstein.senate.gov/public/_cache/files/3/9/3974a291-ddbe-4525-9ed1-22bab43c05ae/934A3562824CACA7BB4D915E97709D2F.simpson-transcript-redacted.pdf, 77.
10. Devlin Barrett, Sari Horowitz, and Adam Entous, "Conservative Website First Paid Fusion GPS for Trump Research That Led to Dossier," *Washington Post*, October 27, 2017.
11. Ibid.
12. Brooke Seipel, "Fusion GPS Paid Ex–British Spy $168,000 for Working on Dossier," The Hill, November 1, 2017; Mark Hosenball, "Ex–British Spy Paid $168,000 for Trump Dossier, U.S. Firm Discloses," Reuters, November 1, 2017.
13. Brooke Singman, "FISA Memo: Steele Fired as an FBI Source for Breaking 'Cardinal Rule'—Leaking to the Media," Fox News, February 2, 2018.
14. Ken Bensinger, Miriam Elder, and Mark Schoofs, "These Reports Allege Trump Has Deep Ties to Russia," BuzzFeed, January 10, 2017; "Company Intelligence Report 2016/080," December 13, 2016, https://www.documentcloud.org/documents/3259984-Trump-Intelligence-Allegations.html.
15. Ibid.
16. Paul Roderick Gregory, "The Trump Dossier Is Fake—and Here are the Reasons Why," *Forbes*, January 13, 2017.
17. Chuck Ross, "Christopher Steele Reportedly Worked for Sanctioned Russian Oligarch," Daily Caller, August 29, 2018; John Solomon, "Russian Oligarch, Justice Department Clear Case of Collusion," The Hill, August 28, 2018; Byron York, "Emails Show 2016 Links Among Steele, Ohr, Simpson—with Russian Oligarch in Background," *Washington Examiner*, August 8, 2018.
18. Kenneth P. Vogel and Matthew Rosenberg, "Agents Tried to Flip Russian Oligarchs. The Fallout Spread to Trump," *New York Times*, September 1, 2018.

19. Virginia Heffernan, "A Close Reading of Glenn Simpson's Trump-Russia Testimony," *Los Angeles Times*, January 14, 2018.

20. Glenn R. Simpson and Mary Jacoby, "How Lobbyists Help Ex-Soviets Woo Washington," *Wall Street Journal*, April 17, 2007.

21. Lee Smith, "Did President Obama Read the 'Steele Dossier' in the White House Last August?," *Tablet Magazine*, December 20, 2017.

22. Senate Judiciary Committee, U.S. Senate, "Interview of: Glenn Simpson."

23. Virginia Heffernan, "A Close Reading of Glenn Simpson's Trump-Russia Testimony," *Los Angeles Times*, January 14, 2018.

24. Chuck Grassley, "Grassley Statement at Hearing on Enforcement of the Foreign Agents Registration Act," July 26, 2017, https://www.grassley.senate.gov/news/news-releases/grassley-statement-hearing-enforcement-foreign-agents-registration-act.

25. Senate Judiciary Committee, U.S. Senate, "Interview of: Glenn Simpson."

26. Ken Dilanian, "Trump Dossier Firm Also Supplied Info Used in Trump Tower Meeting with Russian Lawyer," NBC News, November 10, 2017.

27. Catherine Herridge, Pamela K. Browne, and Cyd Upson, "Russian Lawyer at Center of Trump Tower Meeting Dismisses Dossier Shared with FBI," Fox News, January 19, 2018.

28. "Russian Collusion: It Was Hillary Clinton All Along," *Investor's Business Daily*, August 13, 2018.

29. Michael Doran, "The Real Collusion Story," *National Review*, March 13, 2018.

30. In the High Court of Justice, Queen's Bench Division, Between (1) Aleksej Gubarev, (2) Webzilla B.V., (3) Webzilla Limited, (4) XBT Holding S.A. and (1) Orbis Business Intelligence Limited, (2) Christopher Steele, "Defendants' Response to Claimants' Request for Further Information Pursuant to CPR Part 18," https://assets.documentcloud.org/documents/3892131/Trump-Dossier-Suit.pdf, 7.

31. Christopher Steele, disposition, *Gubarev v. Orbis*, June 18, 2018, https://www.scribd.com/document/401997457/Steele-deposition-Exhibit-66#from_embed?campaign=VigLink&ad_group=xxc1xx&source=hp_affiliate&medium=affiliate.

32. Ashe Schow, "Christopher Steele's Former MI6 Boss Slams Trump Dossier as 'Overrated,'" Information Clearing House, March 19, 2019, http://www.informationclearinghouse.info/51291.htm.

33. Martin Robinson, "Former Spy Chris Steele's Friends Describe a 'Show-Off' 007 Figure but MI6 Bosses Brand Him 'An Idiot' for an 'Appalling Lack of Judgement' over the Trump 'Dirty Dossier,'" *The Daily Mail*, January 13, 2017.

34. Catherine Herridge, Pamela K. Brown, and Cyd Upson, "Clinton Associates Fed Information to Trump Dossier Author Steele, Memo Says," Fox News, February 5, 2018.

35. Michael Isikoff and David Corn, *Russian Roulette: The Inside Story of Putin's War on America and the Election of Donald Trump* (New York: Twelve, 2018).

36. In the High Court of Justice, Queen's Bench Division, "Defendants' Response," 7.

37. Senate Judiciary Committee, U.S. Senate, "Interview of: Glenn Simpson."

38. Executive Session, Permanent Select Committee on Intelligence, U.S. House of Representatives, "Interview of: Glenn Simpson," November 14, 2017, https://docs.house.gov/meetings/IG/IG00/20180118/106796/HMTG-115-IG00-20180118-SD002.pdf, 70.

39. Kimberley A. Strassel, "Who Paid for the 'Trump Dossier'?," *Wall Street Journal*, July 27, 2017; Kimberley A. Strassel, "The Fusion Collusion," *Wall Street Journal*, October 19, 2017.

40. Senate Judiciary Committee, U.S. Senate, "Interview of: Glenn Simpson."

41. Mike Levine, "Trump 'Dossier' Stuck in New York, Didn't Trigger Russia Investigation, Sources Say," ABC News, September 18, 2018; Luke Harding, "How Trump Walked into Putin's Web," *Guardian*, November 15, 2017.

42. Jeff Carlson, "Exclusive: McCabe's FBI Tried to Re-engage Christopher Steele After Comey Was Fired," *The Epoch Times*, January 14, 2019 (updated March 8, 2019); Jane Mayer, "Christopher Steele, the Man Behind the Trump Dossier," *The New Yorker*, March 12, 2018.

43. Rowan Scarborough, "Obama Aide Started Christopher Steele–FBI Alliance," *Washington Times*, March 13, 2018.

44. Jeff Carlson, "Clinton Campaign Relied on Former Spy's Web of Connections to Frame Trump," *The Epoch Times*, January 23, 2019 (updated March 8, 2019).

45. Ken Bensinger, Miriam Elder, and Mark Schoofs, "These Reports Allege Trump Has Deep Ties to Russia," BuzzFeed, January 10, 2017; "Company Intelligence Report 2016/080."

46. Scott Shane, Mark Mazzetti, and Adam Goldman, "Trump Adviser's Visit to Moscow Got the FBI's Attention," *New York Times*, April 19, 2017.

47. Philip Bump, "A Timeline of the Roger Stone–WikiLeaks Question," *Washington Post*, November 27, 2018; Ellen Nakashima, "Russian Government Hackers Penetrated DNC, Stole Opposition Research on Trump," *Washington Post*, June 14, 2016; Mark Tran, "WikiLeaks to Publish More Hillary Clinton Emails—Julian Assange," *The Guardian*, June 12, 2016.

48. Gregg Jarrett, *The Russia Hoax: The Illicit Scheme to Clear Hillary Clinton and Frame Donald Trump* (New York: Broadside Books, 2018), 135–38.

49. Anna Giaritelli, "Carter Page Says He's Never Spoken with Trump in His Life," *Washington Examiner*, February 6, 2018; U.S. House of Representatives, Permanent Select Committee on Intelligence, "Testimony of Carter Page," November 2, 2017, http://www.documentcloud.org/documents/4366245-Carter-Page-Transcript-of-Interview-With-House.html, 157.

50. Ibid., 36.

51. Rosie Gray, "Michael Cohen: 'It Is Fake News Meant to Malign Mr. Trump,'" *The Atlantic*, January 10, 2017.

52. "Full Text of Michael Cohen's Testimony to Congress," *The Guardian*, February 27, 2019; also https://www.scribd.com/document/400649065/Testimony-of-Michael-D-Cohen.

53. Andrew C. McCarthy, "The Strzok-Page Texts and the Origins of the Trump-Russia Investigation," *National Review*, May 14, 2018.

54. Katie Leach, "'White House Is Running This' Mystery Has Top Republican

Squeezing DOJ to Unredact Strzok-Page Texts," *Washington Examiner*, May 23, 2018.

55. Executive Session, Committee on the Judiciary, Joint with the Committee on Government Reform and Oversight, U.S. House of Representatives, "Interview of: Bruce Ohr," August 28, 2018, https://www.scribd.com/document /401389538/Ohr-Interview-Transcript-8-28-18#from_embed.

56. Ibid.

57. House Permanent Select Committee on Intelligence, Minority, memorandum re "Correcting the Record—The Russia Investigations," Unclassified, January 29, 2018 (released to the public on February 24, 2018), https://fas.org/irp /congress/2018_cr/hpsci-dem-memo.pdf, 3.

58. Brooke Singman, "Nellie Ohr, Wife of DOJ Official, Did Extensive Oppo Research on Trump Family, Aides: Transcript," Fox News, March 28, 2019.

59. Chuck Ross, "Nellie Ohr Researched Trump's Kids For Fusion GPS," *The Daily Caller*, January 30, 2019.

60. Executive Session, Committee on the Judiciary, Joint with the Committee on Government Reform and Oversight, "Interview of: Nellie Ohr," October 19, 2018, https://dougcollins.house.gov/sites/dougcollins.house.gov /files/10.19.18%20Nellie%20Ohr%20Interview.pdf?utm_source=Collins +Judiciary+Press+List&utm_campaign=10bc31267f-EMAIL_CAMPAIGN _2019_03_28_01_14&utm_medium=email&utm_term=0_ff92df788e -10bc31267f-168882913; Chuck Ross, "Nellie Ohr: 'I Favored Hillary Clinton,'" Daily Caller, March 28, 2019.

61. Jeremy Herb, "Fusion GPS Contractor Nellie Ohr Doesn't Say Much at House Interview," CNN, October 19, 2018.

62. "Criminal Referral from Rep. Mark Meadows Asking Justice Department to Investigate Whether Nellie Orr Provided False Testimony," May 1, 2019, https://www.scribd.com/document/408347748/Final-Criminal-Referral -Nellie-Ohr-5-1-19; Daniel Chaitin and Jerry Dunleavy, "Mark Meadows Sends Criminal Referral Targeting Nellie Ohr to DOJ," *Washington Examiner*, May 1, 2019.

63. John Solomon, "Nellie Ohr's 'Hi Honey' Emails to DOJ About Russia Collusion Should Alarm Us All," The Hill, May 1, 2019.

64. Executive Session, Committee on the Judiciary, Joint with the Committee on Government Reform and Oversight, U.S. House of Representatives, "Interview of: Bruce Ohr."

65. "FBI Records Show Dossier Author Deemed 'Not Suitable for Use' as Source, Show Several FBI Payments in 2016," Judicial Watch (documents obtained pursuant to a Freedom of Information Act lawsuit filed by Judicial Watch), August 3, 2018; Tom Winter, "FBI Releases Documents Showing Payments to Trump Dossier Author Steele," NBC News, August 3, 2018.

66. Matt Apuzzo, Adam Goldman, and Nicholas Fandos, "Code Name Crossfire Hurricane: The Secret Origins of the Trump Investigation," *New York Times*, May 16, 2018.

67. Laura Jarrett and Evan Perez, "FBI Agent Dismissed from Mueller Probe

Changed Comey's Description of Clinton to 'Extremely Careless,'" CNN, December 4, 2017.

68. Executive Session, Committee on the Judiciary, Joint with the Committee on Government Reform and Oversight, U.S. House of Representatives, "Interview of: Bruce Ohr," 39, available at https://dougcollins.house.gov/sites/dougcollins .house.gov/files/Ohr%20Interview%20Transcript%208.28.18.pdf.

69. Ibid., 93.

70. Ibid., 30–31.

71. Ibid., 125.

72. House Permanent Select Committee on Intelligence Majority Members, memorandum re "Foreign Intelligence Surveillance Act Abuses at the Department of Justice and the Federal Bureau of Investigation," January 18, 2018 (declassified by order of the president, February 2, 2018), https://perry.house.gov/uploaded files/memo_and_white_house_letter.pdf, 3.

73. Executive Session, Committee on the Judiciary, Joint with the Committee on Government Reform and Oversight, U.S. House of Representatives, "Interview of: Bruce Ohr," p. 22.

74. House Permanent Select Committee on Intelligence, Minority, memorandum re "Correcting the Record—The Russia Investigations," 7.

75. Executive Session, Committee on the Judiciary, Joint with the Committee on Government Reform and Oversight, U.S. House of Representatives, "Interview of: Bruce Ohr," 79.

76. Ibid.

77. Ibid, p. 125. See also Gregg Jarrett, "Mueller's Team Knew 'Dossier' Kicking Off Trump Investigation Was Biased and Defective," Fox News, January 17, 2019.

78. Luke Rosiak, "DOJ Official Bruce Ohr Hid Wife's Fusion GPS Payments from Ethics Officials," Daily Caller, February 14, 2018.

79. United States Code, 18 U.S.C. § 1001, "Statements or Entries Generally."

80. United States Code, 18 U.S.C. § 201(b); 18 U.S.C. § 201(c); 18 U.S.C. § 1346.

81. Alex Pappas, "DOJ Official Bruce Ohr Awarded $28G Bonus Amid Russia Probe, Records Indicate," Fox News, June 7, 2019.

82. Executive Session, Committee on the Judiciary, Joint with the Committee on Government Reform and Oversight, U.S. House of Representatives, "Interview of: Bruce Ohr."

83. "FBI Records Show Dossier Author Deemed 'Not Suitable for Use' as Source, Show Several FBI Payments in 2016."

84. Ibid.; Rowan Scarborough, "Christopher Steele Broke FBI Media Rules After Being 'Admonished,' Documents Show," Washington Times, August 4, 2018.

85. Executive Session, Permanent Select Committee on Intelligence, U.S. House of Representatives, "Interview of: Glenn Simpson," November 14, 2017, https:// docs.house.gov/meetings/IG/IG00/20180118/106796/HMTG-115-IG00 -20180118-SD002.pdf, 78.

86. Kelly Cohen, "GPS Founder Glenn Simpson Pleads the Fifth Before House Committees," Washington Examiner, October 16, 2018.

87. Senator Charles Grassley, letter to Senator Richard Blumenthal, December 3, 2018, https://www.judiciary.senate.gov/imo/media/doc/2018-12-03%20 CEG%20to%20Blumenthal%20-%20Trump%20Jr%20and%20Cohen.pdf; Rowan Scarborough, "Why Key Architect of the Anti-Trump Dossier Is Now Accused of Lying to Congress," *Washington Times*, Mary 31, 2018; Lee Smith, "Did Glenn Simpson Lie to Congress?," *Tablet*, January 12, 2018.

88. Charles E. Grassley and Lindsey O. Graham, memorandum to Rod J. Rosenstein, U.S. Department of Justice, and Christopher A. Wray, Federal Bureau of Investigation, re "Referral of Christopher Steele for Potential Violation of 18 U.S.C. § 1001," January 4, 2018, https://www.judiciary.senate.gov/imo /media/doc/2018-02-02%20CEG%20LG%20to%20DOJ%20FBI%20 (Unclassified%20Steele%20Referral).pdf. See also Chuck Grassley, "Senators Grassley, Graham Refer Christopher Steele for Criminal Investigation," January 5, 2018, https://www.grassley.senate.gov/news/news-releases /senators-grassley-graham-refer-christopher-steele-criminal-investigation.

89. Email exchange between Bruce Ohr and Christopher Steele, September 21, 2016; Executive Session, Committee on the Judiciary, Joint with the Committee on Government Reform and Oversight, U.S. House of Representatives, "Interview of: Bruce Ohr."

90. Matt Apuzzo, Michael S. Schmidt, Adam Goldmand, and Eric Lichtblau, "Comey Tried to Shield the FBI from Politics. Then He Shaped the Election," *New York Times*, April 22, 2017; Nick Giampia, "FBI Offered Christopher Steele $50K to Confirm Trump Dossier: Judge Napolitano," Fox Business, December 18, 2017.

91. Jonathan Winer, "Devin Nunes Is Investigating Me. Here's the Truth," *Washington Post*, February 8, 2018.

92. Caitlin Yilek, "Ex–Obama Official Confirms Trump Dossier Was Given to State Department," *Washington Examiner*, February 9, 2018.

93. Winer, "Devin Nunes Is Investigating Me,"; Eric Felton, "The Weird Tales of Jonathan Winer," *The Weekly Standard*, February 10, 2018.

94. Chuck Ross, "Here's How the Steele Dossier Spread Through the Media and Government," Daily Caller, March 18, 2019.

95. Mark Hemingway, "The Other Secret Dossier," *The Weekly Standard*, February 6, 2018. In his story, Hemingway recounts how Clinton aide Sidney Blumental, together with Shearer, ran a "secret spy network" funneling information to Clinton. At the same time, according to Hemingway, they were "trying to leverage their connections to Clinton . . . to get a lucrative contract in Libya." (See also *Slate* article on Shearer: A.O. Scott, "Cody Shearer: If He Didn't Exist, The Vast Right-Wing Conspiracy Would Have Invented Him," *Slate*, May 22, 1999, available at https://slate.com/news-and-politics /1999/05/cody-shearer.html). According to Hemingway, Shearer was also "at the center of the Democrats 1996 fundraising scandal" and the accusations by Kathleen Willey against Bill Clinton. For a digest of Clinton "scandals," see Carrie Johnson, "Clinton Scandals: A Guide From Whitewater To The Clinton Foundation," *NPR*, June 21, 2016, and "A Brief Guide To Clinton

Scandals From Travelgate To Emailgate," *Washington Examiner Staff*, May 17, 2016.

96. Ibid.

97. Jeff Carlson, "Baker Testimony Reveals Perkins Coie Lawyer Provided FBI with Information on Alfa Bank Allegations," *The Epoch Times*, March 8, 2019.

98. Chuck Ross, "Here's How the Steele Dossier Spread Through the Media and Government," Daily Caller, March 18, 2019; Executive Session, Committee on the Judiciary, Joint with the Committee on Government Reform and Oversight, U.S. House of Representatives, "Interview of: James Baker," October 3, 2018, https://dougcollins.house.gov/sites/dougcollins.house.gov/files/Baker Redacted.pdf?utm_source=Collins+Judiciary+Press+List&utm_campaign =2c66dbe45c-EMAIL_CAMPAIGN_2019_04_08_07_36&utm_medium =email&utm_term=0_ff92df788e-2c66dbe45c-169041485, 44–53.

99. Michael Isikoff, "U.S. Intel Officials Probe Ties Between Trump Adviser and Kremlin," Yahoo! News, September 23, 2016.

100. Ibid.

101. "Hillary for America Statement on Bombshell Report About Trump Aide's Chilling Ties to Kremlin," *Milwaukee Courier*, September 24, 2016.

102. David Corn, "A Veteran Spy Has Given the FBI Information Alleging a Russian Operation to Cultivate Donald Trump," *Mother Jones*, October 31, 2016.

103. Ibid.

104. Ibid.

105. Ibid.

106. Executive Session, Committee on the Judiciary, Joint with the Committee on Government Reform and Oversight, U.S. House of Representatives, "Interview of: James Baker," 37–42.

107. Ibid., 36, 43.

108. Michael Isikoff and David Corn, *Russian Roulette: The Inside Story of Putin's War on America and the Election of Donald Trump* (New York: Twelve, 2018).

109. Steven Lee Meyers, "Was the 2016 Election a Game of Russian Roulette?," *New York Times*, March 14, 2018.

110. William Cummings, "Reporter Who Broke Steele Dossier Story Says Ex–British Agent's Claims 'Likely False,'" *USA Today*, December 18, 2018.

111. Ibid.

112. William Barr, Attorney General, letter to the Chairmen and Ranking Members of the House and Senate Judiciary Committees, March 24, 2019, https://eshoo .house.gov/wp-content/uploads/2019/03/Letter-From-AG-Barr-To-Congress.pdf.

113. Luke Harding, Stephanie Kirchgaessner, and Nick Hopkins, "British Spies Were First to Spot Trump Team's Links with Russia," *The Guardian*, April 13, 2017.

114. Ibid.

115. Jonathan Landay, "CIA Unveils New Rules for Collecting Information on Americans," Reuters, January 18, 2017.

116. "Testimony of CIA Director John Brennan Before the House Intelligence Committee," CNN, May 23, 2017.

117. Eric Lichtblau, "CIA Had Evidence of Russia Effort to Help Trump Earlier than Believed," *New York Times*, April 6, 2017.

118. Ibid.

119. Victor Davis Hanson, "John Brennan's Dishonesty: A Long Record," *National Review*, June 4, 2018.

120. "Spying on Trump: Was Obama Behind CIA, FBI Plan to Elect Hillary?," *Investor's Business Daily*, May 22, 2018.

121. Paul Sperry, "Exclusive: CIA Ex-Director Brennan's Perjury Peril," RealClearInvestigations, February 11, 2018.

122. Senator Harry Reid, letter to FBI Director James Comey, August 27, 2016, https://www.documentcloud.org/documents/3035844-Reid-Letter-to-Comey.html.

123. David E. Sanger, "Harry Reid Cites Evidence of Russian Tampering in U.S. Vote, and Seeks FBI Inquiry," *New York Times*, August 29, 2016.

124. Josh Rogan, "Democrats Ask the FBI to Investigate Trump Advisers' Russia Ties," *Washington Post*, August 30, 2016.

125. Senator Harry Reid, letter to FBI Director James Comey, October 30, 2016, available at https://archive.org/details/20161030ReidLetterToComeyexplosive Information.

126. Aaron Blake, "Harry Reid's Incendiary Claim About 'Coordination' Between Donald Trump and Russia," *Washington Post*, October 31, 2016.

127. Rowan Scarborough, "Harry Reid Sent Sensitive Trump Collusion Letter over CIA Objections," *Washington Times*, May 12, 2018.

128. John Brandt, "Reid Repeats Rumor on Senate Floor That Romney Paid No Taxes, Campaign Denies," Fox News, August 2, 2012; Louis Jacobson, "Harry Reid Says Anonymous Source Told Him Mitt Romney Didn't Pay Taxes for 10 Years," PolitiFact, August 6, 2012; Glenn Kessler, "4 Pinocchios for Harry Reid's Claim About Mitt Romney's Taxes," *Washington Post* Fact Checker, August 7, 2012; Ashley Condianni, "Harry Reid Doesn't Regret Accusing Mitt Romney of Not Paying Taxes," CNN, March 31, 2015.

129. Hanson, "John Brennan's Dishonesty"; Becket Adams, "John Brennan, Famous for Lying and Spying on the Senate, Baselessly Accuses Trump of Treason," *Washington Examiner*, July 17, 2018; "Ex-CIA Chief Brennan's Security Clearance Should Have Been Revoked Long Ago," *Investor's Business Daily*, August 17, 2018.

130. Daniel Chaitin, "Trump Never Revoked John Brennan's Security Clearance," *Washington Examiner*, May 25, 2019.

131. Julian E. Zelizer, "Was McCain's Campaign the Worst Ever?," *Newsweek*, November 4, 2008.

132. Jonathan Martin and Alan Rappeport, "Donald Trump Says John McCain Is No War Hero, Setting Off Another Storm," *New York Times*, July 18, 2015.

133. Brent D. Griffiths, "McCain: Trump Never Apologized for Saying I Wasn't a War Hero," The Hill, September 24, 2017.

134. United States District Court, Southern District of Florida, deposition of David Kramer, *Gubarev v. BuzzFeed*, December 13, 2017, https://www.scribd

.com/document/401932342/Kramer-Depositioin#from_embed?campaign
=VigLink&ad_group=xxc1xx&source=hp_affiliate&medium=affiliate.

135. Jerry Dunleavy and Daniel Chaitin, "John McCain Associate Behind Dossier Leak Urged BuzzFeed to Retract Its Story: 'You Are Gonna Get People Killed!,'" *Washington Examiner*, March 14, 2019.

136. John Haltiwanger, "John McCain Described How He Received the Steele Dossier That Contains the Most Salacious Allegations About Trump and Russia," Business Insider, March 22, 2019.

137. Chuck Ross, "John McCain Associate Had Contact with a Dozen Reporters Regarding Steele Dossier," Daily Caller, March 14, 2019.

138. United States District Court, Southern District of Florida, deposition of David Kramer.

139. Ibid.

140. Joe Pompeo, "'The Broad Outline of What Steele Was Writing Is Unquestionably True': BuzzFeed Wins Its Dossier Suit, and Ben Smith Takes a Victory Lap," *Vanity Fair*, December 20, 2018.

141. Andrew O'Hehir, "James Clapper on Donald Trump, Edward Snowden, Torture and 'The Knowability of Truth,'" Salon, May 26, 2018.

142. Ibid.

143. Ibid.; John Bowden, "Clapper: 'More and More' of Steele Dossier Proving to Be True," The Hill, May 26, 2018.

144. Paul Sperry, "Two Colleagues Contradict Brennan's Denial of Reliance on Dossier," RealClearInvestigations, May 15, 2018; Paul Sperry, "Exclusive: CIA Ex-Director Brennan's Perjury Peril," RealClearInvestigations, February 11, 2018; Natasha Bertrand, "Former CIA Director: I Was Concerned About 'Interactions' Between Russians and the Trump Campaign," Business Insider, May 23, 2017.

145. Sperry, "Exclusive: CIA Ex-Director Brennan's Perjury Peril."

146. Ibid.

147. Jack Shafer, "The Spies Who Came In to the TV Studio," Politico, February 6, 2018.

148. Bill Hoffman, "Ex–CIA Chief Brennan Headed to NBC/MSNBC," Newsmax, February 2, 2018.

149. Jonathan Easley, "D Report: Clapper Told CNN Host About Trump Dossier in 2017," The Hill, April 27, 2018.

150. Sean Davis, "Declassified Congressional Report: James Clapper Lied About Dossier Leaks to CNN," The Federalist, April 27, 2018. (Note: The *Washington Post* "fact checker" Glenn Kessler wrote a column on May 3, 2018, claiming that his newspaper's reporters were correct in stating that Clapper's leak was in May 2017, not January. That attempt by Kessler and the *Washington Post* to exonerate Clapper is demonstrably erroneous. Clapper told the Intelligence Committee that he had leaked the "dossier" information to CNN host Jake Tapper "pretty close to when we briefed it." President Obama was briefed on January 5, 2017. President-elect Trump was briefed the next day. Therefore, Clapper's leak to CNN occurred in early January, not May as alleged by the *Washington Post* and Kessler. For

an accurate account of this, see Sean Davis, "Washington Post 'Fact Check' on James Clapper's Leaks Ignores Basic Facts," The Federalist, May 3, 2018.)

151. Rowan Scarborough, "Obama DNI Clapper Leaked Dossier Story on Trump: House Intel Report," *Washington Times*, April 28, 2018.

152. Marshall Cohen and Jeremy Herb, "Revisiting the Trump-Russia Dossier: What's Right, Wrong and Still Unclear?," CNN, January 7, 2019.

153. Office of the Director of National Intelligence, "Background to 'Assessing Russian Activities and Intentions in Recent U.S. Elections': The Analytic Process and Cyber Incident Attribution," January 6, 2017, https://www.dni.gov/files /documents/ICA_2017_01.pdf.

154. Scott Shane, "Russia Isn't the Only One Meddling in Elections. We Do It, Too," *New York Times*, February 17, 2018.

155. Michael S. Schmidt, Mark Mazzetti, and Matt Apuzzo, "Trump Campaign Aides Had Repeated Contacts with Russian Intelligence," *New York Times*, February 14, 2017.

156. Bob Woodward, *Fear: Trump in the White House* (New York: Simon & Schuster, 2018), 84.

157. Ibid. See also Senate Intelligence Committee, Testimony of James Comey, June 8, 2017, "Full Transcript and Video: James Comey's Testimony on Capitol Hill," the *New York Times*, June 8, 2017.

158. Olivia Beavers, "Kremlin Spokesman: Russian Ambassador Met with Advisers to Clinton Campaign Too," The Hill, March 12, 2017.

159. John Solomon, "Ukrainian Embassy Confirms DNC Contractor Solicited Trump Dirt In 2016," The Hill, May 2, 2019; Kenneth P. Vogel and David Stern, "Ukrainian Efforts To Sabotage Trump Backfire," *Politico*, January 11, 2017; Gregg Re, "Clinton-Ukraine Collusion Allegations 'Big' And 'Incredible,' Will Be Reviewed, Trump Says," Fox News, April 25, 2019; Ian Schwartz, "Giuliani: 'Massive Collusion' Between DNC, Obama Admin, Clinton People & Ukraine To Create False Info About Trump," *RealClear Politics*, May 10, 2019.

160. Alan Cullison and Brett Forrest, "Trump Tower Moscow? It Was the End of a Long, Failed Push to Invest in Russia," *Wall Street Journal*, November 29, 2018.

161. Megan Twohey and Steve Eder, "For Trump, Three Decades of Chasing Deals in Russia," *New York Times*, January 16, 2017.

162. Gregg Jarrett, "Big Lebowski, Esq., Takes Up Gen. Flynn's Case," Fox News, April 29, 2017.

163. Robert Windrem, "Guess Who Came to Dinner with Flynn and Putin," NBC News, April 18, 2017.

164. Shannon Bream, interview with Daniel Hoffman, former CIA station chief, *Fox News @ Night with Shannon Bream*, Fox News, April 23, 2019.

165. Attorney General William Barr, letter to Lindsey Graham, Jerrold Nadler, Dianne Feinstein, and Doug Collins.

166. James Clapper, *Facts and Fears: Hard Truths from a Life in Intelligence* (New York: Viking, 2018).

167. Gregg Re, "Rand Paul: 'Source' Says John Brennan Pushed Discredited Steele Dossier," March 27, 2019.

168. Liam Quinn, "Ex–CIA Director John Brennan Admits He May Have Had 'Bad Information' Regarding President Trump and Russia," Fox News, March 26, 2019.

169. Woodward, *Fear*, 64.

170. Andrew C. McCarthy, "Steele's Shoddy Dossier," *National Review*, June 6, 2019.

171. Jan Crawford, "William Barr Interview: Read the Full Transcript," CBS News, May 31, 2019.

CHAPTER 3: LYING AND SPYING

1. "William Barr's Testimony Before the Senate Appropriations Subcommittee on Commerce, Justice, Science, and Related Agencies," CNN, April 10, 2019.

2. Ibid.

3. Gary Abernathy, "Admit It: Fox News Has Been Right All Along," *Washington Post*, April 15, 2019.

4. Justin Wise, "CNN Legal Analyst Knocks Barr Spying Claim: He 'Talks like Sean Hannity,'" The Hill, April 11, 2019.

5. Joseph Wulfsohn, "Media Take Issue with AG Barr for Saying 'Spying Did Occur' on Trump Campaign," Fox News, April 10, 2019.

6. Timothy L. O'Brien, "Bill Barr Is Trying Hard to Be President Trump's Roy Cohen," Bloomberg, April 10, 2019.

7. Jennifer Rubin, "William Barr, Trump's Toady," *Washington Post*, April 10, 2019.

8. Whitney Tipton, "CNN Freaks Out over Barr 'Spying' Test," Daily Caller, April 11, 2019.

9. Definition of the word "spy," Lexico, https://en.oxforddictionaries.com/definition/spy.

10. Byron York, "Barr Is Right, Spying on Trump Campaign Did Occur," *Washington Examiner*, April 10, 2019.

11. Ibid.

12. Zack Budryk, "Clapper: Barr's Spying Claim 'Stunning and Scary,'" The Hill, April 10, 2019.

13. Hans von Spakovsky, "Dems Wrong to Attack Barr for Telling Truth About Fed Spying on Trump Campaign," Fox News, April 10, 2019.

14. "William Barr's Testimony Before the Senate Appropriations Subcommittee on Commerce, Justice, Science, and Related Agencies."

15. Kimberley A. Strassel, "Barr Brings Accountability," *Wall Street Journal*, April 11, 2019.

16. "Read James B. Comey's Opening Statement Ahead of His Testimony," June 7, 2017, NBC News.

17. Ibid.

18. Andrew C. McCarthy, "FBI Russia Investigation Was Always About Trump," Fox News, January 13, 2019.

19. Andrew C. McCarthy, "Behind The Obama Administration's Shady Plan To Spy On The Trump Campaign," *New York Post*, April 15, 2019.

20. Ibid., "Read James B. Comey's Opening Statement."
21. Senator Charles E. Grassley, Chairman, Committee on the Judiciary, letter sent to Michael E. Horowitz, Inspector General, U.S. Department of Justice, February 28, 2018, https://www.grassley.senate.gov/sites/default/files/judiciary/upload/2018-02-28%20CEG%20LG%20to%20DOJ%20OIG%20%28referral%29.pdf, 7.
22. Ibid., 4, 5.
23. Brooke Singman, "Comey Scoffs at Barr Testimony, Claims 'Surveillance' Is Not 'Spying,'" Fox News, April 12, 2019.
24. Ibid.
25. Author's interview with Carter Page, March 12, 2018.
26. R. J. Sharpe, *Law of Habeas Corpus* (Oxford: Clarendon Press, 1989); Paul Halliday, *Habeas Corpus: From England to Empire* (Cambridge: Belknap Press of Harvard University Press, 2010).
27. Constitution of the United States of America, Amendments V and VI (ratified effective December 15, 1791).
28. *Faretta v. California*, 422 U.S. 806 (1975), opinion by Justice Potter Stewart.
29. Foreign Intelligence Surveillance Act, 1978, 50 U.S.C., chapter 36, https://fas.org/irp/agency/doj/fisa/.
30. Conor Clarke, "Is the Foreign Intelligence Surveillance Court Really a Rubber Stamp?," *Stanford Law Review*, February 2014.
31. Larry Abramson, "FISA Court: We Approve 99% of Wiretap Applications," National Public Radio, October 15, 2013.
32. Evan Perez, "Secret Court's Oversight Gets Scrutiny," *Wall Street Journal*, June 9, 2013.
33. Erika Eichelberger, "FISA Court Has Rejected .03 Percent of All Government Surveillance Requests," *Mother Jones*, June 10, 2013.
34. Daniel Chaitin, "Lack of FISA Court Hearings on Carter Page Warrants Sparks Fierce Debate," *Washington Examiner*, September 2, 2018; Dave Boyer, "FISA Court Didn't Hold Hearings Before Granting Warrants on Carter Page, Trump Notes in Tweet," *Washington Times*, September 1, 2018.
35. David Kris, "How the FISA Court Really Works," Lawfare, September 2, 2018.
36. Asha Rangappa, "It Ain't Easy Getting a FISA Warrant: I Was an FBI Agent and Should Know," Just Security, March 6, 2017.
37. "Foreign Intelligence Surveillance Act Court Orders 1979 to 2017," Electronic Privacy Information Center, https://epic.org/privacy/surveillance/fisa/stats/.
38. Rick Moran, "Judicial Watch: No Hearings Held on Carter Page FISA Warrants," American Thinker, September 1, 2018.
39. United States House of Representatives, Permanent Select Committee on Intelligence, "Testimony of Carter Page," November 2, 2017, https://www.documentcloud.org/documents/4176234-Carter-Page-Hpsci-Hearing-Transcript-Nov-2-2017, 11,15.html; Anna Giaritelli, "Carter Page Says He's Never Spoken with Trump in His Life," *Washington Examiner*, February 6, 2018.
40. United States House of Representatives, Permanent Select Committee on Intelligence, "Testimony of Carter Page," 157.

41. "Text: Obama Speech at the New Economic School," *New York Times*, July 7, 2009.

42. Ibid.

43. Scott Shane, Mark Mazzetti, and Adam Goldman, "Trump Adviser's Visit to Moscow Got the FBI's Attention," *New York Times*, April 19, 2017.

44. United States House of Representatives, Permanent Select Committee on Intelligence, "Testimony of Carter Page," 19.

45. Tim Hains, "Carter Page: I Experienced the Trump-Russia Witch Hunt First Hand, Now We're Getting the Real Truth," RealClearPolitics, March 31, 2019.

46. "The Lecture of Trump's Advisor Carter Page in Moscow," YouTube, July 7, 2016.

47. United States House of Representatives, Permanent Select Committee on Intelligence, "Testimony of Carter Page," 40.

48. Eric Felten, "Carter Page Is Mr. Clean," *National Review*, April 19, 2019.

49. "Company Intelligence Report 2016/080," December 13, 2016, https://www.documentcloud.org/documents/3259984-Trump-Intelligence-Allegations.html, 9.

50. Julie Kelly, "Vindication for Carter Page," American Greatness, July 25, 2018.

51. Michael Isikoff, "U.S. Intel Officials Probe Ties Between Trump Adviser and Kremlin," Yahoo! News, September 23, 2016.

52. Ibid.

53. Dylan Stableford, "Yahoo News' Michael Isikoff Describes Crucial Meeting Cited in Nunes Memo," Yahoo! News, February 2, 2018.

54. Chuck Ross, "The Problem with the News Article at Center of Carter Page Spy Warrant," Daily Caller, February 5, 2018.

55. Isikoff, "U.S. Intel Officials Probe Ties Between Trump Adviser and Kremlin."

56. Joe Pompeo, " 'He Was Actually the Paul Revere': As the Steele Dossier's Moment of Truth Arrives, Journalists Argue Its Impact," *Vanity Fair*, April 25, 2019.

57. United States House of Representatives, Permanent Select Committee on Intelligence, "Testimony of Carter Page," attached exhibit, 13, letter from Page to James B. Comey, September 25, 2016.

58. Daniel Chaitin, "Carter Page Says He Consulted State Department, FBI, and CIA for Years," *Washington Examiner*, June 9, 2019.

59. United States House of Representatives, Permanent Select Committee on Intelligence, "Testimony of Carter Page."

60. Ibid., 38.

61. Ibid., 36.

62. United States Code, 18 U.S.C. §§ 1621, 1623, "Perjury"; 19 U.S.C. § 1001, "False and Misleading Statements"; 18 U.S.C. § 1503, et al. "Obstruction of Justice"; 18 U.S.C. § 1031 "Major Fraud Against the U.S."; 18 U.S.C. § 371 "Conspiracy to Defraud the U.S."; 18 U.S.C. § 242 "Deprivation of Rights Under Color of Law," 50 U.S.C. § 1809 "Electronic Surveillance Under Color of Law."

63. Gregg Jarrett, *The Russia Hoax: The Illicit Scheme to Clear Hillary Clinton and Frame Donald Trump* (New York: Broadside Books, 2018), 139–170.

64. United States Foreign Intelligence Surveillance Court, Washington, D.C., warrant applications and corresponding orders dated October 2016, January 2017, April 2017, and June 2017, https://vault.fbi.gov/d1-release/d1-release/view.

65. Ibid., see "Verification" document (54) and "Certification" document (63, 65).

66. Sharyl Attkisson, "Nunes Memo Raises Question: Did FBI Violate Woods Procedures?," The Hill, February 4, 2018.

67. John Solomon, "Mueller Hauled Before Secret FISA Court to Address FBI Abuses in 2002, Congress Told," The Hill, February 6, 2019.

68. Devin Nunes, Chairman, U.S. House of Representatives, Permanent Select Committee on Intelligence, to Jeff Sessions, Attorney General, March 1, 2018, https://www.scribd.com/document/372746970/Nunes-to-Sessions-FBI-may-have-violated-criminal-statutes-in-Carter-Page-FISA-application#from_embed.

69. Ibid.

70. Andrew C. McCarthy, "The Steele Dossier And The 'Verified Application' That Wasn't," National Review, May 18, 2019.

71. Brooke Singman, "FISA Memo: Steele Fired As An FBI Source For Breaking 'Cardinal Rule'—Leaking To The Media," Fox News, February 2, 2018.

72. Andrew C. McCarthy, "FISA Applications Confirm: The FBI Relied On The Unverified Steele Dossier," National Review, July 23, 2018.

73. Catherine Herridge and Pamela Brown, "DOJ Releases FISA Docs That Formed Basis for Surveillance of Ex-Trump Adviser Carter Page," Fox News, July 22, 2018.

74. "Full Transcript and Video of James Comey's Testimony on Capitol Hill," New York Times, June 8, 2017.

75. John Solomon, "Comey's Confession: Dossier Not Verified Before, or After, FISA Warrant," The Hill, December 8, 2018.

76. Andrew C. McCarthy, "FISA Applications Confirm: The FBI Relied on the Unverified Steele Dossier," National Review, July 23, 2018.

77. House Permanent Select Committee on Intelligence Majority Members, memorandum re "Foreign Intelligence Surveillance Act Abuses at the Department of Justice and the Federal Bureau of Investigation," January 18, 2018 (declassified by order of the president, February 2, 2018), https://www.documentcloud.org/documents/4365338-Nunes-memo.html.

78. Byron York, "FISA Warrant Application Supports Nunes Memo," Washington Examiner, July 22, 2018.

79. Ian Schwartz, "Baier to Comey: If Dossier Was 'Salacious,' Why Did You Use It to Get FISA Warrant?," RealClearPolitics, April 26, 2018.

80. "Foreign Intelligence Surveillance Act Court Orders 1979 to 2017"; Gregg Re, "FBI Told FISA Court Steele Wasn't Source of Report Used to Justify Surveilling Trump Team, Docs Show," Fox News, July 22, 2018.

81. "Foreign Intelligence Surveillance Act Court Orders 1979 to 2017," 15.

82. Ibid., 100.

83. Mollie Ziegler Hemingway, "Confirmed: DOJ Use Materially False Information to Secure Wiretaps on Trump Associate," The Federalist, July 23, 2018.

84. For more details on the assistance Carter Page supplied to the FBI, see Gregg Jarrett, *The Russia Hoax: The Illicit Scheme to Clear Hillary Clinton and Frame Donald Trump* (New York: Broadside Books, 2018), 148.

85. United States Code, 50 § U.S.C. 1804, "Applications for Court Orders"; Andrew C. McCarthy, "The Schiff Memo Harms Democrats More Than It Helps Them," *National Review*, February 25, 2018.

86. "Foreign Intelligence Surveillance Act Court Orders 1979 to 2017." See also House Permanent Select Committee on Intelligence Majority Members, memorandum re "Foreign Intelligence Surveillance Act Abuses at the Department of Justice and the Federal Bureau of Investigation."

87. "Foreign Intelligence Surveillance Act Court Orders 1979 to 2017," footnote 8: Source #1 was approached by an identified U.S. Person, who indicated to Source#1 that a U.S.-based law firm had hired the identified U.S. Person to conduct research regarding Candidate #1's ties to Russia. (The identified U.S. Person and Source #1 have a long-standing business relationship.) The identified U.S. person hired Source #1 to conduct this research. The identified U.S. Person never advised Source #1 as to the motivation behind the research into Candidate #1's ties to Russia. The FBI speculates that the identified U.S. Person was likely looking for information that could be used to discredit Candidate #1's campaign.

88. Senator Charles E. Grassley, Senator Lindsey O. Graham, Senator John Cornyn, and Senator Thom Tillis, letter and memorandum re "Referral of Christopher Steele for Potential Violation of 18 U.S.C. 1001," January 4, 2018, available at https://www.grassley.senate.gov/news/news-releases/senators-grassley-graham-refer-christopher-steele-criminal-investigation.

89. Rowan Scarborough, "Trump Legal Team Makes Case to Prosecute Comey, Others for Lying After Cohen Guilty Plea," *Washington Times*, December 2, 2018; Gregg Jarrett, "James Comey May Be the Only One Who Believes the Stories He's Selling," Fox News, December 18, 2018; Catherine Herridge and Judson Berger, "Trump Team Renews Claims Comey Misled Congress in Wake of Cohen Plea," Fox News, November 30, 2018; John Dowd, lawyer for Trump, letter to Rod Rosenstein, Deputy Attorney General, September 1, 2017, https://www.scribd.com/document/394562504/Trump-legal-team-s-letters-to-Mueller-Rosenstein-on-Comey#from_embed.

90. Matt Apuzzo, Michael Schmidt, Adam Goldman, and Eric Lichtblau, "Comey Tried to Shield FBI from Politics. Then He Helped Shape an Election," *New York Times*, April 22, 2017.

91. Michael Doran, "The Real Collusion Story," *National Review*, March 13, 2018.

92. Gregg Re, Catherine Herridge, and Cyd Upson, "FBI Clashed with DOJ over Potential 'Bias' of Source for Surveillance Warrant: McCabe-Page Texts," Fox News, March 22, 2019.

93. Ibid.

94. Ibid.

95. "Foreign Intelligence Surveillance Act Court Orders 1979 to 2017," 2, 68, 164.

96. Tim Hains, "Deputy AG Rod Rosenstein: 'The Department of Justice Is Not Going to Be Extorted,'" RealClearPolitics, May 1, 2018.

97. "Foreign Intelligence Surveillance Act Court Orders 1979 to 2017," 1, 84, 182, 292.

98. Craig Bannister, "Rosenstein Testifies He Doesn't Need to Read FISA Applications He Signs," CNS News, June 29, 2018.

99. Sonam Sheth, "House Investigators Grill Christopher Wray and Rod Rosenstein in Contentious Hearing About the Trump and Clinton Probes," Business Insider, June 28, 2018;

100. Executive Session, Committee on the Judiciary, Joint with the Committee on Government Reform and Oversight, U.S. House of Representatives, "Interview of: James Comey," December 7, 2018, https://www.lawfareblog.com /document-transcript-james-comeys-dec-7-interview-house-committees.

101. Ibid., 124–26.

102. Gregg Jarrett, "Testimony in Russia Probe Shows FBI and Justice Department Misconduct in Effort to Hurt Trump," Fox News, January 24, 2019.

103. Ibid.

104. John Solomon, "Steele's Stunning Pre-FISA Confession: Informant Needed Trump Dirt Before Election," The Hill, May 7, 2019.

105. John Solomon, "FBI's Steele Story Falls Apart: False Intel and Media Contacts Were Flagged Before FISA," The Hill, May 9, 2019.

106. Catherine Herridge and Adam Shaw, "Lawmakers Say FBI May Have Been Warned of Steele's 'Political Motivations' Before Trump Aide Surveillance," Fox News, May 10, 2019; Senators Ron Johnson and Charles E. Grassley, letter to Michael R. Pompeo, Secretary of State, May 9, 2019, https://www.scribd.com /document/409477564/Lawmakers-letter-to-Secretary-of-State-Pompeo#from _embed.

107. Solomon, "FBI's Steele Story Falls Apart."

108. Ibid.

109. John Solomon, "State Department's Red Flag on Steele Went to a Senior FBI Man Well Before FISA Warrant," The Hill, May 14, 2019.

110. Solomon, "FBI's Steele Story Falls Apart."

111. Solomon, "Steele's Stunning Pre-FISA Confession."

112. Ibid.

113. "Spying on Trump: Was Obama Behind CIA, FBI Plan to Elect Hillary?," Investor's Business Daily, May 22, 2018.

114. Jane Mayer, "Christopher Steele, the Man Behind the Trump Dossier," The New Yorker, March 12, 2018; Jasper Fakkert, "The Origins of 'Spygate': 10 Questions," The Epoch Times, May 16, 2019.

115. Letter from Sen. Charles E. Grassley, Chairman of Committee on Finance, to Mark Esper, Acting Secretary of the Department of Defense, July 12, 2019, available at https://www.grassley.senate.gov/sites/default/files/constituents /2019-07-12%20CEG%20to%20DoD%20(Halper%20Contracts)_0.pdf; Sara Carter, "Grassley Pressures DOD For More Information On FBI Spy Stefan Halper," saraacarter.com, July 22, 2019.

116. Gregg Re, "Carter Page Says FBI Informant 'Intensified' Communications Just Prior to FISA Warrant," Fox News, June 9, 2019.

117. Ibid.

118. George Papadopoulos, *Deep State Target: How I Got Caught in the Crosshairs of the Plot to Bring Down President Trump* (New York: Diversion Books, 2019).

119. Ibid., 101.

120. Adam Goldman, Michael S. Schmidt, and Mark Mazzetti, "FBI Sent Investigator Posing as Assistant to Meet with Trump Aide in 2016," *New York Times*, May 2, 2019.

121. Gregg Re and Brooke Singman, "U.S. Informant Reportedly Tried to Probe Papadopoulos on Trump-Russia Ties, 'Seduce Him' During Campaign," Fox News, May 3, 2019.

122. Papadopoulos, *Deep State Target*, 106.

123. Ibid., 107.

124. Jan Crawford, "William Barr Interview: Read the Full Transcript," CBS News, May 31, 2019.

125. Papadopoulos, *Deep State Target*, 67–69.

126. Ibid., 105.

127. Ibid., 176.

128. Ibid., 213.

129. Ibid., 60.

130. Ibid., 61.

131. Ibid., 75.

132. Ibid., 76.

133. Ibid., 77.

134. Sharon LaFraniere, Mark Mazzetti, and Matt Apuzzo, "How the Russia Inquiry Began: a Campaign Aide, Drinks and Talk of Political Dirt," *New York Times*, December 30, 2017.

135. George Papadopoulos, *Deep State Target*, 78.

136. John Solomon, "Australian Diplomat Whose Tip Prompted FBI's Russia-Probe Has Tie to Clintons," The Hill, March 5, 2018.

137. "FBI Domestic Investigations and Operations Guide (DIOG), Part 03 of 03," 2011, https://vault.fbi.gov/FBI%20Domestic%20Investigations%20and%20Operations%20Guide%20%28DIOG%29/fbi-domestic-investigations-and-operations-guide-diog-2011-version/fbi-domestic-investigations-and-operations-guide-diog-october-15-2011-part-03-of-03/view.

138. Robert S. Mueller, *The Mueller Report: The Final Report of the Special Counsel into Donald Trump, Russia, and Collusion as Issued by the Department of Justice* (New York: Skyhorse Publishing, 2019), vol. I, 179.

139. Ibid., 180.

140. Robert Mendick, Alex Luhn, and Ben Riley-Smith, "Revealed: London Professor at Centre of Trump-Russia Collusion Inquiry Says: 'I Have Clear Conscience,'" *The Telegraph*, October 31, 2017.

141. Robert S. Mueller, *The Mueller Report*, 180–181.

142. Jonathan Turley, "The Mysterious Mister Mifsud And Why No One Wants To Discuss Him," The Hill, July 27, 2019.

143. Andrew C. McCarthy, "Fighting the Politicized, Evidence-Free 'Collusion with Russia' Narrative," *National Review*, May 24, 2017.

144. Michael Isikoff and David Corn, *Russian Roulette: The Inside Story of Putin's War on America and the Election of Donald Trump* (New York: Twelve, 2018), 153.

145. John Solomon, "A Convenient Omission? Trump Campaign Adviser Denied 'Collusion' to FBI Source Early On," The Hill, October 23, 2018.

146. John Solomon, "FBI's Spreadsheet Puts A Stake Through The Heart Of Steele's Dossier," The Hill, July 16, 2019.

147. Ibid.

148. "FBI Domestic Investigations and Operations Guide (DIOG), Part 03 of 03," 13.

149. Papadopoulos, *Deep State Target*, 2. Papadopoulos gave a different version of this conversation when interviewed by Congress, stating that the arresting FBI agents told him, "This is what happens when you don't tell us everything about your Russia contacts."

150. Ibid., 179.

151. Glenn Greenwald, "The FBI Informant Who Monitored the Trump Campaign, Stefan Halper, Oversaw a CIA Spying Operation in the 1980 Presidential Election," The Intercept, May 19, 2018.

152. Ibid.

153. Crawford, "William Barr Interview: Read the Full Transcript."

154. Adam Goldman, Charlie Savage, and Michael S. Schmidt, "Barr Assigns U.S. Attorney in Connecticut to Review Origins of Russia Inquiry," *New York Times*, May 13, 2019.

155. Stephen E. Boyd, Assistant Attorney General, letter to Representative Jerrold Nadler, Chairman, Committee on the Judiciary, June 10, 2019, https://www .scribd.com/document/412966458/2019-6-10-DOJ-Review-of-Intelligence -Activities-Nadler#from_embed.

156. Crawford, "William Barr Interview: Read the Full Transcript."

CHAPTER 4: THE ATTEMPTED COUP

1. Laurence J. Peter and Raymond Hull, *The Peter Principle: Why Things Always Go Wrong* (New York: William Morrow, 1969).

2. Anderson Cooper, "Interview with Former Acting FBI Director Andrew McCabe," CNN, February 19, 2019.

3. Andrew G. McCabe, *The Threat: How the FBI Protects America in the Age of Terror and Trump* (New York: St. Martin's Press, 2019), 136; Casey Quackenbush, "Read the Full Transcript of Former FBI Deputy Director Andrew McCabe's *60 Minutes* Interview," *Time*, February 18, 2019.

4. Sarah Left, "Iraq War 'Waged on False Intelligence,'" *The Guardian*, July 9, 2014; Associated Press, "Official: Iraq War Was Both Intel, Policy Failure," NBC News, December 9, 2008.

5. Jenna Lifhits, "Obama Ignored Iranian Transgressions to Preserve Nuclear Deal, Haden Says," *The Weekly Standard*, December 14, 2016.

6. McCabe, *The Threat*, 207.

7. Ibid.

8. Constitution of the United States of America, Article II, Section 3, Clause 4.

9. McCabe, *The Threat*, 239.

10. Joseph Hinks, "Read Former FBI Director James Comey's Farewell Letter to Colleagues," *Time*, May 11, 2017.

11. McCabe, *The Threat*, 239.

12. "Full Transcript: Acting FBI Director McCabe and Others Testify Before Senate Intelligence Committee," *Washington Post*, May 11, 2017.

13. "Read Full Testimony of FBI Director James Comey in Which He Discusses Clinton Email Investigation," *Washington Post*, May 3, 2017.

14. "Rod Rosenstein Full Remarks to Congress on Comey Memo," *Axios*, May 19, 2017; Jessica Taylor, "Rosenstein On Comey Memo: 'I Wrote It. I Believe It. I Stand By It'," *NPR*, May 19, 2017.

15. Gregg Re, "Comey Reveals He Concealed Trump Meeting Memo from DOJ Leaders," Fox News, December 9, 2018.

16. Office of the Inspector General, U.S. Department of Justice, "A Review of Various Actions by the Federal Bureau of Investigation and Department of Justice in Advance of the 2016 Election," p. 405.

17. McCabe, *The Threat*, 219.

18. Ibid., 221.

19. Ibid., 218.

20. Ibid., 217.

21. Ibid., 216–17.

22. Devlin Barrett, "Clinton Ally Aided Campaign of FBI Official's Wife," *Wall Street Journal*, October 24, 2016.

23. McCabe, *The Threat*, 193.

24. Ibid., 218.

25. Andrew C. McCarthy, "McCabe and *60 Minutes* Avoid Discussing *Why* Russia Factored in Comey's Firing," *National Review*, February 18, 2019.

26. McCabe, *The Threat*, 239.

27. Ibid., 225.

28. Catherine Herridge, "Strzok-Page Texts Calling to 'Open' Case in 'Chargeable Way' Under Fresh Scrutiny," Fox News, September 17, 2018.

29. Ibid.

30. "The Attorney General's Guidelines for Domestic FBI Operations," Department of Justice, https://www.justice.gov/archive/opa/docs/guidelines.pdf, 22.

31. Ibid.

32. McCabe, *The Threat*, 225.

33. Quackenbush, "Read the Full Transcript of Former FBI Deputy Director Andrew McCabe's 60 Minutes Interview." See also Representative Doug Collins, Ranking Member, House Judiciary Committee, letter to Representative Jerrold Nadler, Chairman, February 14, 2019, available at https://gallery

.mailchimp.com/0275399506e2bdd8fe2012b77/files/568cf2d2-eb9b-40
a2-92a5-ada5a0bd5db6/02_14_19_Letter_to_Nadler_Re._McCabe.pdf
?utm_source=Collins+Judiciary+Press+List&utm_campaign=3d4d3af9dd
-EMAIL_CAMPAIGN_2019_02_14_08_33&utm_medium=email&utm
_term=0_ff92df788e-3d4d3af9dd-168924225.

34. McCabe, *The Threat*, 234.

35. "Rod Rosenstein Full Remarks to Congress on Comey Memo," Axios, May 19, 2017.

36. Grace Segers, "What Andrew McCabe Told '60 Minutes' About Trump and the 25th Amendment," CBS News, February 17, 2019.

37. Ibid.

38. Kevin R. Brock, "The Embarrassing Return of Andrew McCabe," The Hill, February 15, 2019.

39. Jake Gibson, "Source in Room Says Rosenstein's 'Wire' Comment Was Case of Sarcasm," Fox News, September 21, 2018.

40. Gregg Jarrett, "Rod Rosenstein Should Immediately Stop Overseeing the Mueller 'Collusion' Investigation," Fox News, October 10, 2018; Catherine Herridge, "Rosenstein Threatened to 'Subpoena' GOP-Led Committee in 'Chilling' Clash over Records, Emails Show," Fox News, June 12, 2018.

41. Adam Goldman and Michael S. Schmidt, "Rod Rosenstein Suggested Secretly Recording Trump And Discussed 25th Amendment," *The New York Times*, September 21, 2018.

42. Brooke Singman, "McCabe Details Central Role In Russia Probes, DOJ Meetings On Whether To Oust President," Fox News, February 14, 2019.

43. Catherine Herridge, "Talks on Rosenstein Possibly Recording Trump Unfolded over 'Couple of Days,' Abandoned as 'Too Risky': Former Top FBI Lawyer," Fox News, February 18, 2019.

44. Kevin Breuniger, "Criminal Charges Recommended for Fired FBI Official Andrew McCabe," CNBC, April 19, 2018.

45. Catherine Herridge, "Former Top FBI Lawyer: 2 Trump Cabinet Officials Were 'Ready to Support' 25th Amendment Effort," Fox News, February 17, 2019.

46. Ibid.

47. Adam Goldman and Michael S. Schmidt, "Rod Rosenstein Suggested Secretly Recording Trump and Discussed 25th Amendment," *New York Times*, September 21, 2018.

48. Ibid.

49. Willis L. Krumholz, "Andrew McCabe Can't Keep His Story Straight," The Federalist, February 20, 2019.

50. Constitution of the United States, Amendment XXV, ratified February 10, 1967.

51. Thomas H. Neal, "Presidential Disability Under the Twenty-fifth Amendment: Constitutional Provision and Perspective for Congress," Congressional Research Service, updated November 5, 2018, https://fas.org/sgp/crs/misc/R45394.pdf.

52. Ibid., 7.

53. Ibid.

54. Ibid.

55. Willis L. Krumholtz, "Andrew McCabe Can't Keep His Story Straight," The Federalist, February 20, 2019.

56. Margot Cleveland, "5 Big Takeaways from the '60 Minutes' Interview with Andrew McCabe," The Federalist, February 19, 2019.

57. "Statement by Attorney General on Firing of FBI's McCabe," Reuters, March 17, 2018.

58. Office of the Inspector General, U.S. Department of Justice, "A Report of Investigation of Certain Allegations Relating to Former FBI Deputy Director Andrew McCabe," February 2018, https://www.scribd.com/document/376298359/DOJ-IG-releases-explosive-report-that-led-to-firing-of-ex-FBI-Deputy-Director-Andrew-McCabe#from_embed, 2.

59. Pamela Brown and Laura Jarrett, "Justice Dept. Watchdog Sends McCabe Findings to Federal Prosecutors for Possible Charges," CNN, April 19, 2018.

60. Martina Stewart and Carrie Johnson, "Source: Fired Deputy FBI Director Took Memos, Notes About Interactions with Trump," NPR, March 17, 2018.

61. Byron Tau, "Rod Rosenstein Won't Meet Lawmakers This Week," Wall Street Journal, October 10, 2018.

62. Caitlin Yilek and Kelly Cohen, "GOP Lawmakers File Resolution to Impeach Deputy AG Rod Rosenstein," Washington Examiner, July 25, 2018.

63. Michael D. Shear, Katie Brenner, Maggie Haberman, and Michael S. Schmidt, "Rod Rosenstein's Job Is Safe, for Now: Inside His Dramatic Day," September 24, 2018.

64. Kyle Cheney, "Trump Makes Nice with Rosenstein, but Congress Isn't Letting Up," Politico, October 10, 2018; Associated Press, "Trump Invites Rosenstein on Air Force One, No Plans to Fire Him," NBC News, October 8, 2018; Jennifer Epstein and Justin Sink, "Trump Says He Had 'Great' Meeting with Rosenstein on Air Force One," Bloomberg, October 8, 2018.

65. Sadie Gurman, Michael C. Bender, and Aruna Viswanatha, "Deputy Attorney General Rod Rosenstein to Meet Donald Trump to Consider His Future at DOJ," Wall Street Journal, September 24, 2018.

66. Sadie Gurman, "Trump Says He Has No Plans to Fire Rosenstein," Wall Street Journal, October 8, 2018.

67. Gregg Jarrett, interview with President Donald J. Trump, Oval Office, White House, June 25, 2019.

68. I had several face-to-face meetings with President Trump in the spring, summer, and fall of 2017, as well as follow-up calls in early 2018. This comment is drawn from several of those discussions and was repeated by him during an interview on September 17, 2017, at Trump National Golf Club, Bedminster, New Jersey.

69. Robert S. Mueller, The Mueller Report: The Final Report of the Special Counsel into Donald Trump, Russia, and Collusion as Issued by Department of Justice (New York: Skyhorse Publishing, 2019), 257–58.

70. Gregg Jarrett, interview with John Dowd, lawyer for President Donald Trump, June 13, 2019.

71. Jarrett, "Rod Rosenstein Should Immediately Stop Overseeing the Mueller 'Collusion' Investigation."

72. Herridge, "Rosenstein Threatened to 'Subpoena' GOP-Led Committee in 'Chilling' Clash over Records, Emails Show."

73. Catherine Herridge, "Rosenstein Launched 'Hostile' Attack in May Against Republicans over Russia Records: Congressional Email," Fox News, September 27, 2018.

74. Gregg Jarrett, *The Russia Hoax: The Illicit Scheme to Clear Hillary Clinton and Frame Donald Trump* (New York: Broadside Books, 2018), 210.

75. Code of Federal Regulations, 28 C.F.R § 45.2, "Disqualification Arising from Personal or Political Relationship."

76. *Washington Post* Staff, "Full Transcript: FBI Director James Comey Testifies on Russian Interference in 2016 Election," *Washington Post*, March 2017.

77. Politico Staff, "Transcript: Jeff Sessions' Testimony on Trump and Russia," Politico, June 13, 2017.

78. Code of Federal Regulations, 28 C.F.R. § 600.1, "Grounds for Appointing a Special Counsel."

79. Rod J. Rosenstein, Acting Attorney General, "Appointment of Special Counsel to Investigate Russian Interference with the 2016 Presidential Election and Related Matters," Order No. 3915-2017, May 17, 2017, https://www.document cloud.org/documents/3726408-Rosenstein-letter-appointing-Mueller-special .html.

80. Zachary Basu, "Mueller Referred 14 Criminal Cases for Outside Prosecution," Axios, April 18, 2019.

81. Code of Federal Regulations, 28 C.F.R. § 600.1.

82. Jonathan Turley, "It's High Time Rod Rosenstein Recuse Himself," The Hill, August 7, 2017.

83. Andrew C. McCarthy, "Rosenstein Fails to Defend His Failure to Limit Mueller's Investigation," *National Review*, August 7, 2017.

84. Michael B. Mukasey, "The Memo and the Mueller Probe," *Wall Street Journal*, February 4, 2018.

85. *The Mueller Report*, 39.

86. "Justice Department Defends Russia Probe from GOP Claims of FBI Political Bias," *PBS NewsHour*, PBS, December 13, 2017.

87. Andrew C. McCarthy, "After Mueller's Exoneration of Trump, *Full* Disclosure," *National Review*, March 23, 2019.

88. Code of Federal Regulations, 28 C.F.R. § 600.4, "Jurisdiction."

89. Ibid.

90. United States District Court for the District of Columbia, *United States of America v. Paul J. Manafort, Jr.*, "Government's Response in Opposition to Motion to Dismiss," No. 17-cr-201-1 ABJ, April 2, 2018, https://www.justsecurity .org/wp-content/uploads/2018/04/USG-Govt-Opposition-to-Motion-to-Dis miss.pdf21.

91. Memorandum For The Attorney General From Rod J. Rosenstein, Deputy Attorney General, "Restoring Public Confidence In The FBI," May 9, 2017, available at http://apps.washingtonpost.com/g/documents/politics/fbi-director -james-b-comeys-termination-letters-from-the-white-house-attorney-general /2430/.

92. Sari Horwitz, Karon Demirjian, and Elise Viebeck, "Rosenstein Defends His Controversial Memo Used to Justify Trump's Firing of Comey," *Washington Post*, May 19, 2017.

93. Jonathan Turley, "If Rod Rosenstein Feels Conflicted, He Should Simply Recuse Himself," The Hill, June 30, 2018.

94. Adam Goldman and Michael S. Schmidt, "Rod Rosenstein Suggested Secretly Recording Trump and Discussed 25th Amendment"; Michael S. Schmidt and Adam Goldman, "Shaken Rosenstein Felt Used by White House in Comey Firing," *New York Times*, June 29, 2018.

95. Andrew G. McCabe, *The Threat: How FBI Protects America in the Age of Terror and Trump* (New York: St. Martin's Press, 2019), 234, 242–43.

96. Michael S. Schmidt and Adam Goldman, "Shaken Rosenstein Felt Used by White House in Comey Firing."

97. Code of Federal Regulations, 28 C.F.R. § 45.2, "Disqualification Arising from Personal or Political Relationship."

98. Rules of Professional Conduct, Rule 1.7:—Conflict of Interest, "General Rule," https://www.dcbar.org/bar-resources/legal-ethics/amended-rules/rule1-07 .cfm.

99. Code of Federal Regulations, 28 C.F.R. § 600.7, "Conduct and Accountability."

100. Aruna Viswanatha and Del Quentin Wilber, "Special Counsel's Office Interviewed Deputy Attorney General Rod Rosenstein," *Wall Street Journal*, September 19, 2017.

101. *The Mueller Report*, 449. See footnotes 439–463, which make reference to "302," the term used to identify a summary report following an interview.

102. *The Mueller Report*, 250. See also Erica Orden, "Mueller Report Highlights Rosenstein's Role as Witness in the Investigation He Oversaw," CNN, April 18, 2019.

103. The Trump-Rosenstein conversations are footnoted cryptically in *The Mueller Report*. It is unclear whether a "Ms. Gauhar" took notes of the conversations in real time or if they were recorded and then transcribed. Throughout his report, Mueller carefully avoided stating whether Rosenstein had been interviewed by the special counsel. If the deputy attorney general was interviewed, why not say so? Was Mueller attempting to minimize Rosenstein's conflict of interest by concealing reference to an interview? These are legitimate questions that remain unanswered as of this writing.

104. Laura Jarrett, "Exclusive: Rosenstein Consulted with Ethics Adviser at DOJ on Russia Probe," CNN, April 13, 2018.

105. Sadie Gurman, Eric Tucker, and Jeff Horwitz, "Special Counsel's Trump Investigation Includes Manafort Case," Associated Press, June 2, 2017.

106. Jonathan Turley, "Rod Rosenstein Must Recuse Himself," The Hill, October 12, 2018.

107. Matt Zapotosky and Devlin Barrett, "Rosenstein-McCabe Feud Dates Back to Angry Standoff in Front of Mueller," *Washington Post*, October 10, 2018.

108. Ibid.

109. Eli Lake, "Rod Rosenstein Has Some Serious Explaining to Do," Bloomberg, February 19, 2019.

110. Matt Zapotsky and Devlin Barrett, "Rosenstein-McCabe Fued Dates Back to Angry Standoff in Front of Mueller."

111. Quinta Jurecic, "Document: Justice Department Releases Carter Page FISA Application," Lawfare, July 21, 2018.

112. Craig Bannister, "Rosenstein Testifies He Doesn't Need to Read FISA Applications He Signs," CNS News, June 29, 2018.

113. Ibid.

114. Turley, "It's High Time Rod Rosenstein Recuse Himself."

115. Joseph diGenova, "James 'Cardinal' Comey—the Man Who Destroyed the FBI," Fox News, April 12, 2019.

116. James Comey, *A Higher Loyalty: Truth, Lies, and Leadership* (New York: Flatiron Books, 2018).

117. Jurecic, "Document: Justice Department Releases Carter Page FISA Application."

118. *Washington Post* Staff, "Full Transcript: FBI Director James Comey Testifies on Russian Interference in 2016 Election," *Washington Post*, March 20, 2017.

119. "Vengeance is mine, and recompense, for the time when their foot shall slip; for the day of their calamity is at hand, and their doom comes swiftly," Book of Deuteronomy 32:35.

120. Politico Staff, "Full Text: James Comey Testimony Transcript on Trump and Russia," Politico, June 8, 2017. See also Comey, *A Higher Loyalty*, 270.

121. Gregg Jarrett, *The Russia Hoax: The Illicit Scheme to Clear Hillary Clinton and Frame Donald Trump* (New York: Broadside Books, 2018), ch. 10.

122. Federal Bureau of Investigation, Records Management Division, "FBI's Prepublication Review Policy Guide," 0792PG, June 4, 2015, 1; "FBI Employment Agreement," FD-291, available at https://www.fbi.gov/file-repository/fd-291.pdf/view; see also "Sensitive Compartmented Information NonDisclosure Agreement," Federal Bureau of Investigation (Records Related to the Dismissal of FBI Director Comey), available at https://vault.fbi.gov/records-related-to-the-dismissal-of-fbi-director-comey/Records%20Related%20to%20the%20Dismissal%20of%20FBI%20Director%20Comey%20Part%2002%20of%2002/view.

123. Alex Pappas, "Comey Denies Release of Memo Contents Was a 'Leak,' Addresses Dem-Funded Dossier in Fox News Interview," Fox News, April 26, 2018.

124. Comey, *A Higher Loyalty*, 269.

125. Ibid., 270.

126. Michael S. Schmidt, "Comey Memo Says Trump Asked Him to End Flynn Investigation," *New York Times*, May 16, 2017; David Lauter, "Trump Asked Comey to Shut Down Investigation of Flynn, *New York Times* Reports," *Los Angeles Times*, May 16, 2017.

127. Ibid.

128. Brooke Singman, "Comey Memos Reportedly Had Classified Info; Trump Says 'That Is So Illegal,'" Fox News, July 10, 2017; Jeremy Stahl, "My Theory About the Number of Memos Comey Gave His Friend Was Wrong," Slate, April 20, 2018.

129. James B. Comey, Confidential Presidential Memorandum, FBI, January 28, 2017, available at https://www.documentcloud.org/documents/4442900-Ex -FBI-Director-James-Comey-s-memos.html; Peter Kasperowicz, "James Comey Promised Trump: 'I Don't Leak,'" *Washington Examiner*, April 19, 2018.

130. Byron Tau and Aruna Viswanatha, "Justice Department Watchdog Probes Comey Memos over Classified Information," *Wall Street Journal*, April 20, 2018.

131. Brooke Singman, Twitter, April 20, 2018, https://twitter.com/brookefoxnews /status/987378859510259712.

132. Comey, *A Higher Loyalty*, 270; Daniel Chaitin, "James Comey Denies Leaking Memos, Sees 'No Credible Claim' He Broke Law," *Washington Examiner*, April 25, 2108.

133. Lata Nott, "Leaks and the Media," Freedom Forum Institute.

134. United States Code, 18 U.S.C. § 641, "Public Money, Property or Records." See also *United States v. DiGilio*, 538 F.2d 972, 978 (3d Cir. 1976); *Pfeiffer v. C.I.A.*, 60 F.3d 861, 864 (D.C. Cir. 1995).

135. FBI Employment Agreement, Including Provisions and Prohibited Disclosures, FD-291, https://www.fbi.gov/file-repository/fd-291.pdf/view. This agreement reads, in part, "I will not reveal, by any means, any information or material from or related to FBI files or any other information acquired by virtue of my official employment to any unauthorized recipient without prior official written authorization by the FBI."

136. United States Code, 18 U.S.C. § 793, "Gathering, Transmitting or Losing Defense Information"; United States Code, 18 U.S.C. § 1924, "Unauthorized Removal and Retention of Classified Documents or Material."

137. Politico Staff, "Full Text: James Comey Testimony Transcript on Trump and Russia."

138. James B. Comey, "Statement for the Record," Senate Select Committee on Intelligence, June 8, 2017, https://www.intelligence.senate.gov/sites/default/files /documents/os-jcomey-060817.pdf, 5.

139. Author's interview with President Donald J. Trump, Oval Office, White House, June 25, 2019.

140. *The Mueller Report*, 333.

141. United States Code, 18 U.S.C. § 953, "Private Correspondence with Foreign Governments" (the Logan Act).

142. There have been two indictments under the Logan Act, one in 1803 and the other in 1852. Both cases were dropped without prosecution. For the legal effi-

cacy of the act, see Jacob Frenkel, "Why Michael Flynn's Plea May Not Lead to a Logan Act Violation," *Forbes*, December 12, 2017; Dan McLaughlin, "Repeal the Logan Act," *National Review*, May 5, 2018; Michael V. Seitzinger, "Conducting Foreign Relations Without Authority: The Logan Act," Congressional Research Service, March 11, 2015. See also *Waldron v. British Petroleum Co.*, 231 Fed. Supp. 72 (1964).

143. Alex Pappas, "Comey Admits Decision to Send FBI Agents to Interview Flynn Was Not Standard," Fox News, December 13, 2018.

144. Ibid.

145. Gregg Re, "Flynn Says FBI Pushed Him Not to Have Lawyer Present During Interview," Fox News, December 12, 2018.

146. Charles E. Grassley, Chairman, United States Senate Committee on the Judiciary, letter to Rod J. Rosenstein, Deputy Attorney General, and Christopher A. Wray, Director, Federal Bureau of Investigation, May 11, 2018, https://www.judiciary.senate.gov/imo/media/doc/2018-05-11%20CEG%20to%20DOJ%20FBI%20(Flynn%20Transcript).pdf.

147. Alex Pappas, "Mueller Releases Flynn Files Showing FBI Doubts over 'Lying,' Tensions over Interview," Fox News, December 14, 2018.

148. *Washington Post* Staff, "Full Transcript: Acting FBI Director McCabe and Others Testify Before the Senate Intelligence Committee," *Washington Post*, May 11, 2017.

149. Comey, "Statement for the Record," Senate Select Committee on Intelligence, 5; Politico Staff, "Full Text: James Comey Testimony on Trump and Russia," 66.

150. James B. Comey, "Statement for the Record," 5.

151. Politico Staff, "Full Text: James Comey Testimony on Trump and Russia," 20.

152. Ibid.

153. Dershowitz, "Introduction to the Mueller Report," *The Mueller Report*, 10.

154. John Solomon, "James Comey's Next Reckoning Imminent—This Time For Leaking," The Hill, July 31, 2019.

155. Ibid., see also Brooke Singman and Jake Gibson, "DOJ Will Not Prosecute Comey For Leaking Memos After IG Referral: Sources," Fox News, August 1, 2019.

156. Seamus Bruner, *Compromised: How Money and Politics Drive FBI Corruption* (New York: Bombardier Books, 2018), 18.

157. Ibid., 147.

158. Tessa Berenson, "James Comey Says He 'Never' Leaked Information on Trump or Clinton," *Time*, May 3, 2017.

159. Tim Hains, "Comey Pushes Back Against Trump's 'Leaker' Claim in Full 'Special Report' Interview: 'He's Just Wrong,'" RealClearPolitics, April 26, 2018.

160. Peter Elkind, "James Comey's Testimony on Huma Abedin Forwarding Emails Was Inaccurate," ProPublica, May 8, 2017.

161. Comey, "Statement for the Record," Senate Select Committee on Intelligence.

162. "Department of Justice Issues Statement on Testimony of former FBI Director James Comey," Department of Justice, Office of Public Affairs, June 8, 2017.

163. Quinta Jurecic, "Document: Transcript of James Comey's Dec. 7 Interview

with House Committees," Lawfare, December 8, 2018. See also "Statement by James B. Comey, Director of the FBI, before the House Committee on Oversight and Government Reform," July 7, 2016, available at http://www.thompsontimeline.com/congressional-testimony-of-fbi-director-james-comey/.

164. Tom LoBianco, Pamela Brown, and Mary Kay Mallonee, "Comey Drafted Clinton Exoneration Before Finishing Investigation, GOP Senators Say," CNN, September 1, 2017.

165. Jeff Carlson, "Comey's Testimony: Truth or Lies?," *The Epoch Times*, December 11, 2018.

166. Tim Hains, "Comey Pushes Back Against Trump's 'Leaker' Claim in Full 'Special Report' Interview."

167. Senate Judiciary Committee, U.S. Senate, "Interview of: Glenn Simpson," August 22, 2017, https://www.feinstein.senate.gov/public/_cache/files/3/9/3974a291-ddbe-4525-9ed1-22bab43c05ae/934A3562824CACA7BB4D915E97709D2F.simpson-transcript-redacted.pdf.

168. Steve Peoples and Zeke Miller, "Neoconservative Website Washington Free Beacon Hired Fusion GPS," RealClearPolitics, October 28, 2017.

169. James Comey, "How Trump Co-Opts Leaders like Bill Barr," *New York Times*, May 1, 2019; James Comey, "No 'Treason,' No Coup, Just Lies—and Dumb Lies at That," *Washington Post*, May 28, 2019.

170. James Comey, Twitter, March 24, 2019.

171. James Comey, Twitter, March 23, 2019.

172. Kevin R. Brock, "James Comey Is in Trouble and Knows It," The Hill, May 7, 2019.

173. William McGurn, "The Tales of Parson Comey," *Wall Street Journal*, May 20, 2019.

174. David Alexander and Eric Beech, "Ex–FBI Chief Comey Says Trump Undermines Rule of Law with 'Lies,'" Reuters, December 17, 2018.

175. Catherine Herridge and Cyd Upson, "Lisa Page Testimony: Collusion Still Unproven by Time of Mueller's Special Counsel Appointment," Fox News, September 16, 2018.

CHAPTER 5: THE FOLLY OF MUELLER'S MAGNUM OPUS

1. Code of Federal Regulations, 28 C.F.R. § 600.8, "Notification and Reports by the Special Counsel."

2. Robert S. Mueller, *The Mueller Report: The Final Report of the Special Counsel into Donald Trump, Russia, and Collusion as Issued by the Department of Justice* (New York: Skyhorse Publishing, 2019), 39.

3. Ibid., 195.

4. "Read Attorney General William Barr's Written Testimony Before Senate Judiciary Committee," ABC News, May 1, 2019; full testimony of William Barr available at https://www.judiciary.senate.gov/meetings/the-department-of-justices-investigation-of-russian-interference-with-the-2016-presidential-election.

5. Ibid; see also Margot Cleveland, "Robert Mueller's 10 Most Egregious Missteps During Anti-Trump Russia Probe," The Federalist, May 8, 2019.

6. Ibid.

7. *The Mueller Report*, 194–95.

8. "Rep. Schiff on MSNBC: Obstruction Evidence Laid Out by Mueller Is Damning Enough," YouTube, April 19, 2019.

9. Constitution of the United States, Article II, Section 4.

10. William P. Barr, Attorney General, letter to the Chairmen and Ranking Members of the House and Senate Judiciary Committees, March 24, 2019, https:// assets.documentcloud.org/documents/5779688/AG-March-24-2019-Letter -to-House-and-Senate.pdf.

11. Credit should be given to Andrew C. McCarthy for employing the apt description "magnum opus," which I first heard in a conversation with him. He also used it in a published column, Andrew C. McCarthy, "The Mueller Report Vindicates Bill Barr," *National Review*, April 19, 2019.

12. Alan Dershowitz, "Introduction to the Mueller Report," *The Mueller Report: The Final Report of the Special Counsel into Donald Trump, Russia, and Collusion as Issued by the Department of Justice* (New York: Skyhorse Publishing, 2019), 2–3.

13. Barr, letter to the Chairmen and Ranking Members of the House and Senate Judiciary Committees, March 24, 2019.

14. Ibid.

15. Ibid.

16. Giles Snyder and Brian Naylor, "More 2020 Democrats Call for Impeachment Proceedings Against President Trump," NPR, April 23, 2019.

17. Jason Sattler, "2020 Litmus Test: All Democratic Candidates Should Call for Trump Impeachment Proceedings," *USA Today*, May 1, 2019.

18. Nicholas Fandos, "Pelosi Urges Caution on Impeachment as Some Democrats Push to Begin," *New York Times*, April 22, 2019; Kyle Cheney, Heather Caygle, and Andrew Desiderio, "Pelosi Beats Back Calls for Trump Impeachment," Politico, April 22, 2019.

19. Emmet T. Flood, Special Counsel to the President, letter to William P. Barr, Attorney General, April 19, 2019, https://assets.documentcloud.org/documents /5986068/WHSC-to-AG-4-19-19.pdf, 2.

20. *The Mueller Report*, 204.

21. Ibid., 195, 204, and 347.

22. Flood, letter to William P. Barr, April 19, 2019, 2.

23. "Full Transcript: Mueller Testimony Before House Judiciary, Intelligence Committees," NBC News, July 24, 2019, see questioning by Rep. John Ratcliffe, available at https://www.nbcnews.com/politics/congress/full-transcript-robert -mueller-house-committee-testimony-n1033216.

24. Testimony of Attorney General William Barr, Senate Judiciary Committee, May 1, 2019.

25. Ibid.

26. Flood, letter to William P. Barr, April 19, 2019, 2.

27. *The Mueller Report*, 195.

28. American Bar Association, Rule 3.8(f), "Special Responsibilities of a Prosecutor," https://www.americanbar.org/groups/professional_responsibility/publications /model_rules_of_professional_conduct/rule_3_8_special_responsibilities_of _a_prosecutor/.

29. Ibid.

30. Adam Mill, "5 Times the Mueller Probe Broke Prosecutorial Rules That Ensure Justice," The Federalist, April 25, 2019. (Adam Mill is a pen name for an attorney in Kansas City, Missouri.)

31. Rod J. Rosenstein, Acting Attorney General, "Appointment of Special Counsel to Investigate Russian Interference with the 2016 Presidential Election and Related Matters," Order No. 3915-2017, May 17, 2017, https://www.documentcloud .org/documents/3726408-Rosenstein-letter-appointing-Mueller-special.html.

32. "Full Transcript: Mueller Testimony Before House Judiciary, Intelligence Committees," see questioning by Rep. Steve Chabot.

33. Ibid.

34. Chuck Grassley, "Prepared Floor Statement by U.S. Senator Chuck Grassley of Iowa, Senior Member and Former Chairman, U.S. Senate Judiciary Committee, on the Mueller Investigation," May 6, 2019, https://www.grassley.senate .gov/news/news-releases/grassley-mueller-investigation.

35. John Solomon, "Key Figure That Mueller Report Linked to Russia Was a State Department Intel Source," The Hill, June 6, 2019.

36. Andrew C. McCarthy, "Mueller's Preposterous Rationale for Tainting the President with 'Obstruction' Allegations," *National Review*, May 8, 2019.

37. Brie Stimson, "Devin Nunes Calls 'Fraud,' Citing Difference Between Mueller Report, Dowd Transcript," Fox News, June 1, 2019.

38. Code of Federal Regulations, 28 C.F.R. § 600.7, "Conduct and Accountability." See also Code of Federal Regulations, 28 C.F.R. § 45.2, "Disqualification Arising from Personal or Political Relationship."

39. 28 C.F.R. § 45.2.

40. Peter Holley, "Brothers in Arms: The Long Friendship Between Mueller and Comey," *Washington Post*, May 17, 2017; Garrett M. Graff, "Forged Under Fire—Bob Mueller and James Comey's Unusual Friendship," *Washingtonian*, May 30, 2013.

41. United States Code, 28 U.S.C. § 528, "Disqualification of Officers and Employees of the Department of Justice." See also Ronald D. Rotunda, "Alleged Conflicts of Interest Because of 'Appearance of Impropriety,'" Chapman University, Fowler School of Law, 2005.

42. Politico Staff, "Full Text: James Comey Testimony Transcript on Trump and Russia."

43. 28 C.F.R. § 45.2.

44. Eric Felten, "Does Robert Mueller Have a Conflict of Interest?," *The Weekly Standard*, July 5, 2018.

45. Josh Gerstein, "Justice Department Won't Disclose Details on Mueller Ethics Waiver," Politico, December 12, 2017.

46. Code of Federal Regulations, 28 C.F.R. § 600.1, "Grounds for Appointing a Special Counsel."

47. Carrie Johnson, "Special Counsel Robert Mueller Had Been on White House Shortlist to Run FBI," NPR, June 9, 2017; Dan Merica, "Trump Interviewed Mueller for FBI Job Day Before He Was Tapped for Special Counsel," CNN, June 13, 2017.

48. Jonathan Turley, "The Special Counsel Investigation Needs Attorneys Without Conflicts," The Hill, December 8, 2017.

49. American Bar Association, Rule 3.7, "Lawyer as Witness," https://www.amer icanbar.org/groups/professional_responsibility/publications/model_rules_of _professional_conduct/rule_3_7_lawyer_as_witness/.

50. *The Mueller Report*, page 259.

51. Author's interview with President Donald J. Trump, Oval Office, White House, June 25, 2019.

52. Author's discussion with Madeleine Westerhout, director of Oval Office Operations at the White House (previously personal secretary to the president), Oval Office, White House, June 25, 2019.

53. "Full Transcript: Mueller Testimony Before House Judiciary, Intelligence Committees," see questioning by Rep. Louie Gohmert.

54. Ibid.

55. Author's interview with President Donald J. Trump, Bedminster, New Jersey, September 17, 2017; author's interview with President Donald J. Trump, Oval Office, White House, June 25, 2019.

56. Ibid.

57. Ibid.

58. *The Mueller Report*, page 452.

59. Author's interview with President Donald J. Trump, Oval Office, White House, June 25, 2019.

60. Brooke Singman, "Special Counsel Mueller's Team Has Only One Known Republican," Fox News, February 23, 2018; David Sivak, "Exclusive: Not a Single Lawyer Known to Work for Mueller Is a Republican," Daily Caller, February 21, 2018.

61. Singman, "Special Counsel Mueller's Team Has Only One Known Republican."

62. "Robert S. Mueller III (1990–1993)," United States Department of Justice, https://www.justice.gov/criminal/history/assistant-attorneys-general/robert -s-mueller; CNN Wire Staff, "Obama Requests 2 More Years for FBI Chief," CNN, May 13, 2011.

63. Brooke Singman, "More Clinton Ties on Mueller Team: One Deputy Attended Clinton Party, Another Rep'd Top Aide," Fox News, December 8, 2017.

64. Peter Nicholas, Aruna Viswanatha, and Erica Orden, "Trump's Allies Urge Harder Line as Mueller Probe Heats Up," *Wall Street Journal*, December 8, 2017; David Shortell, "Mueller Attorney Praised Yates as DOJ Official Emails Show," CNN, December 5, 2017.

65. Sidney Powell, "Meet the Very Shady Prosecutor Robert Mueller Has Hired for the Russia Investigation," Daily Caller, November 20, 2017; Sidney Powell,

"Political Prosecution: Mueller's Hit Squad Covered for Clinton and Persecutes Trump Associates," Daily Caller, December 6, 2017; Sidney Powell, "In Andrew Weissmann, the DOJ Makes a Stunningly Bad Choice for Crucial Role," Observer, January 12, 2015.

66. Sidney Powell, "Judging by Mueller's Staffing Choices, He May Not Be Very Interested in Justice," The Hill, October 19, 2017.

67. Josh Gerstein, "Details Emerge on Justice Department Meeting with Reporters on Manafort," Politico, July 8, 2018; Chuck Ross, "Mueller's 'Pit Bull' Arranged Meeting with Reporters to Discuss Manafort Investigation," Daily Caller, July 8, 2018; Sara Carter, "New Texts Reveal FBI Leaked Information to the Press to Damage Trump," September 10, 2018, https://saraacarter.com/new-texts-reveal-fbi-leaked-information-to-the-press-to-damage-trump/.

68. Author's interview with John Dowd, lawyer for President Trump, June 13, 2019.

69. Press Room, "Judicial Watch Obtains DOJ Documents Showing Andrew Weissmann Leading Hiring Effort for Mueller Special Counsel," Judicial Watch, May 14, 2019.

70. Author's interview with Rudy Giuliani, lawyer for President Trump, July 10, 2019.

71. Singman, "More Clinton Ties on Mueller Team."

72. Brooke Singman, "Top Mueller Investigator's Democratic Ties Raise New Bias Questions," Fox News, December 7, 2017.

73. "Full Transcript: Mueller Testimony Before House Judiciary, Intelligence Committees," see questioning by Rep. Kelly Armstrong.

74. Singman, "More Clinton Ties on Mueller Team."

75. Adam Mill, "How Long Has Robert Mueller Been Like This?," The Federalist, July 25, 2019.

76. Josh Gerstein, "Another Prosecutor Joins Trump-Russia Probe," Politico, September 15, 2017.

77. Ibid.

78. Byron York, "Key Justice Department Officials, Including Mueller Deputy, Knew About Dossier," Washington Examiner, January 17, 2019.

79. Gregg Jarrett, "Mueller's Team Knew 'Dossier' Kicking Off Trump Investigation Was Biased and Defective," Fox News, January 17, 2019.

80. Brooke Singman, "McCabe Says He 'Made the Decision' to Remove Strzok from Mueller Team: Transcript," Fox News, May 21, 2019.

81. Bill House, "House Republicans Prepare Contempt Action Against FBI & DOJ," Bloomberg, December 3, 2017; Julia Reinstein, "Mueller Removed an FBI Agent from His Russia Probe Due to Alleged Anti-Trump Texts," BuzzFeed News, December 3, 2017.

82. Jerry Dunleavy and Daniel Chaitin, "Peter Strzok Said Mueller Never Asked if Anti-Trump Bias Influenced Russia Investigation Decisions," Washington Examiner, March 14, 2019.

83. Josh Delk, "GOP Senator: Mueller 'Needs to Clean House of Partisans,'" The

Hill, December 16, 2017; Ken Dilanian, "Republicans Step Up Attacks on Special Counsel Robert Mueller," NBC News, December 13, 2017.

84. Deroy Murdock, "The Mueller Probe: A Year-Old Hyper-partisan Circus," *National Review*, May 18, 2018.

85. Ned Ryun, "Mueller Has a Partisan Pack of Wolves and an Illegitimate Investigation," The Hill, December 17, 2017.

86. Olivia Victoria Gazis, "House Intel Dems Slam GOP Draft Report on Russia Probe," CBS News, March 13, 2018.

87. Camilo Montoya-Galvez, "Schiff Says Trump Faces 'Real Prospect of Jail Time' After Leaving Office," CBS News, December 9, 2018.

88. Norman Eisen, tweet, February 27, 2018, https://twitter.com/NormEisen /status/968473271304941568.

89. Representative Jerrold Nadler, CNN, November 30, 2018; Representative Eric Swalwell, CNN, March 16, 2018; Senator Richard Blumenthal, MSNBC, October 17, 2018; Senator Ron Wyden, CNN, December 14, 2017; Tom Perez, NBC, "Meet the Press," April 22, 2018; Representative Maxine Waters, Town Hall, Black Congressional Caucus Foundation, September 21, 2017; former GOP representative David Jolly, MSNBC, May 17, 2017. See also Tim Murtaugh, Director of Communications, Donald J. Trump for President, Inc., memorandum to Television Producers re "Credibility of Certain Guests," March 25, 2019, https://www.scribd.com/document/403100260/March-25 -2019-Tim-Murtaugh-Trump-Campaign-Memo-to-TV-Producers.

90. Aiden McLaughlin, "Former Acting Solicitor General Neal Katyal: Trump's Future Looks Like It's Behind Bars," Mediaite, December 10, 2018.

91. "Neal Katyal: Mueller's Report Is 'The Beginning of the End,'" *The Late Show with Stephen Colbert*, YouTube, April 19, 2019, https://www.youtube.com /watch?v=49V9_JE4S7A.

92. Zack Beauchamp, "Legal Experts Say Donald Trump Jr. Has Just Confessed to a Federal Crime," Vox, July 11, 2017.

93. "Hardball with Chris Matthews: Trump Defends Son, Transcript 7/17/17 Draws Contrast with Clinton," MSNBC, July 17, 2017. See also Jonathan Turley, "Mueller's End: A Conclusion on Collusion, but Confusion on Obstruction," The Hill, March 24, 2019.

94. Rachel Maddow, MSNBC, January 17, 2017; Rachel Maddow, MSNBC, March 17, 2017; Nicolle Wallace, MSNBC and NBC, July 6, 2017; Mika Brzezinski, MSNBC, March 6, 2017; Mika Brzezinski, MSNBC, December 5, 2017; Lawrence O'Donnell, MSNBC, June 5, 2017; Donny Deutsch, MSNBC, May 12, 2017; Donny Deutsch, MSNBC, July 17, 2018; Carl Bernstein, CNN, July 17, 2018; Anderson Cooper, CNN, November 15, 2018; Joy Reid, MSNBC, April 8, 2018.

95. Gregg Jarrett, *The Russia Hoax: The Illicit Scheme to Clear Hillary Clinton and Frame Donald Trump* (New York: Broadside Books, 2018), xii, 104.

96. *The Mueller Report*, 167.

97. Asha Rangappa, "How Barr and Trump Use a Russian Disinformation Tactic," *New York Times*, April 19, 2019.

98. Jarrett, *The Russia Hoax*, 105–06, 173–79.

99. Kimberley A. Strassel, "Mueller's Report Speaks Volumes," *Wall Street Journal*, April 18, 2019.

100. *The Mueller Report*, 41; see also 79–80.

101. Noah Bierman, "Donald Trump Invites Russia to Hack Hillary Clinton's Emails," *Los Angeles Times*, July 27, 2016.

102. James B. Comey, "Statement by FBI Director James B. Comey on the Investigation of Secretary Hillary Clinton's Use of a Personal E-Mail System," FBI National Press Office, July 5, 2016, https://www.fbi.gov/news/pressrel /press-releases/statement-by-fbi-director-james-b-comey-on-the-investigation -of-secretary-hillary-clinton2019s-use-of-a-personal-e-mail-system.

103. Fred Wertheimer and Norman Eisen, "Trump Illegally Asked Russia to Help Him Win in 2016. He Shouldn't Get Away with It," *USA Today*, January 2, 2019.

104. *The Mueller Report*, 42; see also 101–02.

105. Ibid., 105–06.

106. Ibid., 106.

107. Gregg Jarrett, interview with Rudy Giuliani, July 10, 2019.

108. Nolan D. McCaskill, Alex Isenstadt, and Shane Goldmacher, "Paul Manafort Resigns from Trump Campaign," Politico, August 19, 2016.

109. *The Mueller Report*, 132.

110. Sharon LaFraniere, Kenneth P. Vogel, and Maggie Haberman, "Manafort Accused of Sharing Trump Polling Data with Russian Associate," *New York Times*, January 8, 2019.

111. Matt Naham, "Fmr Fed Prosecutor: It Doesn't Get Much More 'Collusive' than Manafort Sharing Polling Data with Kilimnik," Law & Crime, January 9, 2019.

112. Alan Cullison and Dustin Volz, "Mueller Report Dismisses Many Steele Dossier Claims," *Wall Street Journal*, April 19, 2019.

113. "Company Intelligence Report 2016/080," December 13, 2016, 9, 30, https:// www.documentcloud.org/documents/3259984-Trump-Intelligence-Allegations .html.

114. *The Mueller Report*, 87–92.

115. Ibid., 87.

116. Joe Becker, Matt Apuzzo, and Adam Goldman, "Trump Team Met with Lawyer Linked to Kremlin During Campaign," *New York Times*, July 8, 2017.

117. Tom Porter, "Donald Trump Jr. 'Treason' Emails Prove Russian Collusion: Tim Kaine," *Newsweek*, July 11, 2017; Rachel Stockman, " 'This is Treason': Some Legal Experts Say Trump Jr.'s Clinton Dirt/Russia Meeting Was Illegal," Law & Crime, July 9, 2017.

118. United States Code, 18 U.S.C. § 371, "Conspiracy to Commit Offense or to Defraud United States."

119. *Hass v. Henkel*, 216 U.S. 462 (1910); *Hammerschmidt v. United States*, 265 U.S. 182 (1024); see also "923. 18 U.S.C. § 371—Conspiracy to Defraud the United States," United States Department of Justice, https://www.justice.gov/jm /criminal-resource-manual-923-18-usc-371-conspiracy-defraud-us.

120. *The Mueller Report*, 175.

121. Ibid., 168.

122. United States Code, 18 U.S.C. § 1349, et al., "Attempt and Conspiracy"; *The Mueller Report*, 168.

123. Karoun Demirjaian, "Pelosi Suggests Trump Surrogates Violated Law as Members Try to Force Votes on Matters Related to Russia Probe," *Washington Post*, July 14, 2017.

124. Federal Election Commission, "Volunteer Activity," https://transition.fec.gov/pages/brochures/volact.shtml.

125. United States Code, 52 U.S.C. § 30121, "Contributions and Donations by Foreign Nationals."

126. Code of Federal Regulations, 11 C.F.R. § 110.20, "Prohibition on Contributions, Donations, Expenditures, Independent Expenditures, and Disbursements by Foreign Nationals."

127. Code of Federal Regulations, 11 C.F.R. § 100.74, "Uncompensated Services by Volunteers." See also Federal Election Commission, "Volunteer Activity," available at https://www.fec.gov/help-candidates-and-committees/candidate-taking-receipts/volunteer-activity/.

128. *The Mueller Report*, 175.

129. Ibid.

130. United States Code, 52 U.S.C. § 30109(d)(1)(A), "Federal Election Campaign Act: Enforcement"; Andy Grewal, "If Trump Jr. Didn't Know Campaign Finance Law, He Won't Be Prosecuted," *Yale Journal on Regulation*, July 16, 2017.

131. *The Mueller Report*, 175–76.

132. Byron York, "How Pundits Got Key Part of Trump-Russia Story All Wrong," *Washington Examiner*, March 18, 2017.

133. Byron York, "The Personal Cost of the Trump-Russia Investigation," *Washington Examiner*, April 23, 2019.

134. *The Mueller Report*, 128–30.

135. Kevin G. Hall, "Mueller Report States Cohen Was Not in Prague. It Is Silent on Whether a Cohen Device Pinged There," McClatchy, April 18, 2019.

136. *The Mueller Report*, 52–62.

137. Ibid., 127.

138. Ibid., 159–63, 181–82.

139. Jan Crawford, "William Barr Interview: Read the Full Transcript," CBS News, May 31, 2019.

140. Author's interview with John Dowd, June 13, 2019.

141. Ibid.

142. Byron York, "When Did Mueller Know There Was No Collusion?," *Washington Examiner*, April 26, 2019.

143. Andrew McCarthy, "How Long Has Mueller Known There Was No Trump-Russia Collusion?," Fox News, March 26, 2019.

144. Author's interview with John Dowd, June 13, 2019.

145. Ibid.

146. "The Trump Legal Team's Jan. 29, 2018, Confidential Memo to Mueller," at

"The Trump Lawyers' Confidential Memo to Mueller, Explained," June 2, 2018.

147. Author's interview with John Dowd, June 13, 2019.

148. Author's interview with Jay Sekulow, attorney for President Trump, June 19, 2019.

149. Author's interview with John Dowd, June 13, 2019.

150. Ibid.

151. *In re: Sealed Case (Espy)*, 121 F.3d at 756 (United States Court of Appeals, District of Columbia Circuit), 1997.

152. *The Mueller Report.* See also Associated Press, "Full Text of Trump's Answers to Mueller's Questions," *Chicago Tribune*, April 18, 2019.

153. "Mueller Appears to Have Edited Voicemail Transcript Between John Dowd and Flynn Lawyer," Fox News, June 3, 2019.

154. Ibid.

155. Author's interview with John Dowd, June 13, 2019.

156. Author's interview with Rudy Giuliani, July 10, 2019.

157. "Transcript: Former Trump Attorney John Dowd's Interview on ABC News' the Investigation Podcast," ABC News, February 12, 2019, 5.

158. Ibid.

159. Constitution of the United States of America, Article II, Section 2 (appointments and pardons); Article II, Section 3 ("take care" clause).

160. Constitution of the United States of America, Article II, Section 3, Clause 4.

161. Alan Dershowitz, "Trump Well Within Constitutional Authority on Comey, Flynn—Would This Even Be a Question if Hillary Were President?," Fox News, June 12, 2017.

162. Constitution of the United States of America, Article II, Section 1, Clause 1.

163. *United States v. Nixon*, 418 U.S. 683, 693 (1974); *United States v. Fokker Servs., B.V.*, 818 F. 3d 733, 737 (D.C. Cir. 2016); *Wayte v. United States*, 470 U.S. 598, 607 (1985).

164. *United States v. Armstrong*, 517 U.S. 456, 464 (1996). See also Henry L. Chambers, Jr., "The President, Prosecutorial Discretion, Obstruction of Justice, and Congress," *University of Richmond Law Review, Symposium Book*, vol. 52, February 26, 2018.

165. "Constitutionality of Legislation Extending the Term of the FBI Director," United States Department of Justice, June 20, 2011, https://www.justice.gov/file/18356/download.

166. Joseph Hinks, "Read Former FBI Director James Comey's Farewell Letter to Colleagues," *Time*, May 11, 2017.

167. Testimony of Attorney General William Barr, Senate Judiciary Committee, May 1, 2019.

168. Alan Dershowitz, "Introduction to the Mueller Report," 10.

169. Ibid.

170. "This Day in History: Ford Explains His Pardon to Congress," History, November 16, 2009, updated July 7, 2019, https://www.history.com/this-day-in-history/ford-explains-his-pardon-of-nixon-to-congress.

171. Eric Lichtblau and Davan Maharaj, "Clinton Pardon of Rich a Saga of Power,
      Money," *Los Angeles Times*, February 18, 2001.

172. *The Mueller Report*, 333.

173. *Franklin v. Massachusetts*, 502 U.S. 799 (1992); *Public Citizen v. Department of
      Justice*, 491 U.S. 440 (1989).

174. Jack Goldsmith, "The Mueller Report's Weak Statutory Interpretation Analy-
      sis," Lawfare, May 11, 2019.

175. "The Constitutional Separation of Powers Between the President and Con-
      gress," Department of Justice, May 7, 1996, https://www.justice.gov/file/20061
      /download, 178.

176. *The Mueller Report*, 323.

177. Ibid., 335; Dershowitz, "Introduction to the Mueller Report," 11.

178. Office of Legal Counsel, "A Sitting President's Amenability to Indictment and
      Criminal Prosecution," Department of Justice, 1973 (updated on October 16,
      2000, and December 10, 2018), https://www.justice.gov/sites/default/files/olc
      /opinions/2000/10/31/op-olc-v024-p0222_0.pdf.

179. *The Mueller Report*, 194.

180. Robert Mueller, "Full Transcript of Mueller's Statement on Russia Investiga-
      tion," *New York Times*, May 29, 2019.

181. Testimony of Attorney General William Barr, Senate Judiciary Committee,
      May 1, 2019. See also Meg Wagner, Veronica Rocha, Brian Ries, and Amanda
      Wills, "William Barr Testifies on the Mueller Report," CNN, May 1, 2019;
      Alan Neuhauser, "Barr Surprised Mueller Didn't Decide on Obstruction," *U.S.
      News & World Report*, May 1, 2019.

182. Politico Staff, "Transcript: Attorney General William Barr's Press Conference
      Remarks Ahead of Mueller Report Release," Politico, April 18, 2019; Zachary
      Basu, "Transcript: Bill Barr Answers Questions About Mueller Report," Axios,
      April 18, 2019.

183. Testimony of Attorney General William Barr, Senate Judiciary Committee,
      May 1, 2019. According to Barr, "He [Mueller] said that in the future the facts
      of a case against a president might be such that a special counsel would recom-
      mend abandoning the OLC opinion, but this is not such a case."

184. Mueller, "Full Transcript of Mueller's Statement on Russia Investigation."

185. Ibid.

186. Jonathan Turley, "Mueller Must Testify Publicly to Answer Three Critical
      Questions," The Hill, June 1, 2019.

187. "Full Text of the Starr Report," submitted by the Office of the Independent
      Counsel, reprinted by *Washington Post*, September 9, 1998.

188. Ibid.; see also Joseph diGenova, "Mueller Wants Americans to Believe Trump Is
      a Criminal and It's Up to Congress to Impeach Him," Fox News, May 29, 2019.

189. Mueller, "Full Transcript of Mueller's Statement on Russia Investigation."

190. Jonathan Turley, "Robert Mueller 'No Questions' Routine Is Absolute Non-
      sense," The Hill, May 30, 2019.

191. Ibid.

192. Crawford, "William Barr Interview: Read the Full Transcript."

193. William Barr, Attorney General, letter to Lindsey Graham, Chairman, Committee on the Judiciary, March 24, 2019. https://www.nytimes.com/interactive/2019/03/24/us/politics/barr-letter-mueller-report.html.

194. Ibid.

195. "Full Transcript: Mueller Testimony Before House Judiciary, Intelligence Committees," see questioning by Rep. Ted Lieu.

196. Ibid., see opening statement by Mueller, afternoon session.

197. Ibid.

198. United States Code, 18 U.S.C. §§ 1503, 1505, 1512, 1515. See also *The Mueller Report*, 204–08.

199. United States Code, 18 U.S.C. §§ 1505, 1512.

200. United States Code, 18 U.S.C. § 1512(b).

201. United States Code, §§ 1505, 1515; *United States v. Richardson*, 676 F. 3d 491, 508 (5th Cir. 2012); *United States v. Gordon*, 710 F. 3d 1124, 1151 (10th Cir. 2013).

202. *Arthur Andersen LLP v. United States*, 544 U.S. 696 and 705–706 (2005).

203. "Attorney General Confirmation Hearing, Day 1," C-SPAN, January 15, 2019. See also Matt Zapotosky, Carol D. Leonnig, Rosalind S. Helderman, and Devlin Barrett, "Mueller Report to Be Lightly Redacted," *Washington Post*, April 17, 2019.

204. Politico Staff, "Full Text: James Comey Testimony on Trump and Russia," Politico, June 8, 2017.

205. Ibid.

206. Michael S. Schmidt, "Comey Memo Says Trump Asked Him to End Flynn Investigation," *New York Times*, May 16, 2017; David Lauter, "Trump Asked Comey to Shut Down Investigation of Flynn, *New York Times* Reports," *Los Angeles Times*, May 16, 2017.

207. *The Mueller Report*, 333.

208. Politico Staff, "Full Text: James Comey Testimony on Trump and Russia," 66.

209. Ibid., 20.

210. *Washington Post* Staff, "Read Full Testimony of FBI Director James Comey in Which He Discusses Clinton Email Investigation," *Washington Post*, May 3, 2017.

211. Alex Pappas, "Mueller Releases Flynn Files Showing FBI Doubts over 'Lying,' Tensions over Interview," Fox News, December 14, 2018.

212. Confidential Memo to Mueller from Trump Legal Team, January 29, 2018, 13, available at https://www.nytimes.com/interactive/2018/06/02/us/politics/trump-legal-documents.html.

213. Ibid., 14.

214. *In re Aiken County*, 725 F.3d 255, 262–66, U.S. Court of Appeals, District of Columbia Circuit (August 13, 2013); see also *United States. v. Nixon*, 418 U.S. 683 (1974); *United States v. Goodwin*, 457 U.S. 368 (1982); *Community for Creative Non-Violence v. Pierce*, 786 F.2d 1199, 1201, D.C. Circuit Court (1986).

215. *The Mueller Report*, 299.

216. Ibid., 239.

217. Ibid., 239–40.

218. Ibid., 243.
219. Rod J. Rosenstein, memorandum for the Attorney General re "Restoring Public Confidence in the FBI," May 9, 2017, https://www.realclearpolitics.com/docs /Rosenstein_Memo.pdf.
220. Jeff Sessions, Office of the Attorney General, Correspondence with President Donald Trump, May 9, 2017, available at http://apps.washingtonpost.com /g/documents/politics/fbi-director-james-b-comeys-termination-letters-from -the-white-house-attorney-general/2430/.
221. William Cummings, "Full Text of Trump's Letter Telling Comey He's Fired," *USA Today*, May 9, 2017.
222. *The Mueller Report*, 256.
223. Ibid., 255.
224. Tim Haines, "President Trump's Full Interview with Lester Holt: Firing of James Comey," RealClearPolitics, May 11, 2017.
225. *The Mueller Report*, 4, 5.
226. Ibid., 243.
227. Ibid., 253–54.
228. Ibid., 256.
229. Ibid., 280.
230. Eliza Relman, "Jeff Sessions Explains Why He Recused Himself from Trump Campaign–Related Investigations," Business Insider, June 13, 2017.
231. Fox News Staff, "Transcript: What Is Trump's Endgame with Sessions?," Fox News, July 25, 2017.
232. *The Mueller Report*, 283.
233. Ibid., 272.
234. Ibid., 257.
235. Ibid.
236. Ibid.
237. Ibid., 264.
238. Ibid., 305.
239. Ibid., 316.
240. Ibid., 320.
241. Ibid., 321.
242. Ibid.
243. Ibid., 322.
244. Robert Charles, "Mueller's Been Showboating—Time for Senate Judiciary Committee to Ask Him These Questions," Fox News, May 31, 2019.
245. Louie Gohmert, "Robert Mueller: Unmasked," April 26, 2018, https:// www.hannity.com/wp-content/uploads/2018/04/Gohmert_Mueller_UN MASKED.pdf, 1.
246. Ibid.

CHAPTER 6: THE MEDIA WITCH HUNT

1. David Kramer, deposition, *Aleksej Gubarev et al. v. BuzzFeed, Inc, et al.*, United States District Court for the Southern District of Florida, 17-cv-60426-UU,

December 13, 2017, 22, available at https://www.scribd.com/document/401 932342/Kramer-Deposition#from_embed. The description of Kramer's actions in regard to the Steele Dossier is taken from this deposition.

2. Eric Edelman and David J. Kramer, "How Trump's Victory Could Give Russia Another Win," Politico, November 16, 2016.

3. Allegra Kirkland, "GOP National Security Experts Ask Congress to Investigate DNC Hack," Talking Points Memo, July 28, 2016.

4. David E. Sanger and Maggie Haberman, "50 G.O.P. Officials Warn Donald Trump Would Put Nation's Security at Risk," New York Times, August 8, 2016.

5. Brooke Singman, "FISA Memo: Steele Fired as an FBI Source for Breaking 'Cardinal Rule'—Leaking to the Media," Fox News, February 2, 2018.

6. John Solomon, "FBI's Steele Story Falls Apart: False Intel and Media Contacts Were Flagged Before FISA," The Hill, May 9, 2019.

7. Michael Isikoff, "U.S. Intel Officials Probe Ties Between Trump Adviser and Kremlin," Yahoo! News, September 23, 2016.

8. Julia Ioffe, "Who Is Carter Page? The Mystery of Trump's Man in Moscow," Politico, September 23, 2016.

9. David Corn, "A Veteran Spy Has Given the FBI Information Alleging a Russian Operation to Cultivate Donald Trump," Mother Jones, October 31, 2016.

10. Chuck Ross, "Ex–FBI Official: Fusion GPS Founder Tried to 'Elevate' Dossier by Spreading It All Around Washington," Daily Caller, March 22, 2019.

11. Evan Perez, Jim Sciutto, Jake Tapper, and Carl Bernstein, "Intel Chiefs Presented Trump with Claims of Russian Efforts to Compromise Him," CNN, January 12, 2017.

12. Ken Bensinger, Miriam Elder, and Mark Schoofs, "These Reports Allege Trump Has Deep Ties to Russia," BuzzFeed, January 10, 2017.

13. Jim Rutenberg, "BuzzFeed News in Limbo Land," New York Times, January 20, 2019.

14. Chuck Ross, "Emails: Jake Tapper Tore Into 'Irresponsible' BuzzFeed Editor for Publishing the Steele Dossier," Daily Caller, February 8, 2019.

15. Chuck Ross and Peter Hasson, "CNN's Undisclosed Ties to Fusion GPS," Daily Caller, October 28, 2017.

16. Ross, "Emails."

17. Sean Illing, "Did the Media Botch the Russia Story? A Conversation with Matt Taibbi," Vox, April 1, 2019.

18. Sean Davis, "A Catastrophic Media Failure," Wall Street Journal, March 26, 2019.

19. Paul Farhi, "The Washington Post Wins 2 Pulitzer Prices for Reporting on Russian Interference and the Senate Race in Alabama," Washington Post, April 14, 2018.

20. Greg Norman, "Media's Trump–Russian Collusion Coverage Is the 'Worst Journalistic Debacle of My Lifetime': Brit Hume," Fox News, March 25, 2019.

21. Michael Goodwin, "New York Times Reporter Broke the Biggest Rule in Journalism," New York Post, June 9, 2018.

22. Lisa de Moraes, "Tom Brokaw Declares 'We're at War' After Donald Trump Tweets Another 'Fox & Friends' Love Letter," Deadline, December 21, 2017.

23. Hadas Gold, "Survey: 7 Percent of Reporters Identify as Republicans," Politico, May 6, 2014.

24. Examiner Staff Writer, "Obama, Democrats Got 88 Percent of 2008 Contributions by TV Network Execs, Writers, Reporters," *Washington Examiner*, August 28, 2010.

25. "Exhibit 1-1: The Media Elite," Media Research Center, https://www.mrc.org/media-bias-101/exhibit-1-1-media-elite.

26. "Journalists Denying Liberal Bias, Part One," Media Research Center, https://www.mrc.org/media-bias-101/journalists-denying-liberal-bias-part-one.

27. Ibid.

28. Ibid.

29. Chuck Ross and Peter Hasson, "CNN's Undisclosed Ties to Fusion GPS," Daily Caller, October 28, 2017.

30. Mollie Hemingway, "Politico Hides Fusion GPS Employment of Key Source," The Federalist, November 1, 2017.

31. Editorial Board, "The Rest of the Russia Story," *Wall Street Journal*, October 15, 2017.

32. Jason Schwartz, "Murdoch-Owned Outlets Bash Mueller, Seemingly in Unison," Politico, October 30, 2017.

33. Art Swift, "Americans' Trust in Mass Media Sinks to New Low," Gallup, September 14, 2016.

34. John Kass, "Media Has Been Largely Negative on Trump," *Chicago Tribune*, May 19, 2017.

35. Jennifer Harper, "Media Bias Continues: 90% of Trump Coverage in Last Three Months Has Been Negative, Study Says," *Washington Times*, December 12, 2017.

36. Rich Noyes, "Bias: 1,000 Minutes for Trump/Russia 'Collusion' vs. 20 Seconds for Hillary/Russia Scandal," Media Research Center, NewsBusters, October 25, 2017.

37. Jim Rutenberg, "Trump Is Testing the Norms of Objectivity in Journalism," *New York Times*, August 7, 2016.

38. Ibid.

39. Ibid.

40. Justin Baragona, "James Clapper Questions Trump's 'Fitness' for Office: 'I Just Find This Extremely Disturbing,'" Mediaite, August 23, 2017.

41. Madeline Osburn, "4 Different Lies James Clapper Told About Lying to Congress," The Federalist, March 6, 2019.

42. Jonathan S. Tobin, "When Unfounded Smears Are Treated as Facts," *National Review*, March 23, 2018.

43. "'Fake News' Threat to Media; Editorial Decisions, Outside Actors at Fault," Monmouth University Polling Institute, April 2, 2018.

44. Bill D'Agostino, "'The Walls Are Closing In': Cable Journalists Chant Dems' New Mantra Five Times a Day," NewsBusters, December 20, 2018.

45. Franklin Foer, "Putin's Puppet: If the Russian President Could Design a Candidate to Undermine American Interests—and Advance His Own—He'd Look a Lot like Donald Trump," Slate, July 4, 2016.

46. Jeffrey Goldberg, "It's Official: Hillary Clinton Is Running Against Vladimir Putin," *The Atlantic*, July 21, 2016.

47. David Remnick, "Trump and Putin: A Love Story," *The New Yorker*, August 3, 2016.

48. Abigail Tracy, "Is Donald Trump a Manchurian Candidate?," *Vanity Fair*, November 1, 2016.

49. Ibid.

50. Aiko Stevenson, "President Trump: The Manchurian Candidate?," Huffington Post, January 18, 2017.

51. Ross Douthat, "The 'Manchurian' President?," *New York Times*, May 31, 2017.

52. Ibid.

53. David Nakamura and Debbi Wilgoren, "Caught on Open Mike, Obama Tells Medvedev He Needs Space on Missile Defense," *Washington Post*, March 26, 2012.

54. Jonathan Allen and Amie Parnes, *Shattered: Inside Hillary Clinton's Doomed Campaign* (New York: Crown, 2017), 395.

55. Willis L. Krumholz, "35 Key People Involved in the Russia Hoax Who Need to Be Investigated," The Federalist, March 8, 2019.

56. Franklin Foer, "Was a Trump Server Communicating with Russia?," Slate, October 31, 2016.

57. Hillary Clinton, tweet, October 31, 2016, https://twitter.com/HillaryClinton/status/793250312119263233.

58. Rowan Scarborough, "Democrats Lose Interest in Server Conspiracy Claims Linking Trump to Russia's Alfa Bank," *Washington Times*, March 10, 2019.

59. Jeff Carlson, "Perkins Coie Lawyer Michael Sussmann's Coordinated Leaks to Media and FBI's James Baker, Likely Came from Steele, Fusion GPS," *The Epoch Times*, October 4, 2018.

60. Dan Gainor, "Trump Triumphs—Media's 'Primal Scream' Is Heard Round the World," Fox News, November 9, 2016.

61. Ibid.

62. Howard Kurtz, "Some in the Media Dig in Against 'Normalizing' Donald Trump," Fox News, November 21, 2016.

63. Colby Hall, "George Stephanopoulos Challenges Jay Sekulow: 'Cooperation Is Collusion!,'" Mediaite, October 31, 2017.

64. Rebecca Savransky, "Watergate Reporter Says Current White House 'Potentially More Dangerous Situation,'" The Hill, May 14, 2017.

65. Tim Haines, "Dan Rather: 'Hurricane Vladimir' Is 'Approaching Category Four' for Trump Presidency," RealClearPolitics, August 31, 2017.

66. Paul Krugman, "Days of Greed and Desperation," *New York Times*, November 11, 2017.

67. Aidan McLaughlin, "Jake Tapper Goes Cronkite on Don Jr. Emails: 'Evidence of Willingness to Commit Collusion,'" Mediaite, July 11, 2017.

68. "Former Watergate Prosecutor Nick Akerman on Trump Jr. Meeting," *Hardball with Chris Matthews*, MSNBC, July 11, 2017.

69. Dan Gainor, "Impeach Trump? Liberal Media Profiteering from Anti-Trump Clickbait," Fox News, May 30, 2017.

70. Aidan McLaughlin, "MSNBC's Joy Reid: What if Trump Locked Himself in White House and Refused Arrest?," Mediaite, April 9, 2018.

71. Constitution of the United States of America, Amendments V and XIV.

72. Juliet Eilperin and Adam Entous, "Russia Hackers Penetrated U.S. Electricity Grid Through Utility in Vermont," *Washington Post*, December 30, 2016. See also Warner Todd Huston, "Washington Post's Fake News of Russian Vermont Power Plant Hack," Breitbart, December 31, 2016.

73. Jennifer Palmieri, "The Clinton Campaign Warned You About Russia. But Nobody Listened to Us," *Washington Post*, March 24, 2017.

74. *The Rachel Maddow Show*, MSNBC, July 11, 2017.

75. Tim Haines, "Mika Brzezinski: 'Noose' Tightening on Trump Family; Might Go to Jail for the Rest of Their Lives," RealClearPolitics, December 5, 2017.

76. "Chris Hayes on Trump-Russia Allegations: Why Is Everyone Acting Guilty?," *The Late Show with Stephen Colbert*, YouTube, March 9, 2018.

77. Michael S. Schmidt, Mark Mazzetti, and Matt Apuzzo, "Trump Campaign Had Repeated Contacts with Russian Intelligence," *New York Times*, February 14, 2017.

78. Becket Adams, "Mainstream Media Errors in the Trump Era," *Washington Examiner*, February 9, 2017.

79. Byron York, "From Former Trump Lawyer, Candid Talk About Mueller, Manafort, Sessions, Rosenstein, Collusion, Tweets, Privilege and the Press," *Washington Examiner*, April 3, 2019.

80. Vivian Wang, "ABC Suspends Reporter Brian Ross over Erroneous Report About Trump," *Washington Post*, December 2, 2017.

81. Ibid.

82. Glenn Greenwald, "The U.S. Media Suffered Its Most Humiliating Debacle in Ages and Now Refuses All Transparency over What Happened," The Intercept, December 9, 2017.

83. Michael M. Grynbaum and Sydney Ember, "CNN Corrects a Trump Story, Fueling Claims of 'Fake News,'" *New York Times*, December 8, 2017.

84. Jeff Pegues, "House Intel Investigates Trump Jr. Email Involving Documents Hacked During Campaign," CBS News, December 8, 2017.

85. Allan Smith, "CNN Makes Major Correction on Trump-Russia bombshell After Washington Post Threw Timeline into Question," Business Insider, December 9, 2017.

86. Greenwald, "The U.S. Media Suffered Its Most Humiliating Debacle in Ages and Now Refuses All Transparency over What Happened."

87. Jason Schwartz, "CNN Error Extends Run of Journalistic Mishaps," Politico, December 8, 2017.

88. Amber Athey, "The Media's Russia 'Bombshells' Look Even Worse Now That Mueller Found No Collusion," Daily Caller, March 25, 2019.

89. Jim Rutenberg, "BuzzFeed News in Limbo Land," *New York Times*, January 20, 2019.

90. Colby Hall, "BuzzFeed News Bombshell Reporter: No We Have Not Seen the Evidence Supporting Our Report," Mediaite, January 18, 2019.

91. Dan Gainor, "This Week We Learned That Mainstream Media Won't Tattle on Each Other, No Matter How Badly They Do Journalism," Fox News, January 27, 2019.

92. Gregg Jarrett, "BuzzFeed Report with False Attack on Trump Is Media Malpractice," Fox News, January 19, 2019.

93. David Rutz, "CNN Analyst: Many Americans Will Dismiss Media as 'Leftist Liars' over Disputed BuzzFeed Story," *Washington Free Beacon*, January 18, 2019.

94. Gainor, "This Week We Learned That Mainstream Media Won't Tattle on Each Other, No Matter How Badly They Do Journalism."

95. Greg Miller and Greg Jaffe, "Trump Revealed Highly Classified Information to Russian Foreign Minister and Ambassador," *Washington Post*, May 15, 2017.

96. Todd Shepherd, "Fusion GPS Paid Journalists, Court Papers Confirm," *Washington Examiner*, November 2017.

97. Jerry Dunleavy, "No Collusion: 10 Anonymously Sourced Trump-Russia Bombshells That Look Like Busts," *Washington Examiner*, March 24, 2019.

98. Grynbaum and Ember, "CNN Corrects a Trump Story, Fueling Claims of 'Fake News.'"

99. Brian Flood, "CNN Quietly Backtracks Another Report Tying Trump Campaign to Russia," Fox News, December 11, 2017.

100. Peter Stone and Greg Gordon, "Source: Mueller Has Evidence Cohen Was in Prague in 2016, Confirming Part of Dossier," McClatchy, April 13, 2018.

101. Peter Stone and Greg Gordon, "Cell Signal Puts Cohen Outside Prague Around Time of Purported Russian Meeting," McClatchy, December 27, 2018. An editor's note was added after the release of the Mueller Report: "Robert Mueller's report to the attorney general states that Mr. Cohen was not in Prague. It is silent on whether the investigation received evidence that Mr. Cohen's phone pinged in or near Prague, as McClatchy reported."

102. Ivan Watson and Kocha Olarn, "Jailed Russian 'Sex Coaches' Offer to Trade Election Info for US Asylum," CNN, March 6, 2018.

103. Dunleavy, "No Collusion."

104. Liam Quinn, "Trump Dossier, Michael Flynn Testimony, Michael Cohen in Prague: Stories That Fell Flat During Mueller Probe," Fox News, March 25, 2019.

105. Anna Sanders, "WaPo Reporter Apologizes After Trump Calls Out 'Wrong' Rally Photo," *New York Post*, December 9, 2017.

106. Becket Adams, "Mainstream Media Errors in the Trump Era," *Washington Examiner*, February 9, 2017.

107. Margaret Sullivan, "Jeffrey Toobin Went Ballistic About Trump and Comey. It Was Great TV," *Washington Post*, May 10, 2017.

108. Eli Watkins, "Toobin: 'Three Words: Obstruction of Justice,'" CNN, May 16, 2017.

109. Pete Kasperowicz, "Alan Dershowitz: Comey Firing Was Legal Even if Trump Was Trying to End the Russia Investigation," *Washington Examiner*, April 18, 2018.

110. Derek Hawkins, "Jeffrey Toobin to His Former Professor Alan Dershowitz: 'What's Happened to You?,'" *Washington Post*, March 22, 2018.

111. Ibid.

112. Alan M. Dershowitz, "How CNN Misled Its Viewers," The Hill, March 27, 2019.

113. Ibid.

114. Ibid.

115. Chuck Ross, "Van Jones Responds to O'Keefe Sting Video," *The Daily Caller*, June 29, 2017.

116. James Freeman, "CNN and Avenatti," *Wall Street Journal*, April 29, 2019.

117. Post Editorial Board, "Avenatti's Fall Is Only the Latest Sign of Media Anti-Trump Madness," *New York Post*, March 25, 2019. See also Amy Taxin, "Avenatti Pleads Not Guilty on Charges of Cheating, Lying," AP News, April 29, 2019.

118. Brendan Pierson, "Avenatti Pleads Not Guilty to Extorting Nike, Ripping Off Stormy Daniels," Reuters, May 28, 2019.

119. P. J. Gladnick, "Ted Koppel: Trump's 'Not Mistaken' That Liberal Media Are Blatantly 'Out to Get Him,'" NewsBusters, March 18, 2019.

120. Jill Abramson, *Merchants of Truth: The Business of News and Fight for Facts* (New York: Simon & Schuster, 2019), 390.

121. Ibid., 387.

122. Ibid., 391.

123. Ibid., 389.

124. Ibid., 387.

125. Elizabeth Drew, "The Inevitability of Impeachment," *New York Times*, December 27, 2018.

126. Yoni Applebaum, "Impeach: It's Time for Congress to Judge the President's Fitness to Serve," *The Atlantic*, March 2019.

127. Paul Sperry, "Trump-Russia 2.0: Dossier-Tied Firm Pitching Journalists Daily on 'Collusion,'" RealClearInvestigations, March 20, 2019.

128. Sean Davis, "Confirmed: Former Feinstein Staffer Raised $50 Million, Hired Fusion GPS and Christopher Steele After 2016 Election," The Federalist, April 27, 2018.

129. Ibid.

130. Sperry, "Trump-Russia 2.0: Dossier-Tied Firm Pitching Journalists Daily on Collusion."

131. Aidan McLaughlin, "Lara Logan on Fox News: Media Matters Is the 'Most Powerful Propaganda Organization in This Country,'" Mediaite, April 2, 2019.

132. Ben Smith, "Media Matters' War Against Fox," Politico, March 26, 2011.

133. Lara Logan, "Political Bias Is Destroying People's Faith in Journalism," *New York Post*, February 26, 2019.

134. Ken Meyer, "Ex–CBS Star Lara Logan Calls Media 'Mostly Liberal' in Scorched Earth Interview: I'm Committing Professional Suicide," Mediaite, February 18, 2019.

135. Jonathan Chait, "Has Trump Been an Agent for Russia since 1987?," *New York*, July 9, 2018.

136. John Brennan, "President Trump's Claims of No Collusion Are Hogwash," *New York Times*, August 16, 2018.

137. Guy Benson, "Brennan: Okay, I Didn't Necessarily Mean Trump Committed Treason but I 'Stand Very Much by' Saying So," Townhall, August 20, 2018.

138. Ian Schwartz, "Brinkley: Spirit of Trump-Putin Presser 'Clearly Treasonous,' Will be a 'Battle Cry' for 'Blue Wave,'" RealClearPolitics, July 18, 2018.

139. Constitution of the United States of America, Article III, Section 3; United States Code, 18 U.S.C. § 2381, "Treason." For an extended discussion of how treason is inapplicable, see Gregg Jarrett, *The Russia Hoax: The Illicit Scheme to Clear Hillary Clinton and Frame Donald Trump* (New York: Broadside Books, 2018), 173–76.

140. Marc Thiessen, "The Trump-Russia Collusion Hall of Shame," *Washington Post*, March 29, 2019.

141. "'This Week' Transcript 5-27-18: Sen. Marco Rubio, Rep. Adam Schiff and Former CIA Director Michael Hayden," ABC News, May 27, 2018.

142. Ibid.

143. Christina Zhao, "Donald Trump Jr. Will Be Indicted by Mueller, Former Prosecutor Says, and Will Help to Ensnare His Father," *Newsweek*, February 2, 2019.

144. Jonathan Turley, "Media Bias Seen in BuzzFeed and AG Nominee Barr News Coverage," Fox News, January 20, 2019.

145. Ibid.

146. Kyle Drennan, "Chuck Todd Freaks Out over Peter Strzok Firing: 'Extraordinary and Un-democratic,'" NewsBusters, August 14, 2018.

147. Ibid.

148. Ibid.

149. Byron York, "How Did Peter Strzok's Notorious Text Stay Hidden So Long?," *Washington Examiner*, June 20, 2018.

150. Rich Noyes, "FIZZLE: Nets Gave Whopping 2,284 Minutes to Russia Probe," NewsBusters, March 25, 2019.

151. Caleb Howe, "MSNBC Stunned by Intel Committee News: 'Trump Will Claim Vindication . . . and He'll Be Partially Right,'" Mediaite, February 12, 2019.

152. Thomas Lifson, "Senior National Correspondent Predicts 'Reckoning for Progressives, Democrats . . . and Media' if Mueller Finds No Russia Collusion," American Thinker, March 11, 2019.

153. Joshua Caplan, "Politico: Top Democrats 'Certain' Mueller Report 'Will Be a Dud,'" Breitbart, March 22, 2019.

154. Matt Taibbi, "Taibbi: As the Mueller Probe Ends, New Russiagate Myths Begin," *Rolling Stone*, March 25, 2019.

155. Keith Griffith, "How Pundits Reacted to Conclusion of Mueller Probe: Hannity Celebrates End of 'Witch Hunt' While Maddow Foresees Further Investigations," *Daily Mail*, March 23, 2019.

156. Michael Goodwin, "The New York Times Owes Americans a Big, Fat Apology," *New York Post*, March 25, 2019.

157. Paul Mirengoff, "Mueller Report 'Opens Media Outlets to Mockery,'" Power Line, March 25, 2019.

158. David Brooks, "We've All Just Made Fools of Ourselves—Again," *New York Times*, March 25, 2019.

159. Michael Isikoff and David Corn, *Russian Roulette: The Inside Story of Putin's War on America and the Election of Donald Trump* (New York: Twelve, 2018).

160. Charles Hurt, "List of Trusted Journalistic Sources Shrinks in Post–Mueller Report World," *Washington Times*, March 27, 2019.

161. Chuck Ross, "MSNBC Panel Admits Mueller Report 'Undercuts Almost Everything' in Steele Dossier," Daily Caller, March 26, 2019.

162. David Corn, "Trump Aided and Abetted Russia's Attack. That Was Treachery. Full Stop," *Mother Jones*, March 24, 2019.

163. Joshua Caplan, "John Dean: Mueller Probe Conclusion Doesn't Resolve Whether Trump Russian Agent," Breitbart, March 22, 2019.

164. Sean Davis, "A Catastrophic Media Failure," *Wall Street Journal*, March 26, 2019.

165. Ronn Blitzer, "4 Revelations from Ex–Trump Lawyer's Eye Opening Interview About Mueller Probe," Law & Crime, April 4, 2019.

166. Ibid.

167. Howard Kurtz, "How the Media's Distorted Judgment Kept Hyping the Mueller Probe," Fox News, March 26, 2019.

168. Ibid.

169. Lee Smith, "System Fail: The Mueller Report Is an Unmitigated Disaster for the American Press and the 'Expert' Class That It Promotes," *Tablet*, March 27, 2019.

170. Kurtz, "How the Media's Distorted Judgment Kept Hyping the Mueller Probe."

171. Scott Johnson, "John Brennan: From Spittle to Flop Sweat," Power Line, March 30, 2019.

172. Thomas Lifson, "John Brennan Denies Blame for Calling Trump a 'Traitor,' Says 'I Received Bad Information,'" American Thinker, March 26, 2019.

173. Jason Beale, "In Lester Holt Interview, James Comey Proves He Can't Handle the Russian Hoax's Collapse," The Federalist, March 28, 2019. Jason Beale is a pseudonym for "a retired U.S. Army interrogator and strategic debriefer with 30 years experience in military and intelligence interrogation and human intelligence collection operations."

174. Ibid.

175. Tommy Christopher, "Jake Tapper Defends CNN's Mueller Probe Coverage to

Mike Mulvaney: 'I Don't Know Anybody Who Got Anything Wrong,'" Mediaite, March 31, 2019.

176. John Nolte, "Death Spiral Continues as CNN Loses One-Third of Primetime Audience," Breitbart, June 12, 2019; A. J. Katz, "Basic Cable Ranker: Week of June 3, 2019," TVNewser, June 11, 2019.

177. Janita Kan, "'What Wrong Facts Did We Put Out?': Chris Cuomo Defends Liberal Media for Mueller Coverage," *The Epoch Times*, March 27, 2019.

178. Joshua Caplan, "Jeff Zucker: No Regrets on CNN's Russia Hoax Coverage, 'We Are Not Investigators,'" Breitbart, March 26, 2019.

179. Jeremy Diamond and Kevin Liptak, "Trump Moves to Weaponize Mueller Findings," CNN, March 27, 2019.

180. Pardes Seleh, "MSNBC's Katy Tur: It Doesn't Matter if There Was No Collusion, Mueller Already Has 'Quite a Bit' on Trump," Mediaite, March 21, 2019.

181. Bob Bauer, "Trump's Shamelessness Was Outside Mueller's Jurisdiction," *New York Times*, March 25, 2019.

182. Paul Mirengoff, "Less Than Full Disclosure from the New York Times," Power Line, March 26, 2019.

183. Mirengoff, "Mueller Report 'Opens Media Outlets to Mockery.'"

184. Ibid.

185. Josh Feldman, "Glenn Greenwald Rips MSNBC to Tucker Carlson: They Fed People 'Total Disinformation' and Exploited Fears on Russia," Mediaite, March 25, 2019.

186. Mirengoff, "Mueller Report 'Opens Media Outlets to Mockery.'"

187. Matt Taibbi, "It's Official: Russiagate Is This Generation's WMD," RealClearPolitics, March 23, 2019

188. Sharyl Attkisson, "Apologies to President Trump," The Hill, March 25, 2019.

189. Tim Murtaugh, Director of Communications, Donald J. Trump for President, Inc., memorandum to Television Producers re "Credibility of Certain Guests," March 25, 2019, https://www.scribd.com/document/403100260/March-25-2019-Tim-Murtaugh-Trump-Campaign-Memo-to-TV-Producers.

190. James Freeman, "Schiff and the Media," *Wall Street Journal*, March 27, 2019.

191. Malcolm Nance, *The Plot to Destroy Democracy: How Putin and His Spies Are Undermining America and Disarming the West* (New York: Hachette Books, 2018).

192. David Greene, "Media Outlets Became a Target After Mueller Probe Results Surfaced," NPR, March 24, 2019.

193. Sohrab Ahmari, "And the Winner of the Post's Mueller Madness Bracket Is . . . ," *New York Post*, March 27, 2019.

194. Caleb Howe, "The Press Doesn't Learn Things, Unless Those Things Are About How Great They Are," Mediaite, March 27, 2019.

195. Ibid.

196. Brian Flood, "Evening Newscasts Increased 'Impeachment' Talk Following Mueller Report, Study Says," Fox News, June 4, 2019.

197. Brian Flood, "New York Times Needs to 'Look in the Mirror' if It Thinks MSNBC, CNN Are Too Partisan, Critics Say," Fox News, May 31, 2019.

198. Brian Flood, "Rachel Maddow's Credibility and Ratings at Low Ebb Following Mueller Findings, Critics Say," Fox News, June 3, 2019.

199. John Nolte, "Russia Hoax Queen Rachel Maddow's Ratings Crash to Trump-Era Low," Breitbart, June 4, 2019.

200. Flood, "Rachel Maddow's Credibility and Ratings at Low Ebb Following Mueller Findings, Critics Say."

201. Ibid.

202. Holman W. Jenkins, Jr., "Can the Media Survive Mueller?," *Wall Street Journal*, April 17, 2019.

203. Sean Illing, "Did the Media Botch the Russia Story? A Conversation with Matt Taibbi," Vox, April 1, 2019.

204. Matt Taibbi, "As the Mueller Probe Ends, New Russiagate Myths Begin," *Rolling Stone*, March 26, 2019.

CHAPTER 7: CROOKED COHEN COPS A PLEA

1. Associated Press, "Michael Cohen Pleads Guilty to Lying to Congress," Politico, November 29, 2018; Justin Wise, "Cohen Pleads Guilty for Misstatements to Congress About Contacts with Russians," The Hill, November 29, 2018.

2. Jennifer Mercieca, "Michael Cohen's Verbal Somersault, 'I Lied, but I'm Not a Liar,' Translated by a Rhetoric Expert," The Conversation, February 27, 2019.

3. "Robert Khuzami Statement on Michael Cohen Case," C-SPAN, August 21, 2018.

4. Nicole Hong and Rebecca Davis O'Brien, "Special Counsel's Michael Cohen Probe Dates to 2017, New Documents Show," *Wall Street Journal*, March 19, 2019.

5. Ibid.

6. Geoff Earle, "How Did Cheryl Mills Get Immunity if She Was Also Acting as Clinton's Lawyer? GOP Lodges Formal Complaint with D.C. Bar," *Daily Mail*, October 3, 2016.

7. Hong and O'Brien, "Special Counsel's Michael Cohen Probe Dates to 2017, New Documents Show."

8. Jim Mustian and Larry Neumeister, "Records Show FBI Was Probing Michael Cohen Long Before Raid," Associated Press, as published in the *Chicago Tribune*, March 19, 2019.

9. "These Are the Lawyers on Robert Mueller's Special Counsel Team," CBS News, September 20, 2017. See also Matt Zapotosky, "Trump Said Mueller's Team Has '13 Hardened Democrats.' Here Are the Facts," *Washington Post*, March 18, 2018.

10. Ken Bensinger, Miriam Elder, and Mark Schoofs, "These Reports Allege Trump Has Deep Ties to Russia," BuzzFeed, January 10, 2017; "Company Intelligence Report 2016/080," December 13, 2016, https://www.document cloud.org/documents/3259984-Trump-Intelligence-Allegations.html, page 34. See also Adam Mill, "How the Cohen-Prague Story Helped Expose the Collusion Hoax," American Greatness, April 16, 2019. Adam Mill is a pen name.

He works in Kansas City, Missouri, as an attorney specializing in labor and employment and public administration law.

11. "Report: Czech Intelligence Says No Evidence Trump Lawyer Traveled to Prague," Radio Free Europe/Radio Liberty, January 11, 2017.

12. Mill, "How the Cohen-Prague Story Helped Expose the Collusion Hoax."

13. Matt Apuzzo, "F.B.I. Raids Office of Trump's Longtime Lawyer Michael Cohen; Trump Calls It 'Disgraceful,'" *New York Times*, April 9, 2019.

14. Robert S. Mueller, *The Mueller Report: The Final Report of the Special Counsel into Donald Trump, Russia, and Collusion as Issued by the Department of Justice* (New York: Skyhorse Publishing, 2019), 69.

15. Nicholas Fandos, "Felix Sater, Trump Associate, Skips House Hearing and Now Faces a Subpoena," *New York Times*, June 21, 2019.

16. Johnny Dwyer, "What the Mueller Report Didn't Say About Felix Sater and the Trump Tower in Moscow," The Intercept, April 19, 2019.

17. Alberto Luperon, "Federal Judge: Felix Sater Gave the U.S. Government Osama bin Laden's Phone Number," Law & Crime, May 16, 2019.

18. *United States v. Felix Sater*, 98 cr 1101 (ILG). Sater Cooperation Agreement, signed December 10, 1998, https://assets.documentcloud.org/documents/5972612/Felix-Sater-s-Cooperation-Agreement.pdf.

19. Joseph Tanfani and David S. Cloud, "Trump Associate Led Double Life as FBI Informant—and More, He Says," *Los Angeles Times*, March 2, 2017.

20. Jessica Kwong, "Who Is Michael Cohen's Wife? Laura Shusterman Never Charged Though Prosecutors Had Evidence Implicating Her, Report Says," *Newsweek*, December 3, 2018.

21. Michael Rothfield, Alexandra Berzon, and Joe Palazzolo, "'Boss, I Miss You So Much': The Awkward Exile of Michael Cohen," *Wall Street Journal*, April 26, 2018.

22. *US v. Michael Cohen*, 1:18-cr-00602-WHP, Government's Sentencing Memorandum, filed December 7, 2018.

23. Jonathan Lemire and Jake Pearson, "How Michael Cohen, Trump's 'Image Protector,' Landed in the President's Inner Circle," Business Insider, April 22, 2018.

24. Ibid.

25. Rothfield, Berzon, and Palazzolo, "'Boss, I Miss You So Much.'"

26. Julio Ross, "2016 Michael Cohen Interview Shows He 'Certainly Hoped' He'd Be Offered a White House Job," Mediaite, February 28, 2019.

27. Rothfield, Berzon, and Palazzolo, "'Boss, I Miss You So Much.'"

28. Ibid.

29. Ibid.

30. Rebecca Ballhaus, Peter Nicholas, Michael Rothfield, and Joe Palazzolo, "Michael Cohen's D.C. Consulting Career: Scattershot, with Mixed Success," *Wall Street Journal*, May 13, 2018.

31. Rothfield, Berzon, and Palazzolo, "'Boss, I Miss You So Much.'"

32. Ballhaus, Nicholas, Rothfield, and Palazzolo, "Michael Cohen's D.C. Consulting Career."

33. Ibid.

34. Tom Winter, Adiel Kaplan, and Rich Schapiro, "Michael Cohen Search Warrants Show Federal Probe Began a Year Earlier than Known," NBC News, March 19, 2019.
35. Brian Bennett and Haley Sweetland Edwards, "Michael Cohen's 'Essential Consultants' Business Is a Big Problem for Donald Trump," *Time*, May 10, 2018.
36. Ballhaus, Nicholas, Rothfield, and Palazzolo, "Michael Cohen's D.C. Consulting Career."
37. Jonathan Turley, "Trump Must Beware the Cohen Trap," The Hill, April 10, 2018.
38. Ibid.
39. Apuzzo, "F.B.I. Raids Office of Trump's Longtime Lawyer Michael Cohen."
40. Ibid.
41. Tom Winter, Adiel Kaplan, and Rich Schapiro, "Michael Cohen Search Warrants Show Federal Probe Began a Year Earlier than Known."
42. Ibid.
43. Ewan Palmer, "McDougal Alleged 10-Month Affair That Ended in April 2007," *Newsweek*, July 26, 2018.
44. Michael Rothfield and Joe Palazzolo, "Trump Lawyer Won Order to Silence Stormy Daniels," *Wall Street Journal*, March 7, 2018.
45. Madeline Osburn, "How CNN and MSNBC Made Michael Avenatti a Household Name," The Federalist, March 28, 2019.
46. Rebecca Ballhaus, Michael Rothfield, and Joe Palazzolo, "Trump's Former Lawyer Michael Cohen Recorded Conversation About Stormy Daniels Payment with News Anchor," *Wall Street Journal*, July 25, 2018.
47. Ibid.
48. Rothfield, Berzon, and Palazzolo, " 'Boss, I Miss You So Much.' "
49. Ballhaus, Rothfield, and Palazzolo, "Trump's Former Lawyer Michael Cohen Recorded Conversation About Stormy Daniels Payment with News Anchor."
50. Glenn Greenwald, "The Lanny Davis Disease and America's Health Care Debate," Salon, August 18, 2009.
51. Seth Hettena, "Michael Cohen, Lanny Davis and the Russian Mafia," *Rolling Stone*, August 28, 2018.
52. Greenwald, "The Lanny Davis Disease and America's Health Care Debate."
53. "Search Hillary Clinton's Emails," *Wall Street Journal*, March 1, 2016.
54. Tom Fitton, "Cohen Testimony Against Trump Unethical—Dems Commit Abuse of Power," Fox News, February 28, 2019.
55. Ewan Palmer, "Who Is Lanny Davis? Michael Cohen's Lawyer Is a Washington Insider with a Long List of High-Profile Clients," *Newsweek*, July 26, 2018.
56. *US v. Michael Cohen*, 1:18-cr-00602-WHP, Information.
57. Ibid.
58. Joe Palazzolo, Nicole Hong, Michael Rothfield, Rebecca Davis O'Brien, and Rebecca Ballhaus, "Donald Trump Played Central Role in Hush Payoffs to Stormy Daniels and Karen McDougal," *Wall Street Journal*, November 9, 2018.
59. *US v. Michael Cohen*, 1:18-cr-00602-WHP, Information.

60. Ibid.

61. Michael Rothfield and Joe Palazzolo, "Stormy Daniels Sues Trump over Non-disclosure Agreement," *Wall Street Journal*, March 6, 2018.

62. *US v. Michael Cohen*, 1:18-cr-00602-WHP, Information.

63. Kevin Breuniger and Jacob Pramuk, "National Enquirer Boss and Longtime Trump Friend David Pecker Gets Federal Immunity in Michael Cohen Case," CNBC, August 23, 2018.

64. Allan Smith, "Michael Cohen's Attorney: The 'Final Straw' for Cohen Was Trump's Disastrous Summit with Putin, and He Now Feels 'Liberated' After Making a Deal with Prosecutors," Business Insider, August 22, 2019.

65. "Lanny Davis: Cohen Knows 'Almost Everything' About Trump After 'Many Years' as Lawyer," CNN, August 22, 2018.

66. Bradley Smith, "Stormy Weather for Campaign-Finance Laws," *Wall Street Journal*, April 10, 2018.

67. Maggie Haberman, "Obama 2008 Campaign Fined $375,000," Politico, January 1, 2013.

68. Steven Nelson, "Did Michael Cohen Actually Commit Campaign Finance Violations? Some Legal Experts Aren't Sure," *Washington Examiner*, August 22, 2018.

69. M. J. Lee, Sunlen Serfaty, and Juana Summers, "Congress Paid Out $17 Million in Settlements. Here's Why We Know So Little About That Money," CNN, November 16, 2017.

70. Jon Greenberg, "The $130,000 Stormy Daniels Payoff: Was It a Campaign Expenditure?," PolitiFact, May 3, 2018.

71. Ibid.

72. Ibid.

73. Nelson, "Did Michael Cohen Actually Commit Campaign Finance Violations?"

74. "Michael Cohen's Lawyer Crowdfunds 'Truth About Trump,' Whips Up $120k+," RT, August 23, 2018.

75. Chuck Ross, "Michael Cohen's Congressional Testimony Will Be 'Unsatisfying,' Lanny Davis Told Lawmakers," Daily Caller, January 22, 2019.

76. *Post* Editorial Board, "Why Is CNN Avoiding the Truth About Lanny Davis' Lies?," *New York Post*, August 29, 2018.

77. Jim Sciutto, Carl Bernstein, and Marshall Cohen, "Cohen Claims Trump Knew in Advance of 2016 Trump Tower Meeting," CNN, July 27, 2018.

78. Allan Smith, "Lanny Davis's Walk-back of His Bombshell Claim to CNN Is More Complicated than It Looks. And Experts Say It Causes Michael Cohen Some New Problems," Business Insider, August 28, 2018.

79. *Post* Editorial Board, "Why Is CNN Avoiding the Truth About Lanny Davis' Lies?"

80. Smith, "Lanny Davis's Walk-back of His Bombshell Claim to CNN Is More Complicated than It Looks."

81. *Post* Editorial Board, "Why Is CNN Avoiding the Truth About Lanny Davis' Lies?"

82. Thomas Lifson, "Lanny Davis and Michael Cohen Have Gotten Themselves into a Big Mess Trying to Damage Trump," American Thinker, August 28, 2018.

83. Ibid.

84. *Post* Editorial Board, "Why Is CNN Avoiding the Truth About Lanny Davis' Lies?"

85. Smith, "Lanny Davis's Walk-back of His Bombshell Claim to CNN Is More Complicated than It Looks."

86. Ibid.

87. Chuck Ross, "Lanny Davis: Michael Cohen Never Went to Prague as Steele Dossier Claims," Daily Caller, August 22, 2018.

88. Brooke Singman, "Ex–Trump Attorney Michael Cohen Pleads Guilty to Lying to Congress in Russia Probe," Fox News, November 29, 2018.

89. Ibid.

90. Ibid.

91. Ibid.

92. Ibid.

93. Ken Bensinger, Miriam Elder, and Mark Schoofs, "These Reports Allege Trump Has Deep Ties to Russia," BuzzFeed, January 10, 2017; "Company Intelligence Report 2016/080," December 13, 2016.

94. *US v. Michael Cohen*, 1:18-cr-00602-WHP, Government's Sentencing Memorandum, filed December 7, 2018.

95. *US v. Michael Cohen*, 1:18-cr-00602-WHP, Mueller sentencing memorandum, signed by Jeannie Rhee, Andrew Goldstein, L. Rush Atkinson, Special Counsel's Office, December 7, 2018.

96. Paul Sperry, "For Trump, Cohen Plea Deal's Beginning to Look a Lot like Exoneration," RealClearInvestigations, December 3, 2018.

97. Ibid.

98. Hans von Spakovsky, "Trump's Ex-Laywer Didn't Violate Campaign Finance Law, and Neither Did the President," The Daily Signal, December 11, 2018.

99. Andrew C. McCarthy, "Payoffs to Mistresses as In-Kind Contributions? It's an Open Question," *National Review*, December 14, 2018.

100. Laura Mahmias and Darren Samuelsohn, "Michael Cohen Sentenced to 3 Years in Prison," Politico, December 12, 2018.

101. Ibid.

102. Ibid.

103. Tyler Stone, "Cummings: I Would Like Cohen to Testify Before Congress," RealClearPolitics, December 17, 2018.

104. Peter Stone and Greg Gordon, "Cell Signal Puts Cohen Outside Prague Around Time of Purported Russian Meeting," McClatchy, December 27, 2018.

105. Ashe Schow, "McClatchy Tries to Save Face After Mueller Report Thrashes Cohen-in-Prague Narrative," The Daily Wire, April 20, 2019.

106. Chuck Ross, "Michael Cohen Shakes Up Legal Team for the Second Time," Daily Caller, June 28, 2019.

107. Jason Leopold and Anthony Cormier, "President Trump Directed His Attorney to Lie to Congress About the Moscow Tower Project," BuzzFeed, January 17, 2019.

108. Devlin Barrett, Matt Zapotosky, and Karoun Demirjan, "In a Rare Move, Mueller's Office Denies BuzzFeed Report That Trump Told Cohen to Lie About Moscow Project," *Washington Post*, January 19, 2019.

109. Ken Meyer, "New Trove of Documents Suggests Trump Tower Moscow Negotiations Went Beyond Letter of Intent," Mediaite, February 5, 2019.

110. Ronn Blitzer, "Cohen Blames Postponement of Testimony on Trump 'Threats,' as if We Don't Know the Real Reason," Law & Crime, January 23, 2019.

111. Jeremy Herb, Gloria Borger, and Manu Raju, "Michael Cohen Apologizes to Senate Panel for Lying to Congress," CNN, February 26, 2019.

112. Samuel Chamberlain, "Senate Intel Panel Chair Rips Michael Cohen for Asking to Postpone Hearing," Fox News, February 12, 2019.

113. Maggie Haberman and Nicholas Fandos, "On Eve of Michael Cohen's Testimony, Republican Threatens to Reveal Compromising Information," *New York Times*, February 26, 2019.

114. Larry Neumeister and Michael R. Sisak, "After Surgery, Michael Cohen's Prison Date Postponed to May," Associated Press, February 20, 2019.

115. Aiden McLaughlin, "Michael Cohen's Leaked Testimony Appears to Dispute BuzzFeed's Trump Tower Report," Mediaite, February 27, 2019.

116. Madeline Osburn, "6 Things We Learned from Michael Cohen's Testimony Today," The Federalist, February 27, 2019.

117. Joseph Wulfsohn, "Cohen Testimony Was 'Bombshell That Didn't Explode': Marc Thiessen," Fox News, February 28, 2019.

118. Bruce Golding, "WH-Job Diss & Tell Denied," *New York Post*, February 28, 2019.

119. Colby Hall, "Michael Cohen: Trump Frequently Told Me 'Don Jr. Had the Worst Judgment of Anyone in the World,'" Mediaite, February 27, 2019.

120. John Podhoretz, "Firing Blanks," *New York Post*, February 28, 2019.

121. Rachel Frazin, "WikiLeaks Disputes Cohen, Says Assange Never Talked to Stone," The Hill, February 27, 2018.

122. CNN, "Cohen Gives Documents to House Panel on Trump Attorney Alleged Changes to 2017 Testimony," CNN, March 6, 2019.

123. Ryan Bort, "Did Michael Cohen Lie to Congress About Seeking a Pardon from Trump?," *Rolling Stone*, March 7, 2019.

124. Rebecca Ballhaus, "Cohen Told Lawyer to Seek Trump Pardon," *Wall Street Journal*, March 6, 2019.

125. Wulfsohn, "Cohen Testimony Was 'Bombshell That Didn't Explode': Marc Thiessen."

126. "Cohen: Fears No 'Peaceful Transition' if Trump Loses in 2020," Reuters, February 27, 2019.

127. Brooke Singman, "GOP Reps Refer Michael Cohen to DOJ for Alleged Perjury During Hearing," Fox News, February 28, 2019.

128. "Cohen Gives Documents to House Panel on Trump Attorney Alleged Changes to 2017 Testimony."

129. Maria Bartiromo, *Sunday Morning Futures with Maria Bartiromo*, March 3, 2019.

130. Rebecca Ballhaus, "Michael Cohen Begins Serving Three-Year Sentence," *Wall Street Journal*, May 6, 2019.

CHAPTER 8: COLLATERAL DAMAGE

1. Alex Pappas, "Comey Admits Decision to Send FBI Agents to Interview Flynn Was Not Standard," Fox News, December 13, 2018.

2. Ibid.

3. Olivia Beavers, "GOP Senators Question 'Unusual' Message Susan Rice Sent Herself on Inauguration Day," The Hill, February 12, 2018.

4. George Rasley, "The Mike Flynn Travesty Comes to an End," Conservative HQ, December 6, 2018.

5. Paul Sonne and Michael C. Bender, "Donald Trump Offers Michael Flynn Role as National Security Adviser," *Wall Street Journal*, November 18, 2016.

6. Paul Sonne, "Lt. Gen. Michael Flynn Has Clashed with Intelligence Community, Pentagon," *Wall Street Journal*, November 18, 2016.

7. Frank Hawkins, "Understanding Why the Deep State Had to Take Down General Mike Flynn," American Thinker, January 16, 2019.

8. Ibid.

9. Ibid.

10. Michelle Hackman, "Michael Flynn's Son Has Left Trump Transition Team," *Wall Street Journal*, December 6, 2016.

11. Sonne and Bender, "Donald Trump Offers Michael Flynn Role as National Security Adviser."

12. Editorial Board, "Eavesdropping on Michael Flynn," *Wall Street Journal*, February 13, 2017.

13. Gordon Lubold and Shane Harris, "In Spy-Agency Revamp, Michael Flynn Shows His Influence," *Wall Street Journal*, January 6, 2017.

14. House Permanent Select Committee on Intelligence, "Report on Russian Active Measures," March 22, 2018, https://www.scribd.com/document/377590825/HPSCI-Final-Report-on-Russian-Active-Measures-Redacted-Release#. See also excerpt at https://esmemes.com/i/u-the-committees-investigation-also-reviewed-the-opening-in-summer-56920110520a4899b7ad497bfd5a89f7.

15. Ibid., 26.

16. Ibid.

17. "DiGenova on Mueller Memo: 'This Was a Frame-up of Flynn to Get Donald Trump,'" *Tucker Carlson Tonight*, December 5, 2018. In addition to serving as a US attorney, diGenova has served as independent counsel and special counsel to the House of Representatives.

18. David Sanger, "Obama Strikes Back at Russia for Election Hacking," *New York Times*, December 29, 2016.

19. Devlin Barrett and Carol E. Lee, "Mike Flynn Was Probed by FBI over Calls with Russian Official," *Wall Street Journal*, February 15, 2017.

20. David Ignatius, "Why Did Obama Dawdle on Russia's Hacking?," *Washington Post*, January 12, 2017.

21. Adam Entous, Ellen Makashima, and Philip Rucker, "Justice Department Warned White House That Flynn Could Be Vulnerable to Russian Blackmail, Officials Say," *Washington Post*, February 13, 2017.

22. Ignatius, "Why Did Obama Dawdle on Russia's Hacking?"

23. 18 U.S.C. 798 states, "Whoever knowingly and willfully communicates . . . to an unauthorized person, or publishes . . . any classified information obtained by the processes of communication intelligence from the communications of any foreign government, knowing the same to have been obtained by such processes shall be fined under this title or imprisoned not more than ten years, or both."

24. Brian Cates, "New Evidence Appears to Tie Former FBI Official McCabe to Illegal Leak About Flynn," *The Epoch Times*, October 15, 2018. See also "Deputy Director McCabe Office of Professional Responsibility Investigation Part 01 of 01," Federal Bureau of Investigation, https://vault.fbi.gov/deputy-director-mccabe-office-of-professional-responsibility-investigation/Deputy%20Director%20McCabe%20Office%20of%20Professional%20Responsibility%20Investigation%20Part%2001%20of%2006/view.

25. James Barrett, "Report: FBI May Have Launched Russia Prove to 'Retaliate' Against Mike Flynn," The Daily Wire, June 27, 2017. This story references the story on Circa, which is now defunct.

26. Ibid.

27. Ibid.

28. Ibid.

29. Christopher Wallace, "Acting FBI Boss Andrew McCabe Faces Pressure, Probes, Uncertain Future," Fox News, April 25, 2018.

30. Jordain Carney, "Grassley: Why Hasn't Acting FBI Chief Recused Himself on Flynn?," The Hill, June 29, 2017.

31. Paul Sonne and Shane Harris, "Michael Flynn's Contact with Russian Ambassador Draws Scrutiny," *Wall Street Journal*, January 13, 2017.

32. "Face the Nation Transcript January 15, 2017: Pence, Manchin, Gingrich," CBS News, January 15, 2017.

33. Nick Timiraos, "Donald Trump Asks if CIA Director Was 'Leader of Fake News' About Him," *Wall Street Journal*, January 15, 2017.

34. Entous, Makashima, and Rucker, "Justice Department Warned White House That Flynn Could Be Vulnerable to Russian Blackmail, Officials Say," *Washington Post*, February 13, 2017.

35. Carol E. Lee, Devlin Barrett, and Shane Harris, "U.S. Eyes Michael Flynn's Links to Russia," *Wall Street Journal*, January 22, 2017.

36. Devlin Barrett and Carol E. Lee, "Mike Flynn Was Probed by FBI over Calls to Russia," *Wall Street Journal*, February 15, 2017.

37. James Gordon Meek and Ali Dukakis, "Former Trump Adviser Michael

Flynn Faces $5 Million in Legal Fees amid Pardon Buzz: Source," ABC News, March 26, 2019.

38. Geoff Earle, "Comey Told Lawmakers Flynn Was Feeding Pence Information on Russia That Was 'Starkly at Odds' with Classified Information and He Sent FBI Agents to Interview Him as Part of a 'Counterintelligence Mission,'" *Daily Mail*, December 18, 2018.

39. Ibid.

40. *US v. Michael T. Flynn*, 17 CR 232 (D.D.C.), Memorandum in Aid of Sentencing, filed December 11, 2018, by lawyers Stephen P. Anthony and Robert K. Kelner.

41. Ibid.

42. Ibid.

43. Ibid.

44. Ibid.

45. Ibid.

46. Thomas Lifson, "Senator Grassley Appears to Be Preparing to Bust the Frame-up of General Flynn," American Thinker, May 12, 2018. For background, see Office of the Inspector General, U.S. Department of Justice, "Report of Investigation: Recovery of Text Messages from Certain FBI Mobile Devices," December 2018, https://oig.justice.gov/reports/2018/i-2018-003523.pdf.

47. Ibid.

48. Ibid.

49. Ibid.

50. Ibid.

51. Riley Beggin and Veronica Stracqualursi, "A Timeline of Sally Yates' Warnings to the White House About Mike Flynn," ABC News, May 8, 2017.

52. Entous, Nakashima, and Rucker, "Justice Department Warned White House That Flynn Could Be Vulnerable to Russian Blackmail, Officials Say."

53. Ibid.

54. Byron York, "Comey Told Congress FBI Agents Didn't Think Michael Flynn Lied," *Washington Examiner*, February 12, 2018.

55. Ibid.

56. Ibid.

57. Jason Beale, "How Obama Holdover Sally Yates Helped Sink Michael Flynn," The Federalist, April 8, 2019. Jason Beale is a pseudonym for a retired U.S. Army interrogator.

58. York, "Comey Told Congress FBI Agents Didn't Think Michael Flynn Lied."

59. Josh Blackman, "Why Trump Had to Fire Sally Yates," Politico, January 31, 2017.

60. Kimberly Strassel, "Mueller's Gift to Obama," *Wall Street Journal*, December 6, 2018.

61. Adam Shaw, "Shock Claim About FBI's Michael Flynn Interview Raises Questions," Fox News, February 14, 2018.

62. Barrett and Lee, "Mike Flynn Was Probed by FBI over Calls to Russia."

63. Greg Miller, Adam Entous, and Ellen Nakashima, "National Security Adviser

Flynn Discussed Sanctions with Russian Ambassador, Despite Denials, Officials Say," *Washington Post*, February 9, 2017.

64. Richard Pollack, "Exclusive: In Final Interview, Flynn Insists He Crossed No Lines, Leakers Must Be Prosecuted," Daily Caller, February 14, 2017.

65. Barrett and Lee, "Mike Flynn Was Probed by FBI over Calls to Russia."

66. Ibid.

67. Kaitlyn Schallhorn, "Michael Flynn's Involvement in Russia Investigation: What to Know," Fox News, December 18, 2018.

68. York, "Comey Told Congress FBI Agents Didn't Think Michael Flynn Lied."

69. Ibid.

70. Byron Tau, "Senior Democrats Ask White House for More Details on Michael Flynn," *Wall Street Journal*, February 15, 2017.

71. Evan Perez, "Flynn Changed Story to FBI, No Charges Expected," CNN, February 17, 2017.

72. Shane Harris, Paul Sonne, and Carol E. Lee, "Mike Flynn Worked for Several Russian Companies, Was Paid More than $50,000, Documents Show," *Wall Street Journal*, March 16, 2017.

73. Carol E. Lee, Rob Barry, Shane Harris, and Christopher Stewart, "Mike Flynn Didn't Report 2014 Interaction with Russian-British National," *Wall Street Journal*, March 18, 2017.

74. Michael S. Schmidt, Matthew Rosenberg, and Matt Apuzzo, "Kushner and Flynn Met with Russian Envoy in December, White House Says," *Washington Post*, March 2, 2017.

75. Byron Tau and Natalie Andrews, "Panel Chiefs Say Mike Flynn May Have Violated Law over Payments," *Wall Street Journal*, April 25, 2017.

76. Ibid.

77. *U.S. v. Michael T. Flynn*, Crim. No. 17-232 (EGS), Statement of the Offense.

78. Aruna Viswanatha and Shane Harris, "Mike Flynn Pleads Guilty to Lying About Russian Contacts," *Wall Street Journal*, December 1, 2017.

79. Ibid.

80. Ibid.

81. Ibid.

82. Judson Berger, "Michael Flynn Pleads Guilty to False Statements Charge in Russia Probe," Fox News, December 1, 2017.

83. Anna Dubenko, "Right and Left React to Michael Flynn's Guilty Plea," *New York Times*, December 1, 2017.

84. Miles Parks, "The 10 Events You Need to Know to Understand the Michael Flynn Story," NPR, December 5, 2017.

85. Byron York, "An Unusual Turn in the Michael Flynn Case?," *Washington Examiner*, February 15, 2018.

86. Andrew C. McCarthy, "The Curious Michael Flynn Guilty Plea," *National Review*, February 13, 2018.

87. York, "Comey Told Congress FBI Agents Didn't Think Michael Flynn Lied."

88. Andrew C. McCarthy, "Outrageous Redactions to the Russia Report," *National Review*, May 7, 2018.

89. Ibid.

90. Ibid.

91. Ibid.

92. Ibid.

93. Ibid.

94. U.S. Department of Justice, "Report on the Investigation into Russian Interference in the 2016 Presidential Election," March 2019, https://www.justice.gov/storage/report.pdf.

95. Dion Nissenbaum and Rebecca Ballhaus, "Mueller Probe Looks at Mike Flynn's Work on Documentary Targeting Exiled Turkish Cleric," *Wall Street Journal*, November 24, 2017.

96. Schallhorn, "Michael Flynn's Involvement in Russia Investigation: What to Know."

97. U.S. Department of Justice, "Report on the Investigation into Russian Interference in the 2016 Presidential Election."

98. Nicole Darrah, "Michael Flynn Selling Home to Pay for Legal Fees After Pleading Guilty in Trump Probe," Fox News, March 6, 2018.

99. U.S. Department of Justice, "Report on the Investigation into Russian Interference in the 2016 Presidential Election."

100. Margot Cleveland, "Here's What's Weird About Robert Mueller's Latest Michael Flynn Filing," The Federalist, December 17, 2018.

101. *U.S. v. Michael T. Flynn*, U.S. District Court for the District of Columbia, Crim. No. 17-232 (EGS), Government's Reply to Defendant's Memorandum in Aid of Sentencing, Attachment B, filed December 18, 2018, 4–5. https://assets.documentcloud.org/documents/5628473/12-14-17-Mueller-Reply-Flynn-Sentencing.pdf.

102. Alex Pappas, "DOJ Refusing to Give Grassley Access to Agent Who Interviewed Flynn," Fox News, June 11, 2018.

103. *The Byron York Show*, with Guest Mark Meadows, April 11, 2019. Podcast https://ricochet.com/podcast/byron-york-show/who-was-spying-on-whom/.

104. Gregg Jarrett, "Mueller Strikes Out Trying to Nail Trump—Flynn Sentencing Memo Is a Big Nothing," Fox News, December 5, 2018. The sentencing memo is available at https://www.lawfareblog.com/document-michael-flynn-files-sentencing-memo.

105. Ibid.

106. Gregg Re, "Flynn Says FBI Pushed Him Not to Have Lawyer Present During Interview," Fox News, December 12, 2018.

107. Margot Cleveland, "The Federal Judge Overseeing Michael Flynn's Sentencing Just Dropped a Major Bombshell," The Federalist, December 13, 2018.

108. Gregg Re, "Judge in Flynn Case Orders Mueller to Turn Over Interview Docs After Bombshell Claim of FBI Pressure," Fox News, December 13, 2018.

109. Joel B. Pollak, "Robert Mueller Gives Michael Flynn's '302' to Judge: Agents Thought He Did Not Lie," Breitbart, December 14, 2018.

110. *U.S. v. Michael T. Flynn*, Crim. No. 17-232 (EGS), Government's Reply to Defendant's Memorandum in Aid of Sentencing, December 14, 2018, https://

assets.documentcloud.org/documents/5628473/12-14-17-Mueller-Reply-Flynn -Sentencing.pdf.

111. Paul Mirengoff, "Mueller Makes the Flynn 302 Public," Power Line, December 17, 2018. The Flynn 302 is https://www.documentcloud.org/documents /5633260-12-17-18-Redacted-Flynn-Interview-302.html#document/p1.

112. Cleveland, "The Federalist Judge Overseeing the Flynn Sentencing Just Dropped a Major Bombshell."

113. Robert S. Mueller, III, Letter to Judge Emmet G. Sullivan, December 17, 2018, in *United States v. Michael T. Flynn*, Crim. No. 17-232 (EGS). The letter was drafted by Brandon L. Van Grack and Zainab N. Ahmad, the senior assistant special counsels prosecuting Flynn.

114. Ibid.

115. Alan Dershowitz, "Why Did Mueller Team Distort Trump Attorney's Voice-mail?," The Hill, June 7, 2019. See also *The Mueller Report*, vol. 2, 121–22. The audio of Dowd's voice mail is available at https://lawandcrime.com/high-profile /audio-of-voicemail-ex-trump-attorney-left-flynn-lawyer-released-listen/.

116. Ibid.

117. Ibid.

118. Ibid.

119. Ibid.

120. Sidney Powell, *Licensed to Lie: Exposing Corruption in the Department of Justice* (Dallas: Brown Books Publishing Group, May 1, 2014).

121. Sidney Powell, "Powell: Andrew Weissmann—the Kingpin of Prosecutorial Misconduct—Leaves Mueller's Squad," Daily Caller, March 15, 2019.

122. Complaint filed against Andrew Weissmann by W. William Hodes and Sidney K. Powell with the New York State Unified Court System, First Judicial Attorney Grievance Committee, Department Disciplinary Committee for the First Department, New York & Bronx Counties, 2012, https://cdn.licensed tolie.com/wp-content/uploads/2012_NY_Complaint_Against_Weissmann _0731.pdf.

CHAPTER 9: TARGETED INTIMIDATION

1. Catherine Herridge, "FBI's Manafort Raid Included a Dozen Agents, 'Designed to Intimidate,' Source Says," Fox News, August 24, 2017.

2. "Trump Lawyer Slams Special Counsel for 'Gross Abuse' in Manafort Raid, Challenges Warrant," Fox News, August 10, 2017.

3. Herridge, "FBI's Manafort Raid Included a Dozen Agents, 'Designed to Intimidate,' Source Says."

4. Kaitlyn Schallhorn, "How Paul Manafort Is Connected to the Trump, Russia Investigation," Fox News, August 10, 2017.

5. Victoria Toensing, "Mueller Shouldn't Have Taken the Job," *Wall Street Journal*, April 18, 2019.

6. Michael Rothfield and Craig Karmin, "Trump's Former Campaign Chairman Paul Manafort Thrust Back Into Focus," *Wall Street Journal*, February 20, 2017.

7. Kenneth Vogel and David Stern, "Ukrainian Efforts to Sabotage Trump Backfire," Politico, January 11, 2017.
8. Chuck Ross, "Trump Dossier Source Met with Kremlin Crony at Russian Expo," Daily Caller, March 17, 2018.
9. Ibid. See also WikiLeaks, https://wikileaks.org/dnc-emails/emailid/3962.
10. Ibid.
11. Rothfield and Karmin, "Trump's Former Campaign Chairman Paul Manafort Thrust Back into Focus."
12. Colin Kalmbacher, "Why Manafort Getting an FBI Visit Should Scare Team Trump," Law & Crime, August 9, 2017.
13. Schallhorn, "How Paul Manafort Is Connected to the Trump, Russia Investigation."
14. Ibid.
15. Ibid.
16. Ibid.
17. Josh Dawsey and Darren Samuelsohn, "Feds Sought Cooperation from Manafort's Son-in-Law," Politico, August 9, 2017.
18. Josh Dawsey, "Manafort Switching Legal Team as Feds Crank Up Heat on Him," Politico, August 10, 2017.
19. Gregg Jarrett, "Mueller's Targets Face Financial Strain," Fox News, August 11, 2017.
20. Hallie Detrick, "Former Trump Campaign Chair Paul Manafort Has Been Under FBI Surveillance for Years," Fortune, September 19, 2017.
21. Mollie Hemingway, "Manafort Lawyers Claim Leaky Mueller Probe Has Provided No Evidence Of Contacts With Russian Officials," The Federalist, May 3, 2018.
22. Ibid.
23. Catherine Herridge, "Manafort Filing Unmasks DOJ Meeting with AP Reporters, Questions if 'Grand Jury Secrecy' Violated," Fox News, July 9, 2018.
24. Jack Gillum, Chad Day, and Jeff Horwitz, "AP Exclusive: Manafort Firm Received Ukraine Ledger Payout," Associated Press, April 12, 2017.
25. Reuters, "Who Are Paul Manafort and Rick Gates?," October 30, 2017.
26. Ibid.
27. Anna Palmer, "Tony Podesta Stepping Down from Lobbying Giant amid Mueller Probe," Politico, October 30, 2017.
28. Andrew C. McCarthy, "Paul Manafort Was an Agent of Ukraine, Not Russia," National Review, March 9, 2019.
29. Ibid.
30. Robert S. Mueller, The Mueller Report: The Final Report of the Special Counsel into Donald Trump, Russia, and Collusion as Issued by the Department of Justice (New York: Skyhorse Publishing, 2019), 6.
31. Alan Cullison and Brett Forrest, "Russian Linked to Manafort Is a Shadowy Presence," Wall Street Journal, March 6, 2019.
32. Sharon LaFraniere, Kenneth P. Vogel, and Maggie Haberman, "Manafort Ac-

cused of Sharing Trump Polling Data with Russian Associate," *New York Times*, January 8, 2019.

33. Ibid.

34. Cullison and Forrest, "Russian Linked to Manafort Is a Shadowy Presence."

35. "'A Great Honor': McCain's Historic Reign of International Republican Institute Ends," azcentral, August 4, 2018.

36. Cullison and Forrest, "Russian Linked to Manafort Is a Shadowy Presence."

37. Ibid.

38. John Solomon, "Key Figure That Mueller Report Links to Russia Was a State Department Intel Source," The Hill, June 6, 2019.

39. Ibid.

40. Ibid.

41. Ibid.

42. Del Quentin Wilber and Aruna Viswanatha, "Ex–Trump Adviser Richard Gates Pleads Guilty in Mueller Probe," *Wall Street Journal*, February 23, 2018.

43. Shelby Holliday and Byron Tau, "Roger Stone Trial, Russia Hacking Case Among Mueller Probe's Loose Ends," *Wall Street Journal*, March 25, 2019.

44. Wilber and Viswanatha, "Ex–Trump Adviser Richard Gates Pleads Guilty in Mueller Probe."

45. Ibid.

46. Ibid.

47. Judson Berger, "Cracks in Mueller Probe: Questions over Manafort Charges, Flynn Plea Embolden Trump Allies," Fox News, May 7, 2018.

48. Alan Dershowitz, "Stone Indictment Follows Concerning Mueller Pattern," The Hill, January 25, 2019.

49. Kevin Johnson, "Paul Manafort Trial: Judge T. S. Ellis III Known as Taskmaster, Unafraid to Speak His Mind," *USA Today*, July 31, 2018.

50. Kenneth P. Vogel and David Stern, "Authorities Look into Manafort Protégé," Politico, March 8, 2017.

51. Sonam Sheth, "Mueller Just Hit Paul Manafort and a Russian Intelligence Operative with a New Indictment," Business Insider, June 8, 2018.

52. David K. Ki, "Manafort Listed as 'VIP' in Virginia Jail," *New York Post*, June 16, 2018.

53. Kelly Cohen, "Paul Manafort Spends 23 Hours a Day in Solitary Confinement," *Washington Examiner*, July 6, 2018.

54. Kevin Johnson, "Judge Orders Manafort Moved from 'VIP' Jail to Alexandria Lockup," *USA Today*, July 11, 2018.

55. Kevin Johnson, "Paul Manafort Trial: Prosecutors Accuse Ex–Trump Campaign Chief of Amassing Fortune on Foundation of Lies," *USA Today*, July 31, 2018.

56. Kevin Johnson, "Paul Manafort Trial: Prosecutors Detail Raid on Former Trump Campaign Manager's Luxury Condo, Lavish Spending," *USA Today*, August 1, 2018.

57. Aruna Viswanatha, "Manafort Lied About Contacts with Trump Administration, Mueller Filing Says," *Wall Street Journal*, December 7, 2018.

58. Ibid.

59. Toensing, "Mueller Shouldn't Have Taken the Job."

60. Scott Johnson, "Manafort's Noncooperation," Power Line, December 8, 2018.

61. McCarthy, "Paul Manafort Was an Agent of Ukraine, Not Russia."

62. Jonathan Turley, "Are New York Democrats Handing Paul Manafort a Case for Pardon?," The Hill, February 23, 2019.

63. Ibid.

64. Andrew C. McCarthy, "NY's Political Prosecution of Manafort Should Scare Us All," The Hill, March 14, 2019.

65. William K. Rashbaum and Katie Benner, "Paul Manafort Seemed Headed to Rikers. Then the Justice Department Intervened," *New York Times*, June 17, 2019.

66. Dennis Prager, "Leftism Makes People Meaner," American Greatness, June 10, 2019.

67. Ibid.

68. Ibid.

69. Rashbaum and Benner, "Paul Manafort Seemed Headed to Rikers."

70. Andrew Napolitano, "An American Nightmare," Fox News, January 31, 2019.

71. Representative Doug Collins, House Judiciary Committee, letter to Christopher Wray, FBI Director, January 30, 2019, https://gallery.mailchimp.com /0275399506e2bdd8fe2012b77/files/82d2caa5-c991-43a2-8a83-a2083c5e bcb3/20190130160125191_copy_Merged_.pdf?utm_source=Collins+Ju diciary+Press+List&utm_campaign=421c743f48-EMAIL_CAMPAIGN _2019_01_30_11_31&utm_medium=email&utm_term=0_ff92df788e -421c743f48-169118149. Also available at https://republicans-judiciary.house .gov/press-release/collins-requests-explanation-for-fbis-excessive-use-of-force -in-roger-stone-arrest/.

72. Deana Paul, "The Tactics Behind That Early Morning Raid," *Washington Post*, January 25, 2019.

73. Kenneth Strange, "Roger Stone Raid Raises Questions About Wray Among FBI Rank and File," Daily Caller, January 27, 2019.

74. Ibid.

75. Ibid.

76. Author's interview with Danny Coulson, former deputy assistant FBI director, January 26, 2019.

77. Pete Kasperowicz, "CNN Says It Wasn't Tipped Off About Roger Stone Arrest: 'It's Reporter's Instinct,'" *Washington Examiner*, January 25, 2019.

78. Mark Mazzetti, Eileen Sullivan, and Maggie Haberman, "Indicting Roger Stone, Mueller Shows Link Between Trump Campaign and WikiLeaks," *New York Times*, January 25, 2019.

79. Roger Stone and Robert Morrow, *The Clintons' War on Women* (New York: Skyhorse Publishing, 2015).

80. Roger Stone and Saint John Hunt, *Jeb! And the Bush Crime Family* (New York: Skyhorse Publishing, 2016).

81. Brooke Singman, "FBI's Show of Force in Roger Stone Arrest Spurs Criticism of Mueller Tactics," Fox News, January 25, 2019.

82. Marc Caputo, "Sources: Roger Stone Quit, Wasn't Fired by Trump in Campaign Shakeup," Politico, August 8, 2015.

83. *US v. Roger Jason Stone, Jr.*, U.S. District Court, District of Columbia, 1:19-cr-00018-ABJ, https://www.justice.gov/file/1124706/download.

84. Ibid.

85. Ibid.

86. Andrew C. McCarthy, "Fever Dream: Mueller's Collusion-Free Collusion Indictment of Roger Stone," *National Review*, February 2, 2019.

87. Michael S. Schmidt, Mark Mazzetti, Maggie Haberman, and Sharon LaFraniere, "Read the Emails: The Trump Campaign and Roger Stone," *New York Times*, November 1, 2018.

88. Gregg Jarrett, "Stone Indictment Shows No Evidence of Trump-Russia Collusion," Fox News, January 25, 2019.

89. Maggie Haberman, "Roger Stone Says Text Exchanges Cited in Indictment Were Mischaracterized," *New York Times*, January 27, 2019.

90. Sharon LaFraniere, Michael S. Schmidt, Maggie Haberman, and Danny Hakim, "Roger Stone Sold Himself to Trump's Campaign as a WikiLeaks Pipeline. Was He?," *New York Times*, November 1, 2018.

91. Aidan McLaughlin, "Clapper: Roger Stone Indictment Shows 'Connection, Coordination, Synchronization' Between Trump Campaign and Russia," Mediaite, January 25, 2019.

92. Peter Baker and Maggie Haberman, "Trump, in Interview, Calls Wall Talks 'Waste of Time' and Dismisses Investigations," *New York Times*, January 31, 2019.

93. Rachel Frazin, "Wikileaks Disputes Cohen, Says Assange Never Talked to Stone," The Hill, February 17, 2019.

94. Ibid.

95. Chuck Ross, "Mueller Plea Documents Outline Emails Between Jerome Corsi and Roger Stone," Daily Caller, November 27, 2018.

96. Haberman, "Roger Stone Says Text Exchanges Cited in Indictment Were Mischaracterized."

97. LaFraniere, Schmidt, Haberman, and Hakim, "Roger Stone Sold Himself to Trump's Campaign as a WikiLeaks Pipeline. Was He?"

98. Roger Stone Legal Defense Fund, https://www.stonedefensefund.com.

99. Katelyn Polantz, "DOJ Attorneys Defend Mueller's Ability to Investigate Trump in Roger Stone Filing," CNN, May 5, 2019.

100. Jerome Corsi, *Where's the Birth Certificate?: The Case That Barack Obama Is Not Eligible to Be President* (Washington, DC: WND Books, 2011).

101. Jerome Corsi, *Silent No More: How I Became a Political Prisoner of Mueller's "Witch Hunt"* (New York: Post Hill Press, 2019).

102. Ibid., 17.

103. Ibid., 20.

104. Ibid., 56.

105. Mark Tran, "WikiLeaks to Publish More Hillary Clinton Emails—Julian Assange," *The Guardian*, June 12, 2016.

106. Corsi, *Silent No More*, 72.

107. Ibid., 73.

108. Ibid, 72–75.

109. Ibid., 82.

110. Ibid., 83.

111. Ibid., 85.

112. Ibid., 102.

113. Shelby Holliday and Aruna Viswanatha, "Jerome Corsi, Target of Mueller Probe, Says Stepson Faces Subpoena," *Wall Street Journal*, January 15, 2019.

114. Corsi, *Silent No More*, 93.

115. Ibid., 150.

116. Jerome R. Corsi, memorandum to David Gray, November 25, 2018, provided to Gregg Jarrett by Corsi.

117. Ross, "Mueller Plea Documents Outline Emails Between Jerome Corsi and Roger Stone."

118. Corsi, *Silent No More*, 182.

119. Jerome Corsi, memorandum to David Gray, November 25, 2018.

120. Sara Murray and Katelyn Polantz, "Stone's Efforts to Seek WikiLeaks Documents Detailed in Draft Mueller Document," CNN, November 28, 2018.

121. Jerome Corsi, memorandum to David Gray, November 25, 2018.

122. Holliday and Viswanatha, "Jerome Corsi, Target of Mueller Probe, Says Stepson Faces Subpoena."

123. Kaitlyn Schallhorn, "Who is Jerome Corsi? 3 Things to Know About the Controversial Author," Fox News, November 23, 2018.

124. Ibid.

125. Author's interview with Rudy Giuliani, lawyer for President Trump, July 10, 2019.

126. Ryan Teague Beckwith, "Here Are All the Indictments, Guilty Pleas and Convictions from Robert Mueller's Investigation," *Time*, March 24, 2019.

127. Paul Sperry, "'Scorched Earth': Mueller's Targets Speak Out," RealClearInvestigations, June 6, 2019.

128. Ibid.

129. Ibid.

130. *The Mueller Report*, 61.

131. Manu Raju, "Michael Caputo Says 'It's Clear' Mueller Investigators Focused on Russian Collusion," CNN, May 3, 2018.

132. David Smiley and Glenn Garvin, "Mystery Miamian Tied to Trump Had Many Names, Foul Mouth, 2 DUI Busts," *Miami Herald*, June 9, 2018.

133. Ibid.

134. Sperry, "'Scorched Earth.'"

135. Ibid.

136. Ibid.

137. Ibid.

138. Ibid.

139. Jacob Carozza, "Former Trump Adviser Says He Was Detained at Logan as Part of the Mueller Investigation," *Boston Globe*, March 31, 2018.

140. Theodore Roosevelt Malloch, *The Plot to Destroy Trump: How the Deep State Fabricated the Russian Dossier to Subvert the President* (New York: Skyhorse Publishing, 2018).

141. Sperry, "'Scorched Earth.'" See also Jenna McLaughlin, Jim Sciutto, and Carl Bernstein, "Exclusive: Trump Adviser Played Key Role in Pursuit of Possible Clinton Emails from Dark Web Before Election," CNN, April 7, 2018.

142. *The Mueller Report*, 1.

143. Margot Cleveland, "Why Did the Obama Administration Ignore Reports of Russian Election Meddling?," The Federalist, June 4, 2019.

144. Justin Baragona, "Susan Rice Reportedly Told Staffers Planning Cyber Counterstrike on Putin in 2016 to 'Stand Down,'" Mediaite, March 9, 2018; According to the story, Rice didn't want to box the president in and force him to confront Putin, nor did she want to send a signal calling the integrity of the election into question. "Don't get ahead of us," she told the head of the division; see also Christian Datoc, "Obama's Cybersecurity Coordinator Confirms Susan Rice Ordered Him to 'Stand Down' on Russian Meddling," *Washington Examiner*, June 20, 2018.

145. *US v. Internet Research Agency et al.*, 1:18-cr-00032-DLF, Indictment, filed February 16, 2018.

146. Dele Quentin Wilbur and Aruna Viswanatha, "Russians Charged with Interfering in U.S. Election," *Wall Street Journal*, February 16, 2018.

147. *US v. Internet Research Agency et al.*, 1:18-cr-00032-DLF, Indictment.

148. Michael Crowley and Louis Nelson, "Mueller: Russians Entered U.S. to Plot Election Meddling," Politico, February 16, 2018.

149. Scott Johnson, "Notes on the Indictment," Power Line, February 17, 2018.

150. *US v. Internet Research Agency et al.*, 1:18-cr-00032-DLF, 18.

151. Shelby Holliday and Rob Barry, "Russian Operation Targeted U.S. Business Owners," *Wall Street Journal*, December 20, 2018.

152. Crowley and Nelson, "Mueller: Russians Entered U.S. to Plot Election Meddling."

153. Ibid.

154. Sean Illing, "Did the Media Botch the Russia Story? A Conversation with Matt Taibbi," Vox, April 1, 2019.

155. Ryan Saavedra, "CNN Doxes Elderly Trump Supporter, Harrasses Her, Accuses Her of Working for Russian Trolls," The Daily Wire, February 2, 2018.

156. Ibid.

157. Aruna Viswanatha, "Lawyer for Russian Firm Hits Back at Mueller's Probe," *Wall Street Journal*, May 16, 2018.

158. Tim Ryan, "Mueller Asks Court to Push Back Arraignment of Russian Business," Courthouse News Service, May 4, 2018.

159. Scott Johnson, "Mueller Indicts a Ham Sandwich," Power Line, May 12, 2018.

160. Ibid.

161. Ibid.

162. Viswanatha, "Lawyer for Russian Firm Hits Back at Mueller's Probe."

163. Aruna Viswanatha, "Lawyer for Russian Firm Says Mueller Is Prosecuting a 'Made-Up Crime,'" *Wall Street Journal*, October 15, 2018.

164. Jake Gibson, "Mueller Team Wants to Withhold 3.2 Million 'Sensitive' Docs from Indicted Russian Company," Fox News, March 7, 2019.

165. *US v. Concord Management and Consulting LLC*, CR 1:18-cr-00032-2-DLF, Defendant Concord Management and Consulting LLC's Opposition to Government's Motion for Leave to File an Ex Parte, in Camera, Classified Addendum to Its Opposition to Defendant's Motion to Disclose Discovery Pursuant to Protective Order, filed December 27, 2018.

166. Andrew C. McCarthy, "Mueller's Tough Week in Court," *National Review*, May 7, 2018.

167. Aruna Viswanatha, Sadie Gurman, and Del Quentin Wilber, "Mueller Probe Indicts 12 Russians in Hacking of DNC and Clinton Campaign," *Wall Street Journal*, July 14, 2018.

168. Peter Nicholas and Vivian Salama, "Russian Agents' Indictment Raises Stakes Ahead of Trump-Putin Summit," *Wall Street Journal*, July 13, 2018.

169. Andrew C. McCarthy, "Why Isn't Julian Assange Charged with 'Collusion with Russia'?," *Wall Street Journal*, April 13, 2019.

170. Ibid.

AFTERWORD: THE RECKONING

1. House Permanent Select Committee on Intelligence, "Report on Russian Active Measures," March 22, 2018, https://www.scribd.com/document/377590825/HPSCI-Final-Report-on-Russian-Active-Measures-Redacted-Release#. See also excerpt at https://esmemes.com/i/u-the-committees-investigation-also-reviewed-the-opening-in-summer-56920110520a4899b7ad497bfd5a89f7. See also Paul Wood, "Trump 'Compromising' Claims: How and Why Did We Get Here?," BBC News, January 12, 2017; George Neumayr, "Crossfire Hurricane: Category Five Political Espionage," The American Spectator, May 18, 2018; George Neumayr, "John Brennan's Plot to Infiltrate the Trump Campaign," The American Spectator, May 22, 2018.

2. In the High Court of Justice Queen's Bench Division, *Gubarev v. Orbis*, "Defendants' Response to Claimants' Request for Further Information Pursuant to CPR Part 18," https://assets.documentcloud.org/documents/3892131/Trump-Dossier-Suit.pdf, 7.

3. John Solomon, "FBI's Spreadsheet Puts A Stake Through The Heart Of Steele's Dossier," The Hill, July 16, 2019.

4. John Solomon, "FISA Shocker: DOJ Official Warned Steele Dossier Was Connected to Clinton, Might Be Biased," The Hill, January 16, 2019; Gregg Re and Catherine Herridge, "State Department Official Cited Steele in Emails with Ohr After Flagging Credibility Issues to FBI, Docs Reveal," Fox News, May 15, 2019.

5. United States Foreign Intelligence Surveillance Court, "Verified Application," *In re Carter W. Page*, a U.S. Person, October 2016, https://assets.document

cloud.org/documents/4614708/Carter-Page-FISA-Application.pdf, 15, footnote 8.

6. Jeff Carlson, "Exclusive: Transcripts of Former Top FBI Lawyer Detail Pervasive Abnormalities in Trump Probe," *The Epoch Times*, January 18, 2019 (updated March 8, 2019); Brian Cates, "Why FBI Special Agent Joseph Pientka Is the DOJ's Invisible Man," *The Epoch Times*, January 23, 2019 (updated February 6, 2019).

7. Brooke Singman, "FISA Memo: Steele Fired as an FBI Source for Breaking 'Cardinal Rule'—Leaking to the Media," Fox News, February 2, 2018; Cates, "Why FBI Special Agent Joseph Pietnka Is the DOJ's Invisible Man."

8. Gregg Jarrett, "The Scheme from Bruce Ohr and Comey's Confederates to Clear Clinton, Damage Trump," Fox News, August 28, 2018; Paul Sperry, "Days After Comey Firing, McCabe's Team Re-engaged Fired Dossier Author," RealClearPolitics, January 15, 2019.

9. Judicial Watch, "FBI 302 Interviews With Bruce Ohr On Spygate Released To Judicial Watch," August 8, 2019, available at https://www.judicialwatch.org /press-releases/judicial-watch-fbi-302-interviews-with-bruce-ohr-on-spygate -released-to-judicial-watch/; Gregg Re and Catherine Herridge, "FBI Kept Using Steele Dossier For FISA Applications Despite Documenting Ex-Spy's Bias, Documents Show," Fox News, August 9, 2019; See FBI documents entitled "Ohr 302s FOIA Release 080819," available at https://www.scribd.com /document/421250005/Ohr-302s-FOIA-Release-080819.

10. Charles Creitz, "Graham: New Bruce Ohr Docs Show FISA Warrant Against Ex-Trump Campaign Aide A 'Fraud,'" Fox News, August 8, 2019.

11. Mollie Hemingway, "Comey's Memos Indicate Dossier Briefing of Trump Was a Setup," The Federalist, April 20, 2018; Andrew C. McCarthy, "The Steele Dossier and the 'Verified Application' That Wasn't," *National Review*, May 18, 2019.

12. Andrew C. McCarthy, "Behind The Obama Administration's Shady Plan To Spy On The Trump Campaign," *New York Post*, April 15, 2019.

13. United States Code, 18 U.S.C. § 798, "Disclosure of Classified Information."

14. United States Code, 18 U.S.C. § 953, "Private Correspondence with Foreign Governments" (the Logan Act).

15. *U.S. v. Michael T. Flynn*, U.S. District Court for the District of Columbia, Crim. No. 17-232 (EGS), Government's Reply to Defendant's Memorandum in Aid of Sentencing, Attachment B, filed December 18, 2018, https://assets .documentcloud.org/documents/5628473/12-14-17-Mueller-Reply-Flynn-Sentencing.pdf, 4–5.

16. Politico Staff, "Full Text: James Comey Testimony on Trump and Russia," Politico, June 8, 2017.

17. Adam Goldman and Michael S. Schmidt, "Rod Rosenstein Suggested Secretly Recording Trump and Discussed 25th Amendment," *New York Times*, September 21, 2018.

18. Jonathan Turley, "Mueller Must Testify Publicly to Answer Three Critical Questions," The Hill, June 1, 2019.

19. Testimony of Attorney General William Barr, Senate Judiciary Committee, May 1, 2019; Jan Crawford, "William Barr Interview: Read the Full Transcript," CBS News, May 31, 2019.

20. William P. Barr, Attorney General, letter to the Chairmen and Ranking Members of the House and Senate Judiciary Committees, March 24, 2019, https://assets.documentcloud.org/documents/5779688/AG-March-24-2019-Letter-to-House-and-Senate.pdf.

21. Robert S. Mueller, *The Mueller Report: The Final Report of the Special Counsel into Donald Trump, Russia, and Collusion as Issued by the Department of Justice* (New York: Skyhorse Publishing, 2019).

22. Ben Riley-Smith, "Ex–British Spy Christopher Steele 'Interviewed by Russia Election Investigators for Two Days,'" *The Telegraph*, February 7, 2018; Jerry Dunleavy, "DOJ Inspector General Interviewed Ex-Spy Steele," *Washington Examiner*, July 9, 2019; Chuck Ross, "Report: Christopher Steele Will Not Cooperate with U.S. Attorney's Investigation," Daily Caller, May 28, 2019.

23. *The Mueller Report*, 107.

24. Caitlin Yilek, "Mueller Responds to Trump Allegation He Had Disputed Golf Club Fees," *Washington Examiner*, April 18, 2019.

25. Gregg Jarrett, "Mueller's Team Knew 'Dossier' Kicking Off Trump Investigation Was Biased and Defective," Fox News, January 17, 2019.

26. Author's interview with Rudy Giuliani, lawyer for President Trump, July 10, 2019; Gregg Jarrett, "Cohen Guilty Plea Does Absolutely Nothing to Show Wrongdoing by Trump," Fox News, November 29, 2019.

27. *The Mueller Report*, 39.

28. Letter from Special Counsel Robert S. Mueller, III to Attorney General William P. Barr, March 27, 2019, available at https://apps.npr.org/documents/document.html?id=5984399-Mueller-Letter-to-Barr.

29. "Read Robert Mueller's Full Remarks On The Conclusion Of His Russia Investigation," CBS News, May 29, 2019, available at https://www.cbsnews.com/news/read-robert-muellers-full-remarks-on-the-conclusion-of-russia-investigation-today-2019-05-29/.

30. "Full Transcript: Mueller Testimony Before House Judiciary, Intelligence Committees," NBC News, July 24, 2019, see questioning by Rep. Martha Roby, available at https://www.nbcnews.com/politics/congress/full-transcript-robert-mueller-house-committee-testimony-n1033216.

31. Ibid., questioning by Rep. John Ratcliffe.

32. Ibid.

33. Ibid., questioning by Rep. Louie Gohmert.

34. Ibid., questioning by Rep. Steve Chabot.

35. Editorial, "Beyond Mueller's 'Purview,'" *The Wall Street Journal*, July 26, 2019; Kimberley A. Strassel, "What Mueller Was Trying To Hide," *The Wall Street Journal*, July 26, 2019.

36. Adam Mill, "How Long Has Robert Mueller Been Like This?," *The Federalist*, July 25, 2019 (Adam Mill is a pen name for a Kansas City, Missouri lawyer).

37. "Read Mueller's Full Remarks," see questioning by Rep. Adam Schiff.

38. Sam Dorman, " 'Optics' Of Mueller Hearings Were A 'Disaster' For Democrats, NBC's Chuck Todd Admits," Fox News, July 24, 2019; Katie Pavlich, "Democrats And Media Admit: Mueller's Testimony Was A Total Disaster," Townhall, July 24, 2019.

39. Fox News Staff, "Ken Starr Says Mueller Did A 'Grave Disservice To Our Country,' Did Not Ensure Staff Was 'Fair And Balanced,'" Fox News, July 24, 2019.

40. Jan Crawford, "William Barr Interview: Read the Full Transcript."

41. Adam Goldman, Charlie Savage, and Michael S. Schmidt, "Barr Assigns U.S. Attorney in Connecticut to Review Origins of Russia Inquiry," *New York Times*, May 13, 2019; Brooke Singman and Jake Gibson, "U.S. Attorney John Durham Has Been Reviewing Origins of Russia Probe 'for Weeks': Source," Fox News, Mary 14, 2019.

42. "William Barr's Testimony Before the Senate Appropriations Subcommittee on Commerce, Justice, Science, and related Agencies," CNN, April 10, 2019.

43. Catherine Herridge and Cyd Upson, "Reluctant Witnesses in FISA Abuse Probe Agree to Talk to DOJ Inspector General," Fox News, July 5, 2019.

44. Mark Hosenball, "Trump 'Dossier' Author Grilled by Justice Department Watchdogs: Sources," Reuters, July 9, 2019.

45. Tim Hains, "Rep. Jim Jordan: Comey Said 245 Times 'Don't Remember, Don't Recall, Don't Know,'" RealClearPolitics, December 9, 2018; James Freeman, "The Unbelievable James Comey," *Wall Street Journal*, December 10, 2018.

46. Michael C. Bender and Rebecca Ballhaus, "Trump Gives Barr Authority to Declassify Information About Russia Probe's Origins," *Wall Street Journal*, May 23, 2019; Devlin Barrett, Carol D. Leonnig, and Colby Itkowitz, "Trump Gives Barr Power to Declassify Intelligence Related to Russia Probe," *Washington Post*, May 23, 2019.

ABOUT THE AUTHOR

———

GREGG JARRETT is the *New York Times* bestselling author of *The Russia Hoax*. He is a legal and political analyst for Fox News and was an anchor at the network for fifteen years. Before joining Fox News, he was an anchor and correspondent for MSNBC and an anchor for Court TV. He is a former trial attorney and lives in Connecticut.